Introduction to Modeling and Simulation

Introduction to Modeling and Simulation

A Systems Approach

Mark W. Spong
The University of Texas at Dallas
USA

WILEY

This edition first published 2023

Registered Offices
John Wiley & Sons, Inc., 111 River Street, Hoboken, NJ 07030, USA
John Wiley & Sons Ltd, The Atrium, Southern Gate, Chichester, West Sussex, PO19 8SQ, UK

For details of our global editorial offices, customer services, and more information about Wiley products visit us at www.wiley.com.

Wiley also publishes its books in a variety of electronic formats and by print-on-demand. Some content that appears in standard print versions of this book may not be available in other formats.

Library of Congress Cataloging-in-Publication Data is Applied for:

Hardback ISBN: 9781119982883

Cover image: © billnoll/Getty Images
Cover design: Wiley

Set in 9.5/12.5pt STIXTwoText by Straive, Chennai, India
Printed in Singapore

M WEP273154 250324

To Lila

Contents

Preface

This book grew out of a one-semester course, Engineering Modeling and Simulation, in the Systems Engineering and Management Master of Science degree program at the University of Texas at Dallas. In the spring 2020 semester I was asked to teach this course after the previous instructor left the university, and it has been an enjoyable experience both to teach topics familiar to me and to prepare topics that I had never taught before. My vision in developing this text is to introduce students to a broad range of topics in mathematical modeling that can prepare them for deeper studies of one or more of these topics in subsequent courses. The individual topics in the text were chosen with this in mind, partly based on the availability of more advanced courses in our own curriculum. At the same time, I wanted to include enough material so that the students acquire a basic understanding of each topic whether or not they continue their studies beyond the present text. The emphasis is on systems concepts rather than proofs. I include proofs if doing so adds some particular insight. Otherwise, proofs are either omitted, with pointers to the literature, or else included in the appendix.

The intended audience for this text includes advanced undergraduates and beginning graduate students in science, engineering, management, or mathematics. I assume that the students have a basic knowledge of calculus, linear algebra, probability theory, and differential equations, as well as some familiarity with Matlab, although instructors may wish to supplement this text with an introductory Matlab text.

Organization of the Text

Each chapter of the text is intended to give a more or less self-contained overview of an important topic in systems modeling and simulation. After an introductory chapter, Chapter 2 deals with second-order systems. Second-order systems are an important class of systems because many physical laws, like Newton's second law, are stated in terms of second-order differential equations.

Chapter 3 treats the time response of continuous-time systems of any order. We state sufficient conditions that guarantee a unique solution for nonlinear differential equations. We also discuss linear approximations of nonlinear systems and introduce the matrix exponential, with which we can characterize the solutions of linear systems. We present the Hartman–Grobman theorem which, in effect, says that near a hyperbolic equilibrium point, the response of a nonlinear system and its linear approximation share the same qualitative behavior. We also discuss singularly perturbed systems, which are used to model systems with very large or very small parameters that can pose particular challenges for simulation methods.

Chapter 4 introduces compartmental modeling with applications to population dynamics, the spread of epidemics, and predator–prey models. The basis for these models is the so-called logistic equation, which we treat in some detail.

Chapter 5 deals with stability theory for linear and nonlinear systems. We discuss stability in the sense of Lyapunov and LaSalle's invariance principle.

In Chapter 6 we discuss discrete-time systems, including the discrete-time logistic equation, which shows important differences between continuous- and discrete-time equations, and we present cobweb diagrams as a way to visualize solutions of one-dimensional discrete maps.

Chapter 7 deals with numerical methods. We discuss numerical differentiation, numerical integration, and the most common methods for numerical solution of ordinary differential equations using Matlab and Simulink.

Chapter 8 is an introduction to optimization theory. We discuss necessary and sufficient conditions for solutions of nonlinear programming problems. We discuss gradient search methods and Newton's method for optimization and root finding. For optimization subject to constraints we introduce Lagrange multipliers and we present the so-called Karush–Kuhn–Tucker conditions for optimality. We also discuss the importance of convexity in optimization.

Chapter 9 treats system identification. We discuss linear and nonlinear regression, and the method of recursive least squares. We also present a brief introduction to logistic regression and neural networks.

Chapter 10 treats stochastic systems. In particular, we focus on Markov chain models and Monte Carlo methods for simulation and optimization.

Chapter 11 is an introduction to feedback, which is a fundamental property in natural and human-made systems. We give examples of feedback systems and present an introduction to feedback control.

Chapter 12 treats partial differential equation (PDE) models of systems. We discuss the three most important examples of linear PDE models, namely, the wave equation, heat equation, and Laplace's equation.

Finally Chapter 13 treats complex networks. Many of the most important systems in the modern world, such as the internet, World Wide Web, airline networks, social networks, and others, are modeled as complex networks. We give an introduction to the most relevant concepts in graph theory and network theory applied to both deterministic and random networks.

A Note to the Instructor

My original goal was to produce a text that could be covered in one semester. The book has grown to where it is unlikely that any single course can cover all the topics herein in a single semester. There is, however, a somewhat natural division of the text into chapters dealing primarily with dynamical systems and chapters dealing primarily with computational methods. Below I have outlined suggested courses for both 10 -week quarters and 16 -week semesters. Instructors may also pick and choose topics of interest in order to customize a specific course or a course sequence.

Course 1: Introduction to Modeling and Simulation
Level: Senior undergraduate/graduate
For a one-quarter course (10 weeks):

Chapter 1: Introduction
Chapter 2: Second-Order Systems
Chapter 3: System Fundamentals
Chapter 4: Compartmental Models
Chapter 5: Stability

For a one-semester course (16 weeks) add:

Chapter 11: Feedback Systems
Chapter 12: PDE Models
Chapter 13: Complex Networks

Course 2: Computational Methods in Modeling and Simulation
Level: Senior undergraduate/graduate
For a one-quarter course (10 weeks):

Chapter 1: Introduction
Chapter 2: Second-Order Systems
Chapter 4: Compartmental Models
Chapter 6: Discrete-Time Systems
Chapter 7: Numerical Methods

For a one-semester course (16 weeks) add:

Chapter 8: Optimization
Chapter 9: System Identification
Chapter 10: Stochastic Systems

In addition, the instructor will want to supplement the text with either an introductory Matlab text or tutorials available online.

Acknowledgements

I am indebted to Jim Primbs, who developed the first version of the modeling and simulation course at UT-Dallas. Later, Bobby Gregg took over responsibility for the course and added additional material. Their course outlines and lecture notes were the starting point for my subsequent development of the course. Any errors or omissions in the present text are entirely due to me. I also want to thank Reza Moheimani and Steve Yurkovich for constructive comments and support. Thanks also to Aykut Satici and Alireza Mohammadi for an interesting discussion on variable-mass systems. The students in my class also helped by asking insightful questions and finding errors in early drafts of the text. Several comments from anonymous reviewers also helped considerably in making the text more readable and accessible to students. A special thanks to Richard Leigh for an exceptional editing job, which greatly improved the final manuscript. Finally, thanks to Sandra Grayson and Juliet Booker at John Wiley for their support and encouragement.

August 2022 *Mark W. Spong*

About the Companion Website

This book is accompanied by a companion website:

www.wiley.com/go/spong/engineeringmodeling

From the website you can find the following online materials:

- Solutions Manual
- Case-study Exercises
- Lecture Slides
- MATLAB code

1

Introduction

1.1 Introduction

This text is concerned with *modeling and simulation of systems*, both natural and engineered. We treat both analytical and computational methods for a range of applications to electrical, mechanical, biological, financial, social, and other systems.

The notion of what constitutes a system is very broad and application dependent. Generally, by a system, one means *a combination of interrelated components or parts that works in synergy to collectively perform a desired function*. In business, for example, a system can mean a set of processes or procedures working together, such as a payroll system or inventory system. In physiology, the circulatory system is composed of the heart, blood, arteries, veins, and capillaries and delivers oxygen and nutrients to cells and removes waste products, such as carbon dioxide. In computer science, an operating system is the software that manages hardware and software resources and provides services to application programs. In biology, an ecosystem refers to a community of interacting species and their physical environment.

The distinction between component and system is application dependent. For example, in the semiconductor industry a microprocessor is a highly complex system, whereas in the automotive industry a microprocessor is a component in systems that control engine emissions, braking, cruise control, and other functions.

1.1.1 Systems Engineering

Systems engineering is an interdisciplinary field of engineering that focuses on how to design and manage systems over their life cycles [3, 15, 17]. A key problem in systems engineering is how to deal with the increasing complexity of modern systems. Modeling and simulation are important tools to help design and

Introduction to Modeling and Simulation: A Systems Approach, First Edition. Mark W. Spong.

Companion Website: www.wiley.com/go/spong/engineeringmodeling

analyze modern complex systems. Good models allow meaningful simulation for testing, design validation, hardware-in-the-loop testing and iterative methods. Simulations are less expensive than prototype testing and can be more rapidly developed and modified.

In this text we take the viewpoint that modeling is the more important part of modeling and simulation. Simulations based on poor models are of relatively little use. Thus, we focus on the modeling aspects of systems and use simulations to illustrate their performance and gain insight into their structure.

1.1.2 The Input/Output Viewpoint

At an abstract level, we can view a system as an object or process Σ that transforms *inputs* u to *outputs* y, which we write as $y = \Sigma(u)$; see Figure 1.1. Within this input/output system paradigm various assumptions on the nature of the system must be made to derive concrete models that can be used for analysis, simulation, and prediction. How to do this is the principle subject of this text.

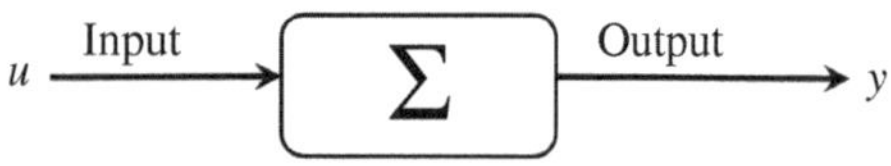

Figure 1.1 A general characterization of a system as an object or process that transforms inputs u to outputs y.

1.1.3 Some Examples

Examples of systems as objects that transform inputs to output are found in numerous fields of engineering and nature. Some representative examples include the following:

(1) **An electric circuit.** In the circuit shown in Figure 1.2 we can take the voltage V_1 as input and the voltage V_2 as output. The system then transforms the input to the output according to Kirchhoff's laws.

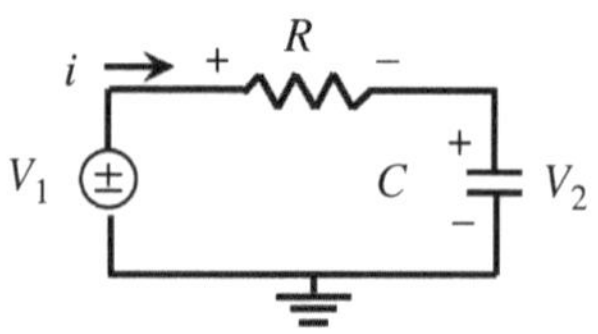

Figure 1.2 An electric circuit with input voltage V_1 and output voltage V_2. The output voltage is determined by Kirchhoff's Laws.

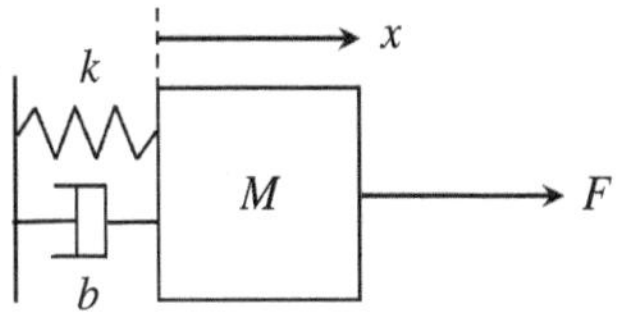

Figure 1.3 A mass–spring–damper system with input force *F* and output position *x*. The motion of the mass as a function of time is governed by Newton's laws.

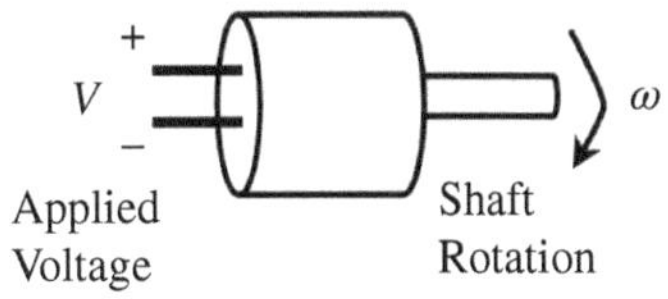

Figure 1.4 A DC motor with input voltage *V* and output speed ω. Both electrical and mechanical models govern the input/output behavior of this system.

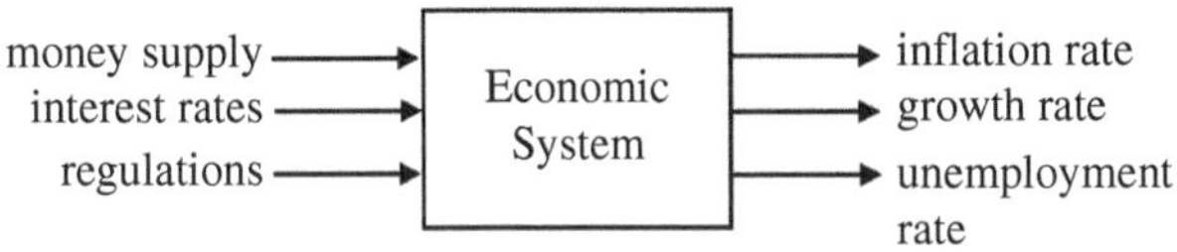

Figure 1.5 An economic system with multiple inputs and outputs. The choice of what variables to choose as inputs and what variables to choose as outputs is also an important part of the modeling process.

(2) **A mechanical system.** The system in Figure 1.3 shows an interconnected mass, spring, and damper. If an input force F acts on the mass, then the output position x of the mass is governed by Newton's laws of motion.
(3) **A DC motor.** In a DC motor, as shown in Figure 1.4, we can take the applied voltage V as input and the shaft angular velocity ω as output. The system then transforms electrical energy into mechanical energy to perform useful work.
(4) **An economic system.** A system may have multiple inputs and/or multiple outputs. For example, in an economic system, variables such as money supply, interest rates, and regulations can be taken as inputs and variables such as inflation rate, growth rate, unemployment rate can be taken as outputs (Figure 1.5).
(5) **A manufacturing plant.** A factory (Figure 1.6) can be thought of as a system that transforms raw materials (the inputs) to finished products (the outputs). The manufacturing process is composed of many individual processes that occur as discrete events, such as metal forming, welding, painting, final assembly, and other processes.

Figure 1.6 A manufacturing plant transforms input raw materials into output finished products. The system can be modeled as the interconnection of many subsystems, such as metal forming, welding, painting, assembly, and so on.

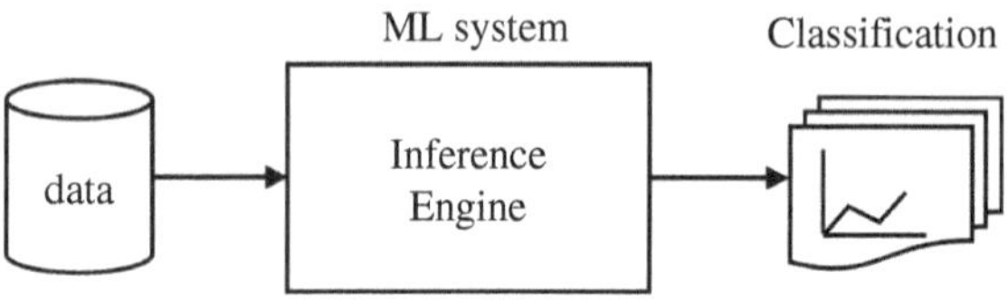

Figure 1.7 A machine learning (ML) system takes input data and produces output decisions. The generation of the input data and the algorithms used for the inference engine are key features of a learning system.

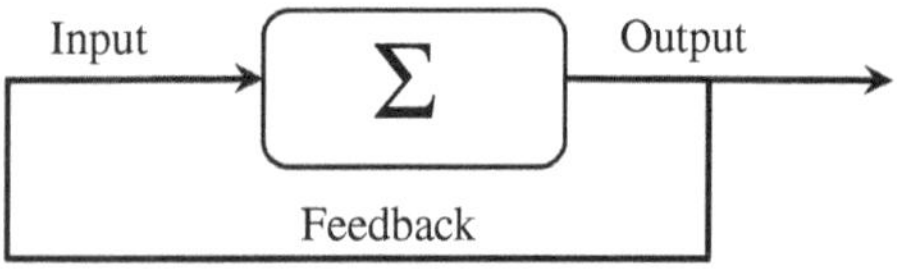

Figure 1.8 A feedback system is one where the output affects future inputs.

(6) **Machine learning.** A machine learning system can be thought of as a computer program that takes data as input and produces estimates (classifications, decisions, suggestions) as outputs (Figure 1.7).
(7) **Feedback systems.** Feedback is a fundamental property in both natural and engineered systems. By a *feedback system* we mean "a system where the input is influenced by the output."

We refer to the system in Figure 1.8 as a *closed-loop system* and the system in Figure 1.1 as an *open-loop system*. Feedback can be positive, which tends to have a destabilizing effect, as in a nuclear chain reaction, or negative, which tends to have a stabilizing effect, for example, when we sweat in hot weather to reduce our body temperature.

Feedback control is indispensable in modern engineering systems, such as automotive systems, aircraft, robotics, chemical and oil refineries, manufacturing, energy and power systems, and a host of other applications. In biological systems, feedback is present in regulating many processes, from body temperature to cell metabolism, gene expression, hormone production, as well as balance and locomotion. In weather and climate dynamics, feedback is an important mechanism affecting global temperatures, weather patterns, ocean currents, and so on.

1.2 Model Classification

A model is a mathematical representation and is both an abstraction and an approximation of physical reality. Developing models of systems takes several forms.

Models from first principles. These are physics-based models, typically differential or difference equations derived from Newton's laws, Maxwell's equations, Kirchhoff's laws, and so on.

Black box models. These are models based on measured data to identify an input/output structure. Such data-driven methods include neural networks, genetic algorithms, frequency-response methods, machine learning, or optimization methods.

Gray box models. These are a combination of the above two using a-priori assumptions on the model together with measured data for parameter identification, systems identification, regression, or other methods.

System models may be classified in several ways, which can be useful to determine the best tools for their analysis. Some of the most useful classifications are listed below.

1.2.1 Static and Dynamic Systems

In a static system, there is an algebraic or functional relationship between input and output, as in a resistor where $V(t) = RI(t)$. Static models will be used, for example, when we discuss optimization, regression, and graph theory. In a dynamic system the output depends on the input together with the current state of the system, which typically evolves according to a differential or difference equation.

1.2.2 Linear and Nonlinear Systems

All systems possess nonlinearities. Despite this fact, linear systems are an important class of systems for two primary reasons. First, nonlinear systems are approximately linear near their equilibrium states. Second, linear systems are well understood and their behaviors are easily characterized. Nonlinear systems, on the other hand, are difficult to analyze and can often only be studied using numerical methods. At the same time, nonlinear systems have extremely rich behavior not possessed by linear systems.

The key to linearity is the principle of superposition, with which the output response to an arbitrary input can be understood in relation to outputs generated by simpler inputs. Specifically, we have the following definition.

Definition 1.1 ***(Linearity)*** A system Σ is linear if and only if it satisfies the principle of superposition: If $y_1 = \Sigma(u_1)$ and $y_2 = \Sigma(u_2)$, then the output y corresponding to the input $u = a_1u_1 + a_2u_2$, where a_1 and a_2 are constant, is given by $y = a_1y_1 + a_2y_2$. In other words,

$$\Sigma(a_1u_1 + a_2u_2) = a_1\Sigma(u_1) + a_2\Sigma(u_2).$$

The principle of superposition says that if y_1 is the output corresponding to the input u_1 and y_2 is the output corresponding to the input u_2, then a linear combination of the inputs u_1 and u_2 produces the same linear combination of the outputs y_1 and y_2. Using superposition, one can gain an understanding of systems by analyzing their constituent components.

This approach is related to the principle of *reductionism*, which refers to a scientific approach in which a system is decomposed into subsystems, each of which is more amenable to analysis. The increasing complexity of modern engineering systems limits the usefulness of the reductionist approach. Complexity leads to *emergent behavior*, which means roughly that the whole is greater than the sum of the parts. Thus, we need to develop methods that are directly applicable to nonlinear, interconnected systems. We will discuss emergent behavior in complex networks in the final chapter of this text.

1.2.3 Distributed-Parameter Systems

By a distributed-parameter system, we mean one that is described by a partial differential equation. Examples of distributed-parameter systems are vibrating strings and membranes, the temperature distribution along a bar, and quantum mechanical systems.

1.2.4 Hybrid and Discrete-Event Systems

Hybrid systems are systems that combine both continuous-time and discrete-time components. A simple example of a hybrid system is a bouncing ball, whose motion is governed by continuous dynamics together with a discrete *reset map* when the ball contacts the ground.

Discrete-event systems are logic-based rather than time-based. For example, in a manufacturing system an assembly operation may take place in discrete stages as parts are transferred between machines.

1.2.5 Deterministic and Stochastic Systems

In a deterministic system the output is uniquely determined from the model and is repeatable. In a stochastic system, the output is determined according to a probability, and multiple repeated trials will generally produce different outputs.

1.2.6 Large-Scale Systems

By a *large-scale system* one typically means a system with numerous components and with many variables and parameters needed to model them. These systems are also complex and difficult to model accurately. A few example of large-scale engineered systems are given below in Figures 1.9 to 1.15.

Figure 1.9 Warehouse robots for order retrieval are an example of factory automation, where the robots act with little or no human supervision.

Figure 1.10 Oil and chemical refineries are large-scale, complex systems that are made more efficient through modeling, simulation, and systems engineering principles for process control and optimization.

Figure 1.11 The electric power grid consists of an interconnected network of power generation stations, transmission lines, substations, and power distribution stations. China is currently the largest producer of electric energy followed by the United States.

Figure 1.12 A wind farm is a collection of wind turbines connected to the power grid. Each turbine is a complicated engineered system. However, in order to optimize energy production, it is not sufficient to study the individual turbines. Modeling the interactions among the turbines is also important.

Figure 1.13 Supply chains and logistics systems are responsible for moving manufactured goods on a global scale. These systems are interconnected networks of ships, airplanes, trucks, and railroads, as well as pipeline networks, loading docks, warehouses and a host of other systems that must function efficiently and economically.

Figure 1.14 Systems of autonomous vehicles are already changing the trucking industry, the shipping industry, and personal transportation. As in the wind-farm example, one must model both the individual vehicles and the interconnections among multiple vehicles.

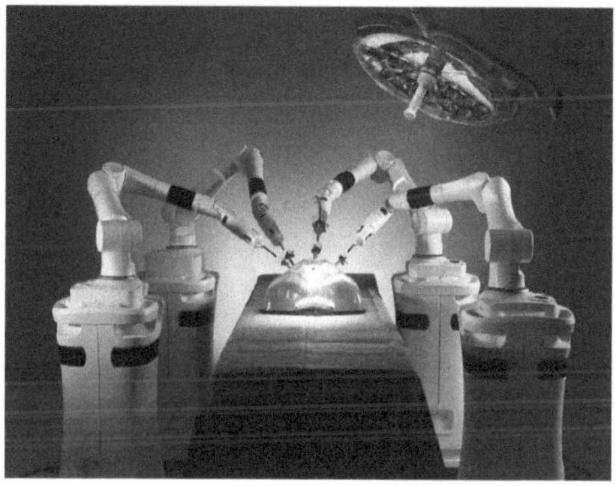

Figure 1.15 Surgical robots are examples of human/machine/computer interfaces and require new methods for modeling human dynamics in addition to robot dynamics.

1.3 Simulation Languages

There are several general purpose and special purpose simulation software programs and packages available, for example:

- **Matlab/Simulink** is a programming language and numerical computing environment. Matlab also allows symbolic computation and graphical representation of data.
- **SysML** is a graphical modeling language for analyzing complex systems. SysML is useful for modeling system requirements, system interconnections, and other features.
- **AnyLogic** is a simulation tool designed primarily for business applications like supply chains, logistics, markets and other applications.
- **FlexSim** is a simulation package designed for modeling and simulation of discrete-event systems.
- **Mathematica** is a general-purpose programming language for symbolic and numerical computation.
- **Modelica** is an object-oriented modeling language, as opposed to a programming language, designed to model systems containing electrical, mechanical, hydraulic, thermal, and other components.

- **ANSYS** is a finite-element modeling package for numerical analysis of mechanical systems.
- **SolidWorks** is a computer-aided design and computer-aided engineering package that runs primarily on Microsoft Windows.
- **Arena** is a discrete-event simulation and automation package for business and production processes.

A discussion of specialized programs for topics such as project management is outside the scope of this text. Instead, we will use Matlab/Simulink as a general purpose programming language to illustrate simulation methods for system models.

1.4 Outline of the Text

Each chapter of the text is intended to give an overview of a specific topic in systems modeling and simulation. After studying the chapters in this text the reader should be prepared for more in-depth treatment of each topic.

Chapter 2 begins with a treatment of second-order systems. Second-order systems form an important class of systems because many physical laws, like Newton's second law, are stated in terms of second-order differential equations. We characterize the behavior of linear second-order systems in terms of the eigenvalues and eigenvectors of the system coefficient matrices. We also discuss the existence of periodic solutions for nonlinear systems and present the Poincaré–Bendixson theorem, which gives conditions for the existence of periodic solutions of nonlinear systems.

Chapter 3 treats the time response of general continuous-time systems. We state sufficient conditions that guarantee a unique solution for nonlinear differential equations. We also discuss linear approximations of nonlinear systems and we introduce the matrix exponential, with which we can characterize the solutions of nth-order linear systems. We also discuss the Hartman–Grobman theorem which, in effect, says that near a hyperbolic equilibrium point, the response of a nonlinear system and its linear approximation have the same qualitative behavior. We also discuss singularly perturbed systems, which are used to model systems with very large or very small parameters that can pose particular challenges for simulation methods.

In Chapter 4 we introduce compartmental modeling and discuss examples and applications, including models of population dynamics, the spread of epidemics, and predator–prey models. The basis for these models is the so-called logistic equation, which we treat in some detail.

Chapter 5 deals with stability theory for linear and nonlinear systems. For systems operating in equilibrium configurations, it is important to know how the system will respond to perturbations and disturbances that move the system state away from equilibrium. We discuss two main results, namely stability in the sense of Lyapunov and LaSalle's invariance principle.

In Chapter 6 we discuss discrete-time systems. We present results that mimic the results for continuous-time systems, including the existence and uniqueness of solutions and stability. We discuss the discrete-time logistic equation, which shows important differences between the continuous-time and discrete-time equations, and we present cobweb diagrams as a convenient way to visualize solutions of one-dimensional discrete-time maps.

In Chapter 7 we discuss numerical methods to simulate system models. We discuss numerical differentiation, numerical integration and the most common methods for numerical solution of ordinary differential equations using Matlab and Simulink.

Chapter 8 is an introduction to optimization theory. We will discuss general necessary and sufficient conditions for solutions of nonlinear programming problems. We discuss gradient search methods, including the method of steepest descent. We then treat Newton's method for optimization and root finding. For optimization subject to constraints we introduce Lagrange multipliers and we present the so-called Karush–Kuhn–Tucker conditions for optimality. We also discuss the importance of convexity in optimization.

Chapter 9 treats system identification. We discuss linear and nonlinear regression, and the method of recursive least squares. We also present a brief introduction to logistic regression and perceptron neural networks for binary classification.

Chapter 10 treats stochastic systems. In particular, we focus on Markov chain models and Monte Carlo methods for simulation and optimization.

In Chapter 11 we discuss feedback systems. We give examples of feedback systems and present an introduction to feedback control.

In Chapter 12 we discuss partial differential equation (PDE) models of systems. We discuss the three most important examples of PDE models, namely, the wave equation, heat equation, and Laplace's equation.

Chapter 13 treats complex networks. Many of the most important systems in the modern world are modeled as complex networks, such as the internet, World Wide Web, airline networks, social networks, and others. We give an introduction to the most relevant concepts in graph theory applied to both deterministic and random networks.

Problems

1.1 For each of the systems below, what would you choose as appropriate inputs and outputs to model the system?

1. Roof-top solar cells
2. A home thermostat

3. An insurance company
4. The respiratory system

Linear Algebra Background

1.2 Consider the matrix

$$A = \begin{bmatrix} -5 & 0 & 0 \\ 0 & 0 & 1 \\ 0 & -2 & -3 \end{bmatrix}.$$

1. Compute the determinant of A.
2. Compute the principal minor determinants of A.
3. Compute the trace of A.
4. Compute the inverse of A.
5. Compute $A + A^T$.
6. Compute AA^T.

1.3 Compute the range and null space of the matrix

$$A = \begin{bmatrix} -2 & 0 & 0 \\ 0 & 0 & 0 \\ 0 & -2 & -2 \end{bmatrix}.$$

1.4 Consider the matrix

$$A = \begin{bmatrix} 0 & 1 \\ -2 & -3 \end{bmatrix}.$$

1. Compute the eigenvalues λ_1 and λ_2 of A.
2. Compute unit eigenvectors v_1 and v_2 corresponding λ_1 and λ_2, respectively.
3. Define the matrix $T = [v_1\ v_2]$ and compute the matrix $T^{-1}AT$.

1.5 Consider the vectors $x = [1, -1]^T$ and $y = [2, 1]^T$.
1. Compute the inner product $x^T y$.
2. Computer the norms $||x||$ and $||y||$.
3. Verify that the triangle inequality and Cauchy–Schwarz inequality hold for these vectors.

Real Analysis Background

1.6 Consider the function $f(x_1, x_2) = 2x_1^2 + x_1x_2 + 4x_2^2$. Compute the gradient and Hessian of f. Show that the Hessian is a positive-definite matrix.

1.7 Compute the Jacobian of the vector field

$$f(x_1, x_2) = \begin{bmatrix} 3x_1^3 + 2x_2 \\ x_1 x_2^2 \end{bmatrix}.$$

1.8 Verify the Taylor series expansions of sin(x) and cos(x) given in Appendix B by direct calculation.

Probability Background

1.9 Suppose that you have 100 tickets in a hat numbered 1 to 100. What is the probability of drawing a ticket number divisible by 2 or 5?

1.10 Suppose that you have an exam with 10 multiple-choice questions. Each question has four choices, only one of which is correct. What is the probability of getting five or more correct answers by choosing answers randomly?

1.11 Here is a list of ten students' grades on an exam:

$$98, 93, 89, 77, 68, 83, 94, 88, 65, 77.$$

What are the mean, median, and standard deviation of the set of grades?

Matlab

1.12 Write a Matlab script to input and multiply two matrices. Your script should check that the matrices have compatible dimensions and print an error message if the dimensions are not compatible.

1.13 Given the two matrices,

$$A = \begin{bmatrix} 0 & 1 & 2 \\ -1 & 3 & 4 \\ 0 & 3 & -1 \end{bmatrix}, \quad I = \begin{bmatrix} 1 & 0 & 0 \\ 0 & 1 & 0 \\ 0 & 0 & 1 \end{bmatrix},$$

verify that the Matlab commands `inv(A)`, $A\backslash I$, and I/A return the same matrix, A^{-1}. Check that `A*inv(A)=I`.

1.14 Write a Matlab script to input an $n \times n$ matrix and compute its determinant and its eigenvalues using the Matlab functions `det` and `eig`.

1.15 Write a Matlab script to input an $n \times n$ matrix A and compute the square of each entry of A. Note that this is different from computing A^2.

1.16 Write a Matlab script to input an anonymous function $f(x)$ of a single variable x and plot the graph of $f(x)$ on an interval $[-a, a]$.

1.17 Review the documentation on Matlab symbolic variables and functions, `sym` and `syms`. Compute the symbolic derivative and integrals of the following functions:

1. $\sin(3x)$ and $\cos(3x)$
2. $x^3 - 3x^2 + 4$
3. e^{-2t}

1.18 Write a Matlab script to input two polynomials, multiply them together, and find the roots of the resulting polynomial.

1.19 A particle of mass $m = 1$ is dropped in a gravitational field $g = 9.8$ m/s^2, starting at a height $h = 100$ meters above the ground. Plot its position and velocity for 3 seconds.

1.20 Repeat the above problem assuming now that the particle is initially thrown upward with a velocity of 10 m/s.

2

Second-Order Systems

2.1 Introduction

We begin our study of modeling and simulation with a treatment of second-order, continuous-time systems. Second-order systems form an important class of systems for a couple of reasons. First, the state space is two-dimensional and so graphical methods can be used for visualization and analysis. Second, many physical laws, such as Newton's second law, $f = ma$, give rise to second-order systems [29].

We first discuss trajectories and phase portraits for autonomous second-order systems. We then discuss methods for sketching phase portraits of second-order systems based on the concept of the direction field, and show how to classify second-order systems in terms of the eigenvalues of the system coefficient matrix. We also discuss the existence of limit cycles, which are isolated periodic solutions. We present a fundamental result, known as the Poincaré–Bendixson theorem, which can be used to determine the existence of periodic solutions to second-order systems. Finally, we discuss coupled second-order systems that frequently arise from physical models of mechanical and electrical systems.

To start, suppose that the input/output system model in Figure 1.1 is described by a second-order linear differential equation of the form

$$\frac{d^2y}{dt^2} + a_1\frac{dy}{dt} + a_0y = u, \tag{2.1}$$

with input $u(t)$, output $y(t)$, and constant coefficients a_1 and a_0. The equation is *homogeneous* if $u(t) = 0$; otherwise it is *inhomogeneous*. Classic examples of this model are a mass–spring–damper system and an *RLC* circuit as shown in Figure 2.1.

Introduction to Modeling and Simulation: A Systems Approach, First Edition. Mark W. Spong.

Companion Website: www.wiley.com/go/spong/engineeringmodeling

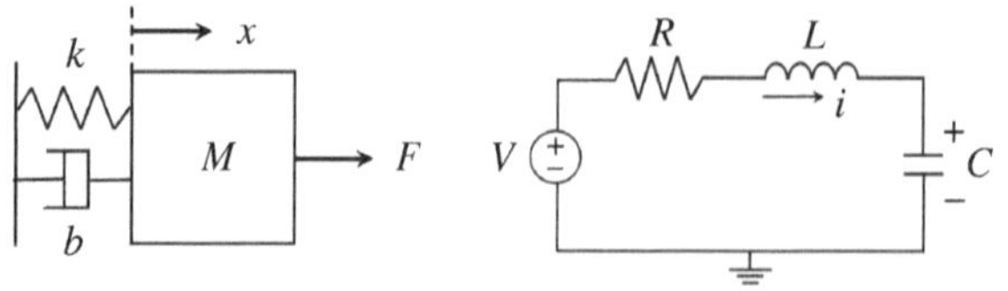

Figure 2.1 A mass–spring–damper system (left) and an *RLC* circuit (right). Both the electrical and mechanical system are governed by the second-order differential Equation (2.1).

Using basic circuit theory (Problem 2.1), it is easy to show that the equation for the electrical circuit is

$$\frac{d^2i}{dt^2} + \frac{R}{L}\frac{di}{dt} + \frac{1}{LC}i = \frac{V}{L}, \tag{2.2}$$

where i is the current and V is an applied voltage. Likewise, from Newton's laws (Problem 2.4), the equation for the mass–spring–damper system is

$$\frac{d^2x}{dt^2} + \frac{b}{M}\frac{dx}{dt} + \frac{k}{M}x = \frac{F}{M}, \tag{2.3}$$

where x is the position of the mass and F is an applied force. Both systems (2.2) and (2.3) are in the form of Equation (2.1) for suitable assignments of the variables y and u and coefficients a_0 and a_1.

Remark 2.1 The above two examples illustrate the important fact that systems in diverse fields, such as electrical, mechanical, chemical, and biological systems, are often governed by the same underlying dynamical equations, which makes abstraction and analysis of generic (or canonical) models advantageous. We will, in fact, study the linear system (2.1) in great detail in the present chapter as a canonical representation for applications of second-order linear systems.

In particular, there is a strong analogy between electrical and mechanical systems and their governing equations. One of the first people to utilize this analogy was James Clerk Maxwell, who used mechanical analogies to explain less familiar (at that time) electrical phenomena. The reverse is now more common, where electrical circuit models are used to model mechanical systems, including concepts such as impedance and admittance.

In the remainder of this chapter we will primarily treat the homogeneous case $u = 0$ and return to the inhomogeneous case in Chapter 11 where we discuss feedback systems. For simplicity, we will also use the "dot" notation for the derivatives with respect to time, $\dot{y} = \frac{dy}{dt}$, $\ddot{y} = \frac{d^2y}{dt^2}$, and rewrite the system (2.1) as

$$\ddot{y} + a_1\dot{y} + a_0y = 0. \tag{2.4}$$

The Harmonic Oscillator

An important special case of Equation (2.4) is the *harmonic oscillator*

$$\ddot{y} + \omega^2 y = 0, \tag{2.5}$$

which models systems that undergo pure sinusoidal motion. The harmonic oscillator arises in the study of elasticity, acoustics, circuits, molecular vibrations, electromagnetic fields, vibrating strings, and optics. The parameter ω is called the *natural frequency* of the system.

We can illustrate the harmonic oscillator using the mass–spring system shown in Figure 2.2. The mass at position x is acted upon by the restoring force $F = -kx$ and gravitational force Mg, where k is the spring stiffness constant and g is the acceleration due to gravity. This leads to the equation

$$M\ddot{x} + kx = Mg. \tag{2.6}$$

The equilibrium position of the mass is equal to $x_0 = Mg/k$. If we define $y = x - x_0$, then Equation (2.6) is of the form (2.5) with natural frequency $\omega = \sqrt{k/M}$.

The general solution of the harmonic oscillator equation is

$$y(t) = C_1 \cos(\omega t) + C_2 \sin(\omega t),$$

where C_1 and C_2 are constants determined by the initial conditions,

$$C_1 = y(0), \quad C_2 = \frac{\dot{y}(0)}{\omega}.$$

The solution $y(t)$ is periodic with period $\frac{2\pi}{\omega}$. For example, with $y(0) = r$, $\dot{y}(0) = 0$, the solution is $y(t) = r\cos(\omega t)$.

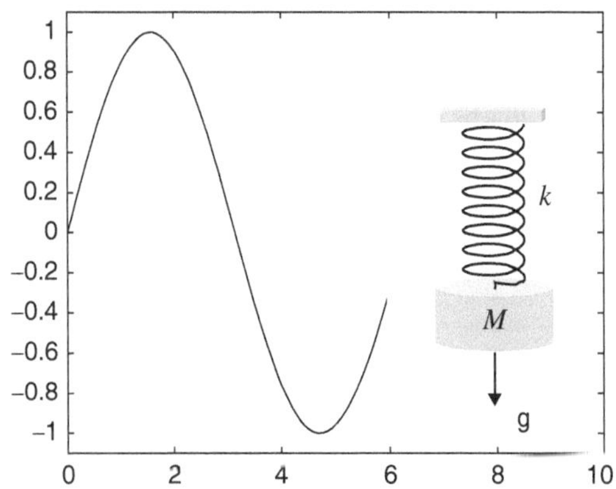

Figure 2.2 Mass–spring system that undergoes harmonic motion.

Symbolic Solution of Differential Equations

Matlab's `dsolve` command can be used for symbolic solutions of differential equations like the harmonic oscillator.

Symbolic solution of the harmonic oscillator

```
omega=2;           % the natural frequency
syms t y(t);       % define t and y as symbolic variables
harmonic = diff(y,t,2) + omega^2*y == 0;
y = dsolve(harmonic)
```

Running the script returns

Symbolic solution of the harmonic oscillator

```
y = C1*cos(2*t) - C2*sin(2*t).
```

We include initial conditions as follows.

Solution of the harmonic oscillator with initial conditions

```
omega=2;
syms t y(t);
harmonic = diff(y,t,2) + omega^2*y == 0;
Dy = diff(y,t);
C1 = 3; C2=0;
y = dsolve(harmonic, [y(0) == C1, Dy(0) == -C2/omega]);
```

Running this script returns

Solution of the harmonic oscillator with initial conditions

```
y = 3*cos(2*t).
```

Example 2.1 ***(Hyperbolic Functions)***
If we change the sign in the second-order system (2.5) to give

$$\ddot{y} - \lambda^2 y = 0, \tag{2.7}$$

then the solution is given in terms of the hyperbolic trigonometric functions as

$$y(t) = C_0 \cosh(\lambda t) + C_1 \sinh(\lambda t). \tag{2.8}$$

The hyperbolic functions cosh and sinh are related to the exponential functions as

$$\cosh(t) = \frac{1}{2}(\exp(t) + \exp(-t)), \quad \sinh(t) = \frac{1}{2}(\exp(t) - \exp(-t)).$$

The Matlab symbolic differential equation function `dsolve` returns the solution in terms of the exponential function. In order to express the solution in terms of cosh and sinh, the Matlab function `rewrite` should be used as shown below.

Solution of the modified harmonic oscillator equation

```
lambda=1;
syms t y(t);
hyperbolic = diff(y,t,2) - lambda^2*y == 0;
Dy = diff(y,t);
y = dsolve(hyperbolic, [y(0) == -2 Dy(0) == 1]);
y = rewrite(y,'cosh');
```

Running this script returns

Solution of the modified harmonic oscillator equation

```
y = sinh(t) - 2*cosh(t).
```

Equation (2.7) and its solution in terms of hyperbolic trigonometric functions are important in the solution of partial differential equations. We will return to this equation in Chapter 12 when we discuss series solutions of partial differential equations.

2.2 State-Space Representation

In this section we introduce the *state-space representation* of the system (2.4). The state-space representation is useful for, among other things, obtaining qualitative information about the behavior of a system without having to explicitly solve the differential equations. This is particularly important in the case of nonlinear systems which, in many cases, cannot be solved in closed form.

Given the system (2.4), we define variables $x_1 = y$ and $x_2 = \dot{y}$. From this we can see that x_1 and x_2 satisfy the equations

$$\dot{x}_1 = x_2 \tag{2.9}$$

$$\dot{x}_2 = -a_0 x_1 - a_1 x_2. \tag{2.10}$$

Thus, we have transformed the single (scalar) second-order equation in y into a system of two first-order equations in x_1 and x_2. The vector $x = [x_1, x_2]^T$ is called the *state* of the system or *state vector*, and $\mathbb{R}^2$ is referred to as the *state space*. In terms of the state vector x, the equations (2.9)–(2.10) can be written compactly as

$$\dot{x} = Ax, \quad \text{where } A = \begin{bmatrix} 0 & 1 \\ -a_0 & -a_1 \end{bmatrix}.$$

The matrix A is called the *state matrix* or *coefficient matrix*. For two-dimensional systems the state space is also called *phase space* and x_1 and x_2 are called *phase variables.*

The Vector Field

Second-order linear systems are special cases of more general systems in the form

$$\begin{aligned}\dot{x}_1 &= f_1(t, x_1, x_2)\\ \dot{x}_2 &= f_2(t, x_1, x_2).\end{aligned}$$

The function $f(t, x) = \begin{bmatrix} f_1(t, x_1, x_2) \\ f_2(t, x_1, x_2) \end{bmatrix}$ is called a *vector field*. The vector field f is called *autonomous* or *time invariant* if $f(x) : \mathbb{R}^2 \to \mathbb{R}^2$ does not depend explicitly on time t. Otherwise, it is *nonautonomous* or *time varying*.

Example 2.2 The following system is a second-order, time-varying nonlinear system:

$$\begin{aligned}\dot{x}_1 &= x_2\\ \dot{x}_2 &= -3tx_1 - e^{-t}x_2^2.\end{aligned}$$

It is nonlinear because of the term x_2^2 and time varying because the coefficients of x_1 and x_2 are explicit functions of time. On the other hand, the system

$$\begin{aligned}\dot{x}_1 &= x_2\\ \dot{x}_2 &= -3x_1 - 2x_2^2\end{aligned}$$

is an autonomous nonlinear system since the right-hand sides of the equations do not depend explicitly on t.

Example 2.3 ***(Variable-Mass System)***
A physical example of a nonautonomous system is a *variable-mass system*, such as a rocket whose mass decreases as it burns fuel. In this case, Newton's second law, $f = ma$, must be modified to account for the momentum of the mass ejected from the rocket [56]. The modified equation is of the form

$$f = ma + \frac{dm}{dt}v_{\text{rel}},$$

where $v_{\text{rel}} = v - u$ is the relative velocity of the ejected mass with respect to the center of mass of the object.

We can illustrate a variable-mass system using the harmonic oscillator. Suppose that the mass M in Equation (2.6) is a "leaky bucket," such as a bucket of water with a hole in it. Following [58], we model the variable mass $M(t)$ as

$$M(t) = \begin{cases} M_b + M_w(1 - at)^2, & \text{if } 0 < t < 1/a, \\ M_b, & \text{if } t \geq 1/a, \end{cases}$$

where M_b is the mass of the empty bucket, M_w is the initial mass of the water at time $t = 0$, and

$$a = f\sqrt{\frac{g}{2h_0}},$$

where $f = A_h/A_b$ is the ratio of the cross-sectional area A_h of the hole and the cross-sectional area A_b of the bucket, and h_0 is the initial height of the water in the bucket. The modified harmonic oscillator equation is now

$$M(t)\ddot{y} + ky = -\frac{dm}{dt}v_{\text{rel}} + M(t)g, \tag{2.11}$$

where

$$y = x - M(0)g/k$$

and $M(0)g/k$ is the initial equilibrium position. Equation (2.11) is now a nonautonomous system due to the dependence of the mass on time.

Figure 2.3 shows a typical response of the system. The amplitude and frequency of oscillation change as the water drains from the bucket. When all the water has drained from the bucket the solution is a pure oscillation with natural frequency $\sqrt{k/M_b}$.

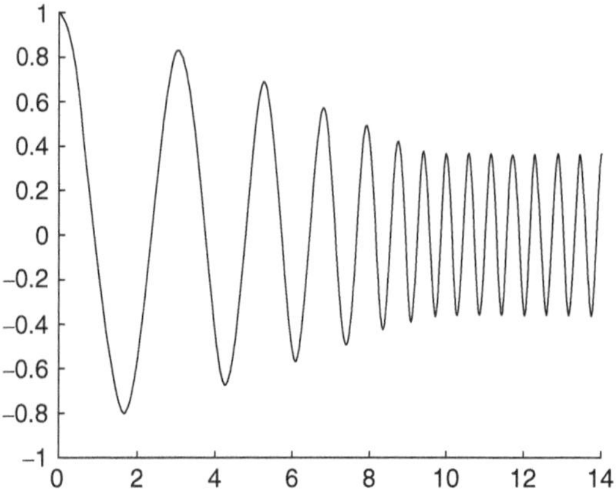

Figure 2.3 Oscillation of a leaky bucket. As the mass of the water in the bucket decreases, the frequency of oscillation increases while the amplitude decreases. After time $t = 1/a$, the bucket is empty and the oscillations have constant amplitude with natural frequency equal to $\sqrt{k/M_b}$.

2.3 Trajectories and Phase Portraits

Initial-Value Problem

Consider a second-order system of the form

$$\dot{x} = f(t, x), \quad x \in \mathbb{R}^2. \tag{2.12}$$

Equation (2.12), together with an initial condition $x(t_0) = x_0$, is called an *initial-value problem.*

Definition 2.1 ***(Trajectory)*** A trajectory of the system (2.12) with initial condition $x(t_0) = x_0$ at the initial time t_0, is a differentiable function $x(t) : \mathcal{J} \to \mathbb{R}^2$ with $\mathcal{J} \subset \mathbb{R}$, a time interval, such that for all $t \in \mathcal{J}$,

$$\frac{d}{dt}x(t) = f(t, x(t)), \quad \text{and } x(t_0) = x_0.$$

A trajectory is also called an *orbit*. Other notation for the solution trajectory, starting at x_0 at time t_0, include $x(t, t_0, x_0)$, or $x_t(t_0, x_0)$ if we want to emphasize the initial conditions. In this chapter we will assume that the time interval $\mathcal{J} = [t_0, \infty)$. The case where the interval $\mathcal{J}$ is restricted to a finite time interval will be discussed in the next chapter. A trajectory is therefore a curve in $\mathbb{R}^2$, parameterized by time, such that the vector field f is tangent to the curve at each point, as shown in Figure 2.4.

In the remainder of this chapter we restrict our attention to autonomous systems, which means that the vector $f(x)$ is a constant vector at each point $x \in \mathbb{R}^2$.

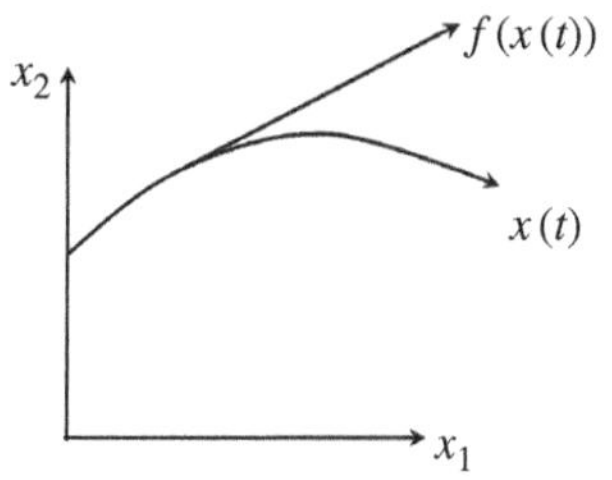

Figure 2.4 A solution trajectory in $\mathbb{R}^2$. The vector field f is tangent to the trajectory at each point.

Definition 2.2 ***(Phase Portrait)*** The phase portrait (or flow) is the collection of all trajectories in $\mathbb{R}^2$, in other words, the solutions starting at all possible initial conditions in $\mathbb{R}^2$. Phase portraits give qualitative, geometric information to visualize the behavior of the system.

Example 2.4 ***(Harmonic Oscillator Revisited)***
The harmonic oscillator system in state space is given in terms of the state variables $x_1 = y, x_2 = \dot{y}$ as

$$\begin{aligned}\dot{x}_1 &= x_2\\ \dot{x}_2 &= -\omega^2 x_1.\end{aligned}$$

In this case, the vector field f is given by

$$f = \begin{bmatrix} x_2 \\ -\omega^2 x_1 \end{bmatrix} = \begin{bmatrix} 0 & 1 \\ -\omega^2 & 0 \end{bmatrix}\begin{bmatrix} x_1 \\ x_2 \end{bmatrix}.$$

As an example, the solution with the initial conditions $x_1(0) = r, x_2(0) = 0$ is then

$$x(t) = \begin{bmatrix} r\cos(t) \\ -r\sin(t) \end{bmatrix}.$$

Since $x_1^2(t) + x_2^2(t) = r^2$ for all t, the solution trajectory in state space lies on the circle of radius r in $\mathbb{R}^2$. This is illustrated in Figure 2.5.

If we graph the solutions for various initial conditions, we can construct the *phase portrait* of the harmonic oscillator. The trajectories lie on concentric circles in $\mathbb{R}^2$ as shown in Figure 2.6.

Example 2.5 ***(Simple Pendulum)***
Next, we consider the simple pendulum shown in Figure 2.7, which is a nonlinear system with periodic solutions. The equation of motion of a simple pendulum of length ℓ is given by [34]

$$\ddot{\theta} + \frac{g}{\ell}\sin\theta = 0. \tag{2.13}$$

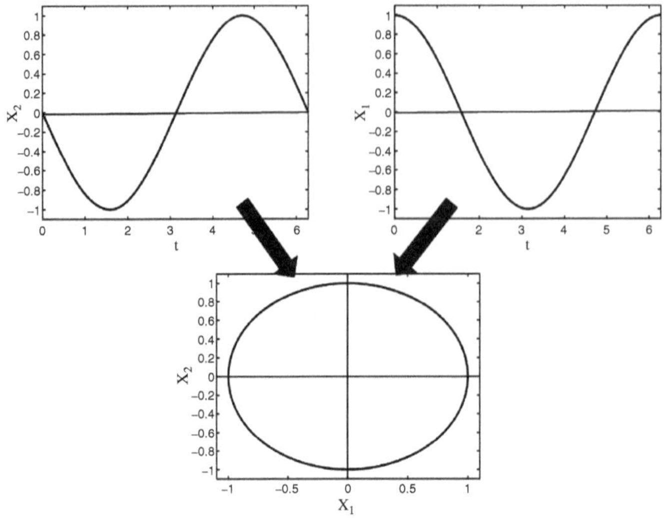

Figure 2.5 A single trajectory for the harmonic oscillator. The two state variable solutions $x_1(t) = r\cos(t)$, $x_2(t) = -r\sin(t)$ can be graphed in state space as points on the circle of radius r parameterized by t.

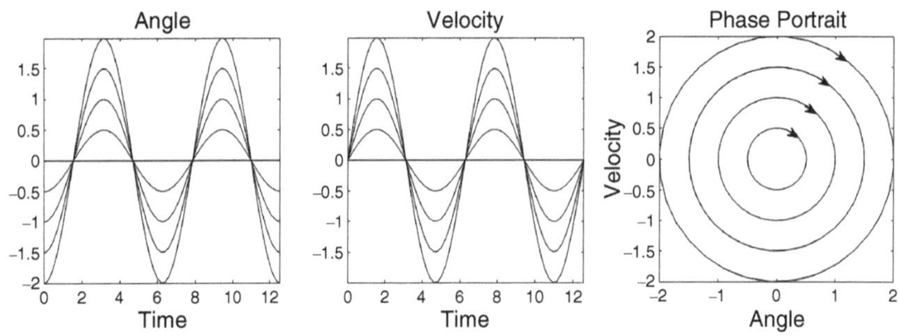

Figure 2.6 Multiple trajectories in time (left, center) and a portion of the phase portrait for the harmonic oscillator (right).

The term "simple" refers to the fact that the simple pendulum has a single degree of freedom. Its importance as a model is significant. First, note that Equation (2.13) does not depend on the mass of the pendulum bob. In Aristotelian physics, prior to the work of Galileo, it was thought that heavier objects fall faster than lighter objects. Galileo's experimental study of the pendulum, beginning in the early seventeenth century, showed correctly that the acceleration due to gravity is independent of the mass. On a deeper level, this fact is related to the

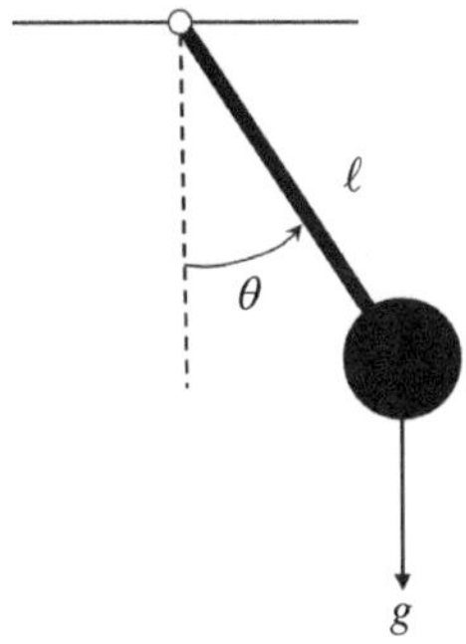

Figure 2.7 A simple pendulum with length ℓ rotating in a vertical plane under the influence of gravity.

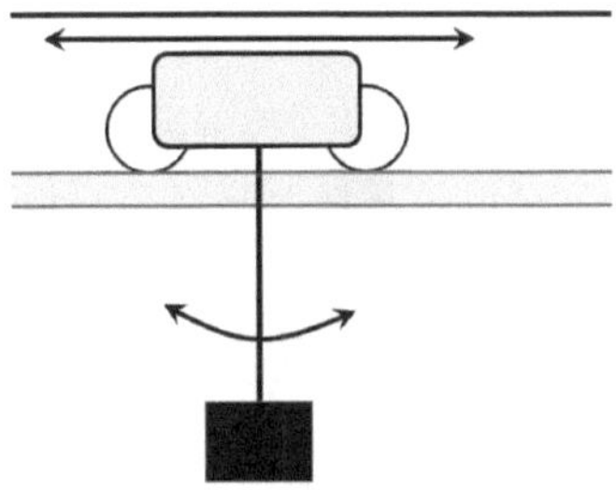

Figure 2.8 The load transported by an overhead crane is modeled as a hanging pendulum. Reprinted from [64] with permission.

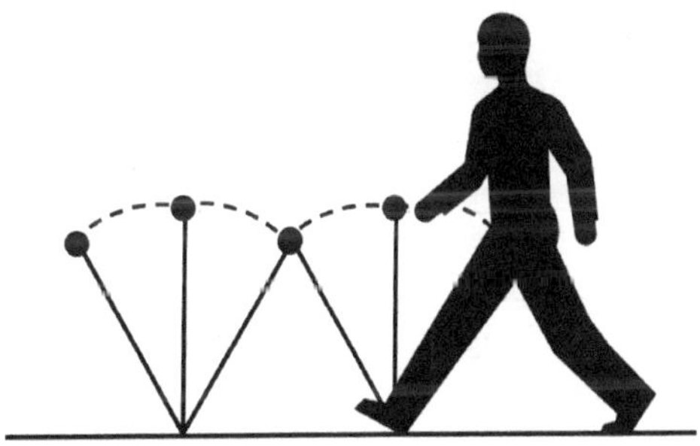

Figure 2.9 Inverted pendulum model for bipedal locomotion. The stance leg behaves as an inverted pendulum and the position of the hip traces the endpoint motion represented by the dotted line. Reprinted from [64] with permission.

question of the equivalence of inertial and gravitational mass, which was not fully resolved until the work of Einstein in the twentieth century.

The pendulum is used to model many physical phenomena. For example, the load transported by an overhead crane, shown in Figure 2.8, swings back and forth as a hanging pendulum. The simple pendulum can also be used as a model for fuel slosh in rockets and other vehicles, and the rolling motion of ships in waves.

An inverted pendulum, as opposed to a hanging pendulum, is widely used to model bipedal locomotion as shown in Figure 2.9.

Another example of an inverted pendulum is a rocket ascending vertically, where the direction of the thrust produced by the rocket engine is controlled to

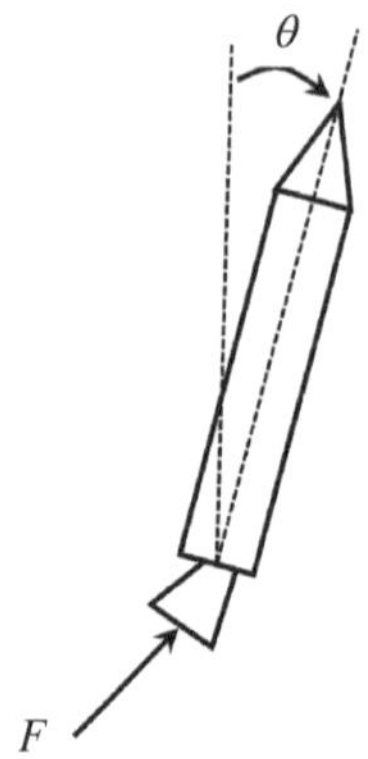

Figure 2.10 A rocket ascending vertically is modeled as an inverted pendulum. The vectored thrust of the rocket motor powers the ascent and also maintains balance about the vertical position. Reprinted from [64] with permission.

Figure 2.11 The Segway Human Transporter. The controller for the Segway tries to keep the vehicle balanced as an inverted pendulum.

keep the proper orientation of the rocket (Figure 2.10). The recent achievement by the company SpaceX to land its rockets after a flight is a remarkable achievement requiring an understanding of the dynamics and control of the inverted pendulum.

Finally, the control of the Segway Human Transporter (Figure 2.11) is based on an inverted pendulum model. The controller attempts to keep the Segway vertical so, as the rider leans forward, the Segway must move forward to try to keep the pendulum pivot directly below the handlebar.

With state variables $x_1 = \theta, x_2 = \dot{\theta}$, the state-space equations for the simple pendulum are

$$\begin{aligned}\dot{x}_1 &= x_2 \\ \dot{x}_2 &= -\frac{g}{\ell}\sin(x_1).\end{aligned}$$

The vector field defining this system is therefore

$$f(x) = \begin{bmatrix} x_2 \\ -\frac{g}{\ell}\sin(x_1) \end{bmatrix}. \tag{2.14}$$

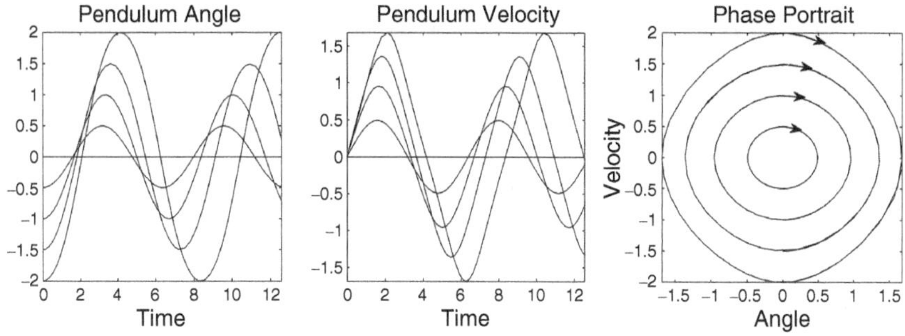

Figure 2.12 Solution trajectories and a portion of the phase portrait for the simple pendulum. The period of oscillation depends on the amplitude, in other words, on the initial conditions.

Because the simple pendulum is a nonlinear system, the vector field $f(x)$ in Equation (2.14) cannot be expressed as a matrix times x. The solution of the simple pendulum equation involves elliptic integrals (which we will not study here) and therefore cannot be expressed in closed form. The Matlab function `dsolve` will not return a solution to an initial-value problem for this system. In these cases, we can compute the trajectories (Figure 2.12) by numerical integration using, for example `ode45`, which is a Matlab numerical integration function. We will treat numerical solutions of differential equations in detail in Chapter 7.

For the harmonic oscillator, the period of oscillation is independent of the amplitude, i.e., of the initial conditions. For the simple pendulum, the period depends on the initial conditions. Starting the pendulum at a larger angle will produce a longer period oscillation.

2.4 The Direction Field

In this section we introduce the direction field, which provides a way to visualize the flow of a system and sketch the phase portrait. The significance of the direction field is that we can gain qualitative information about the behavior of a system based only on the defining differential equations, i.e., without having to solve these equations.

Definition 2.3 ***(Direction Field)*** Given a vector field $f = \begin{bmatrix} f_1 \\ f_2 \end{bmatrix} : \mathbb{R}^2 \to \mathbb{R}^2$, the direction field assigns a unit vector $\bar{f}$ at each point $x \in \mathbb{R}^2$ with direction

$$\theta_f = \tan^{-1}\left(\frac{f_2(x_1, x_2)}{f_1(x_1, x_2)}\right).$$

The vector $\bar{f}$ is just a scaled version of the vector field at each point. The arrows in the direction field are tangent to the trajectories of the differential equation. Note that it is not necessary that we assign a unit vector at each point. We may scale the direction field to any convenient value at each point in order to facilitate visualization of the flow.

Example 2.6 ***(Sketching the Direction Field)***
The Matlab command `quiver` can be used to sketch direction fields. We first create a grid of points in the plane and then the `quiver` function attaches an arrow at each point. The directions of the arrows are determined by the vector field defining the differential equation.

Direction Field for the Harmonic Oscillator

```
[x,y]=meshgrid(-2:0.25:2,-2:0.25:2);
u = y;         % state equations for the harmonic oscillator
v =-x;
L=sqrt(u.^2 + v.^2); % normalize the length of the arrows
q=quiver(x,y,u./L,v./L,0.4);
hold on
q.Color='black';
q.AutoScale='on';
q.MaxHeadSize=0.2;
axis([-2 2 -2 2],'square');
```

The direction field for the harmonic oscillator produced by the above script is shown in Figure 2.13.

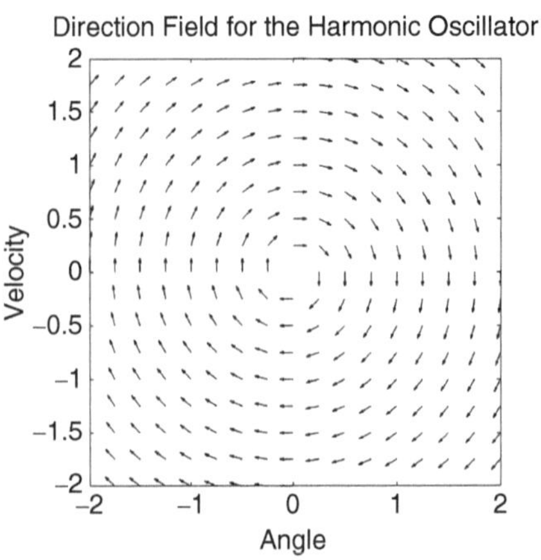

Figure 2.13 Direction field for the harmonic oscillator, which shows qualitatively the oscillatory nature of the system. Individual solution trajectories can be drawn so that the arrows are tangent to trajectories.

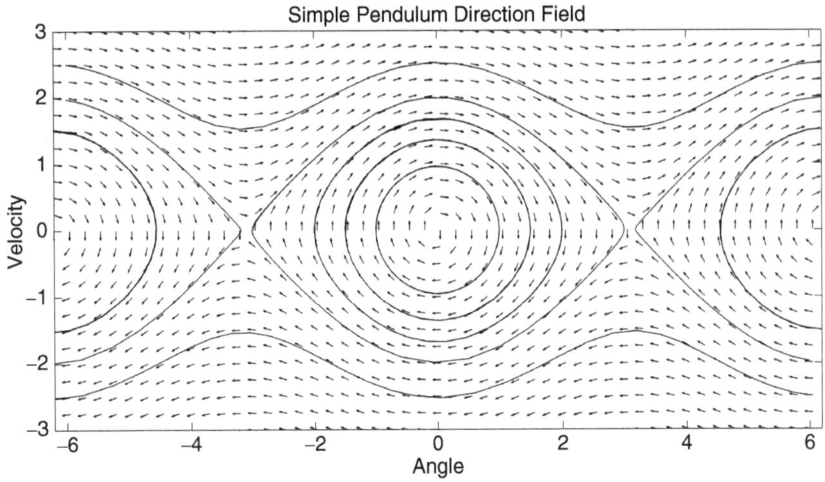

Figure 2.14 Direction field and a portion of the phase portrait for the simple pendulum generated by the above Matlab script. The individual trajectories are generated using the Matlab ordinary differential equation solver `ode45`.

We can see the general trend of the harmonic oscillator behavior as oscillating clockwise around the origin without solving the equation.

Example 2.7 ***(Direction Field for the Simple Pendulum)***
Using the Matlab script below, we plot the direction field and a few trajectories for the simple pendulum in Figure 2.14.

Direction Field for the Simple Pendulum

```
[x,y]=meshgrid(-6.2:0.25:6.2,-3:0.25:3);
u=y;
v=-sin(x);
L = sqrt(u.^2 + v.^2);
q=quiver(x,y,u./L,v./L,0.4);
hold on
q.Color='black';
q.AutoScale='on';
q.MaxHeadSize=0.2;
axis([-3.2 3.2 -2 2]);
%
f=@pend;
```

(Continued)

(Continued)

```
xic = [-6.0 -6.0 -6.0 -6 -2.98 -1 -1.5 6 6 6 -6 -2];
yic = [2.5 1.5 1.98 -2.5 0 0 0 -1.5 -1.98 -2.5 -1.5 0];
for ic = [xic;yic]
[t,z] = ode45(f,[0,22],[ic]);
plot(z(:,1),z(:,2),'k');
hold on
end
function dxdt = pend(t,z) % Simple Pendulum Equation
x1=z(1);
x2=z(2);
dxdt = [x2;-sin(x1)];
end
```

From the direction field we can deduce that for initial conditions close to the origin, the simple pendulum behaves as a harmonic oscillator. For initial conditions farther away from the origin, the behavior of the simple pendulum is much different than that of the harmonic oscillator.

2.5 Equilibria

A large part of systems analysis deals with studying systems in *equilibrium*. For example, in a power system the goal is to maintain a constant voltage and frequency of transmission. In aircraft control the goal is to maintain a constant altitude and speed. Equilibrium solutions of nonlinear systems take several forms, including:

- Constant solutions (fixed points)
- Periodic solutions (oscillatory solutions)
- Limit cycles (isolated periodic solutions)
- Strange attractors (chaotic solutions).

The definition of an equilibrium point or fixed point for a dynamical system that we give next is stated for two-dimensional systems, but the notion of equilibrium for n-dimensional systems is identical.

Definition 2.4 ***(Equilibrium Point)*** A vector $\overline{x} \in \mathbb{R}^2$ is an *equilibrium point* (also called a *stationary point* or *fixed point*) for the dynamical system $\dot{x} = f(x)$ if and only if $f(\overline{x}) = 0$.

The significance of an equilibrium point is that, if $\bar{x} \in \mathbb{R}^2$ is an equilibrium, then the initial-value problem,

$$\dot{x} = f(x), \quad x(0) = \bar{x},$$

has the constant solution

$$x(t) = \bar{x}, \quad \text{for } t \geq 0.$$

In order to see this, a simple calculation shows that if $x(t) = \bar{x}$, then

$$\frac{dx(t)}{dt} = \frac{d\bar{x}}{dt} = 0 = f(\bar{x}) = f(x(t)).$$

Intuitively, if the system "starts at an equilibrium $\bar{x}$", it remains at $\bar{x}$ for all time. The solution $x(t) = \bar{x}$ is called the *equilibrium solution* or *null solution* of the system.

Definition 2.5 ***(Isolated Equilibrium)*** A vector $\bar{x}$ is an *isolated equilibrium* if and only if

1. $\bar{x}$ is an equilibrium and
2. there exists $\epsilon > 0$ such that $f(x) \neq 0$ for $||x - \bar{x}|| < \epsilon$.

The above definition means that, if an equilibrium point $\bar{x}$ is not isolated, then there are other equilibrium points that are arbitrarily close to $\bar{x}$. For a linear system, $\dot{x} = Ax$, the origin $\bar{x} = 0$ is always an equilibrium point. If the matrix A is nonsingular, then $\bar{x} = 0$ is the only equilibrium point (Problem 2.10).

On the other hand, if A is singular, there are infinitely many, nonisolated, equilibrium points forming a subspace, namely the *null space* of the matrix A. Thus, for linear systems, there are only two possibilities, a unique isolated fixed point, or infinitely many nonisolated fixed points.

Example 2.8 ***(Nonisolated Equilibria)***
Consider the second-order linear system

$$\begin{aligned}\dot{x}_1 &= 2x_1\\ \dot{x}_2 &= -x_1.\end{aligned}$$

Then points of the form $(x_1, x_2) = (0, c)$, for any constant c are equilibrium points. Thus the entire x_2-axis in $\mathbb{R}^2$ consists of equilibrium points, and hence they are nonisolated. Note that the A matrix for the above system is $A = \begin{bmatrix} 2 & 0 \\ -1 & 0 \end{bmatrix}$, which is easily seen to be a singular matrix.

In contrast to the case of linear systems, nonlinear systems may have multiple isolated equilibrium points, either a finite number or infinitely many.

Example 2.9 Consider the nonlinear system

$$\dot{x}_1 = x_2$$
$$\dot{x}_2 = -x_2 + x_1(x_1 - 1)(x_1 + 1).$$

From the first equation we see that any equilibrium $\bar{x}$ must satisfy $\bar{x}_2 = 0$ and, therefore, from the second equation, we obtain $\bar{x}_1 = 0, 1, -1$. Therefore, there are exactly three isolated equilibrium points, namely, $(0, 0), (1, 0), (-1, 0)$.

Example 2.10 ***(Equilibria for the Simple Pendulum)***
Recall the simple pendulum equations in state space

$$\dot{x}_1 = x_2$$
$$\dot{x}_2 = -\frac{g}{\ell}\sin(x_1).$$

Setting the right-hand side of the equations to zero, we see that all points with $\bar{x}_2 = 0, \bar{x}_1 = k\pi,\ k = 0, \pm 1, \ldots$, are isolated equilibrium points (Figure 2.15).

If we identify 0 with 2π, 4π, etc., then there are two isolated fixed points, namely

$$x_0 = \begin{bmatrix} 0 \\ 0 \end{bmatrix}, \begin{bmatrix} \pi \\ 0 \end{bmatrix}.$$

Physically, these fixed points correspond to the pendulum at rest in the downward and inverted positions, respectively (Figure 2.16).

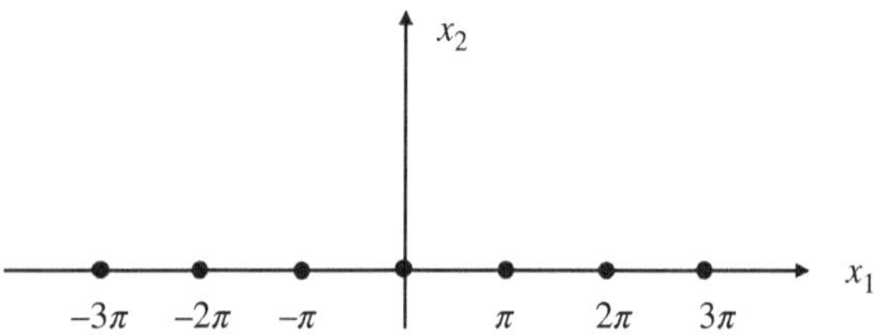

Figure 2.15 The simple pendulum has infinitely many isolated equilibrium points in $\mathbb{R}^2$.

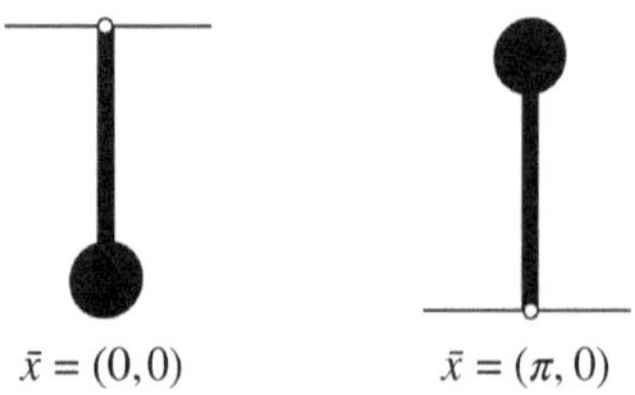

Figure 2.16 Unique equilibrium states of the simple pendulum by identifying 0 and 2π.

2.6 Linear Systems

For second-order linear systems, we can give a complete characterization of the phase portraits by examining the eigenvalues and eigenvectors of the system matrices. To this end, given a second-order linear system

$$\dot{x} = Ax, \quad x \in \mathbb{R}^2, \tag{2.15}$$

we write this as

$$\dot{x}_1 = ax_1 + bx_2$$
$$\dot{x}_2 = cx_1 + dx_2,$$

so that the coefficient matrix A is given by

$$A = \begin{bmatrix} a & b \\ c & d \end{bmatrix}. \tag{2.16}$$

The eigenvalues of the matrix A are easily shown to be $\lambda = \frac{1}{2}p \pm \frac{1}{2}\sqrt{p^2 - 4q}$, where $p = a + d$ is the trace of the coefficient matrix A and $q = ad - bc$ is the determinant of A (Problem 2.7). The term $p^2 - 4q$ is called the *discriminant* and its sign determines whether or not the eigenvalues of A are real or complex.

Definition 2.6 ***(Hyperbolic Equilibrium)*** The equilibrium $\bar{x} = 0$ for the linear system (2.15) is called *hyperbolic* if and only if the eigenvalues of A do not have real part equal to zero.

In particular, if the equilibrium $\bar{x}$ is hyperbolic, then $\lambda = 0$ is not an eigenvalue, which implies that A is nonsingular and $\bar{x} = 0$ is the unique equilibrium point. Hyperbolic equilibria play a crucial role in the study of both linear and nonlinear systems, as we shall see.

Theorem 2.1 *Given any 2×2 matrix A, there is a nonsingular matrix*

$$T = \begin{bmatrix} T_{11} & T_{12} \\ T_{21} & T_{22} \end{bmatrix}$$

such that the matrix $\bar{A} = T^{-1}AT$ has one of the following forms:

$$\bar{A}_1 = \begin{bmatrix} \lambda_1 & 0 \\ 0 & \lambda_2 \end{bmatrix}, \quad \bar{A}_2 = \begin{bmatrix} \lambda & 1 \\ 0 & \lambda \end{bmatrix}, \quad \bar{A}_3 = \begin{bmatrix} \alpha & \beta \\ -\beta & \alpha \end{bmatrix},$$

where λ_1, λ_2, λ, and $\alpha \pm j\beta$ are, respectively, the eigenvalues of the matrix $\bar{A}_1, \bar{A}_2, \bar{A}_3$. In the matrix A_1 the eigenvalues are real and distinct, $\lambda_1 \neq \lambda_2$, or else $A = \lambda I$, a multiple of the identity matrix. In the matrix A_2 the eigenvalues are real and repeated. In the matrix A_3 the eigenvalues are complex conjugate, $\alpha \pm j\beta$, where $j = \sqrt{-1}$.

The columns of the matrix T are determined by the eigenvectors of the matrix A, corresponding to the respective eigenvalues. The proof of this theorem can be found in many introductory linear algebra texts. We will illustrate it using examples.

If we now change variables by setting $x = Tz$, then, from $\dot{x} = Ax$, we have

$$\dot{z} = T^{-1}\dot{x} = T^{-1}ATz = \bar{A}_i z, \quad i = 1, 2, \text{or } 3.$$

The components of the vector z are called *modal coordinates*. We can classify all hyperbolic fixed points of the system (2.15) in terms of the eigenvalues of $\bar{A}_i$ and use the modal coordinate representation to sketch the corresponding phase portraits.

Case 1. Suppose the eigenvalues λ_1 and λ_2 of A are nonzero, real, and distinct. Then, in the modal coordinates, the system equations are determined from $\bar{A}_1$ as

$$\dot{z}_1 = \lambda_1 z_1$$
$$\dot{z}_2 = \lambda_2 z_2,$$

which have solutions

$$z_1(t) = e^{\lambda_1 t} z_1(0),$$
$$z_2(t) = e^{\lambda_2 t} z_2(0).$$

The fixed point $\overline{x} = 0$ is called a

1. *stable node* if λ_1 and λ_2 are both negative,
2. *unstable node* if λ_1 and λ_2 are both positive,
3. *saddle node* or *saddle point* if λ_1 and λ_2 have opposite signs.

Representative phase portraits of these cases are shown in modal coordinates in Figure 2.17.

The directions of the z_1 and z_2 axes in the x_1–x_2 plane are determined by eigenvectors corresponding to λ_1 and λ_2, respectively. The z_1 and z_2 axes are called *eigensubspaces* and are invariant subspaces, meaning that solutions starting on the eigensubspaces at $t = 0$ remain on them for all time $t > 0$.

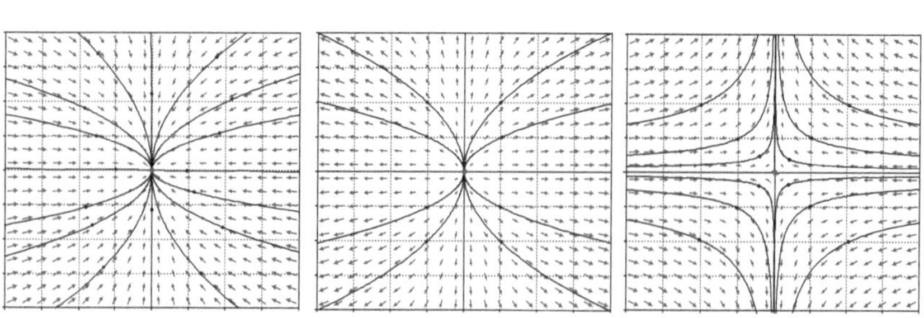

Figure 2.17 Phase portraits in modal coordinates for a stable node (left); unstable node (center); saddle node (right).

The solution of the equation in the original x-coordinates is determined from the modal coordinates as

$$x = Tz = \begin{bmatrix} T_{11}z_1 + T_{12}z_2 \\ T_{21}z_1 + T_{22}z_2 \end{bmatrix}.$$

Example 2.11 Consider the system

$$\begin{aligned} \dot{x}_1 &= x_2 \\ \dot{x}_2 &= -2x_1 - 3x_2. \end{aligned}$$

One can check that the eigenvalues are $\lambda_1 = -1$, $\lambda_2 = -2$, and, hence, the fixed point $x = 0$ is a stable node. To find the eigenvectors, we solve the equation

$$(A - \lambda_i I)x = 0, \quad i = 1, 2.$$

For $\lambda_1 = -1$, we have

$$\begin{bmatrix} 1 & 1 \\ -2 & -2 \end{bmatrix} \begin{bmatrix} x_1 \\ x_2 \end{bmatrix} = \begin{bmatrix} 0 \\ 0 \end{bmatrix},$$

and so $(x_1, x_2) = (1, -1)$ is a solution. A similar calculation shows that $(1, -2)$ is an eigenvector for $\lambda_2 = -2$. The transformation matrix T is now formed from these eigenvectors as

$$T = \begin{bmatrix} 1 & 1 \\ -1 & -2 \end{bmatrix},$$

and a direct calculation shows that

$$T^{-1}AT = \begin{bmatrix} -1 & 0 \\ 0 & -2 \end{bmatrix}.$$

The phase portraits in the modal coordinates and in the original coordinates are shown in Figure 2.18.

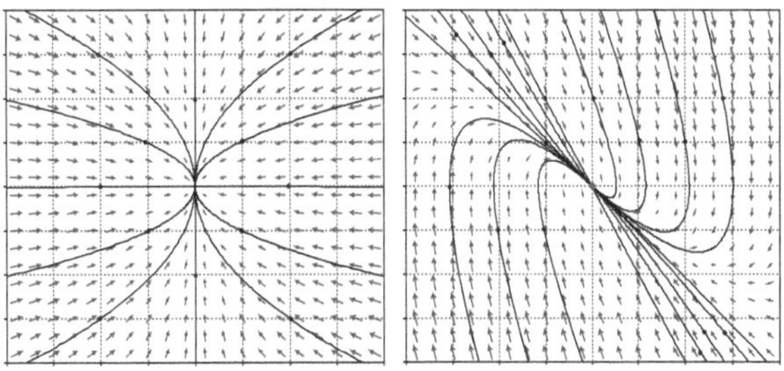

Figure 2.18 Phase portraits of the stable node in modal coordinates (left) and in the original coordinates (right).

Example 2.12 Next, consider the system

$$\dot{x}_1 = 2x_1 + 2x_2$$
$$\dot{x}_2 = -2x_1 - 3x_2.$$

One can check that the eigenvalues are $\lambda_1 = -2$, $\lambda_2 = +1$, and therefore the fixed point $x = 0$ is a saddle node. It is straightforward to show that

$$T = \begin{bmatrix} 1 & -2 \\ -2 & 1 \end{bmatrix}, \quad T^{-1}AT = \begin{bmatrix} -2 & 0 \\ 0 & 1 \end{bmatrix}.$$

The phase portraits of this system in the original coordinates and modal coordinates are shown in Figure 2.19.

Note that all solutions of the saddle node equation are unstable except those with $z_2(0) = 0$. The z_1-axis, corresponding to the eigenvalue -2, is called the *stable eigensubspace*. The z_2-axis, corresponding to the eigenvalue $+1$, is called the *unstable eigensubspace.*

Case 2. Suppose $\lambda_1 = \lambda_2 = \lambda$ are real, repeated eigenvalues. In this case there are two possibilities. Either $A = \lambda I$, in which case the fixed point is a stable or unstable node depending on the sign of λ, or else there exists a nonsingular matrix T such that

$$\bar{A} = T^{-1}AT = \begin{bmatrix} \lambda & 1 \\ 0 & \lambda \end{bmatrix}. \tag{2.17}$$

The matrix $\bar{A}$ in Equation (2.17) is called the *Jordan form*. The fixed point is a *Jordan node* and is stable or unstable, depending on the sign of λ. In this case there is only one invariant (eigen)subspace, $z_2 = 0$.

Example 2.13 Consider the system

$$\dot{x}_1 = -2x_1 + x_2$$
$$\dot{x}_2 = -x_1.$$

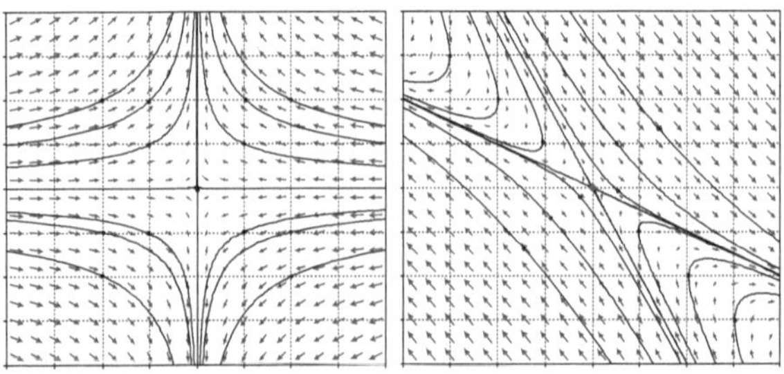

Figure 2.19 Phase portraits of the saddle node in modal coordinates (left) and in the original coordinates (right).

In this case $A = \begin{bmatrix} -2 & 1 \\ -1 & 0 \end{bmatrix}$ and the eigenvalues are repeated at $\lambda = -1$. It is easily seen that $x = \begin{bmatrix} x_1 \\ x_2 \end{bmatrix} = \begin{bmatrix} 1 \\ 1 \end{bmatrix}$ is an eigenvector for $\lambda = -1$. We cannot find another eigenvector for λ independent of x and so we look for a so-called *generalized eigenvector*. This is a vector y satisfying

$$(A - \lambda I)y = x,$$

where x is an eigenvector, in this case

$$\begin{bmatrix} -1 & 1 \\ -1 & 1 \end{bmatrix} \begin{bmatrix} y_1 \\ y_2 \end{bmatrix} = \begin{bmatrix} 1 \\ 1 \end{bmatrix},$$

from which we see that $\begin{bmatrix} y_1 \\ y_2 \end{bmatrix} = \begin{bmatrix} -1 \\ 0 \end{bmatrix}$ is a generalized eigenvector for $\lambda = -1$. So we take

$$T = \begin{bmatrix} 1 & -1 \\ 1 & 0 \end{bmatrix}$$

and compute

$$T^{-1}AT = \begin{bmatrix} -1 & 1 \\ 0 & -1 \end{bmatrix} = \bar{A}.$$

The phase portraits for this example in the original and modal coordinates are shown in Figure 2.20.

Case 3. Suppose λ_1 and λ_2 are complex conjugate eigenvalues,

$$\lambda_1 = \alpha + j\beta, \quad \lambda_2 = \alpha - j\beta.$$

In this case there exists a nonsingular matrix T such that

$$T^{-1}AT = \begin{bmatrix} \alpha & \beta \\ -\beta & \alpha \end{bmatrix}.$$

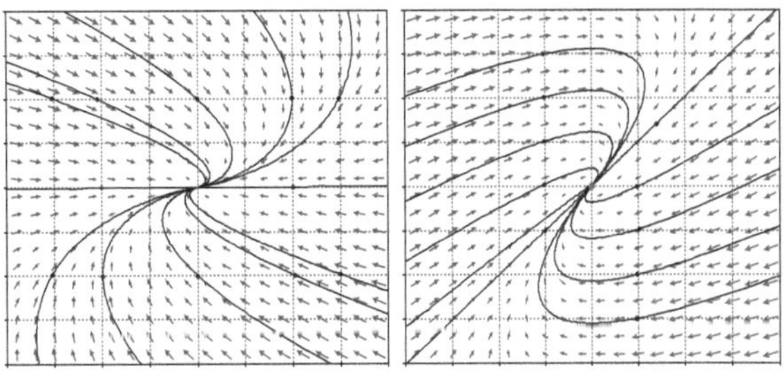

Figure 2.20 Jordan node phase portraits in modal coordinates (left) and in the original coordinates (right).

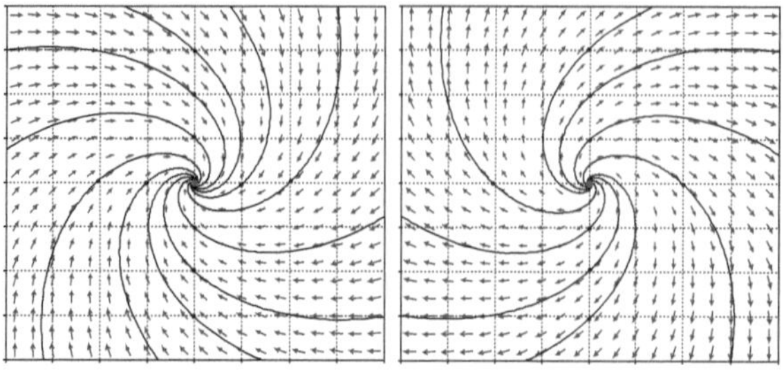

Figure 2.21 Stable focus (sink) left; unstable focus (source) right

The solution trajectories are composed of linear combinations of terms of the form $e^{\alpha t}\cos(\beta t)$ and $e^{\alpha t}\sin(\beta t)$. The fixed point is called a *stable focus* or *sink* if $\alpha < 0$ and an *unstable focus* or *source* if $\alpha > 0$ (Figure 2.21).

Example 2.14 Consider the system

$$\begin{aligned}\dot{x}_1 &= x_2\\ \dot{x}_2 &= -2x_1 - 2x_2.\end{aligned}$$

One can check that the eigenvalues of this system are $\lambda_1 = -1 + j$, $\lambda_2 = -1 - j$, so the fixed point is a sink. With complex eigenvalues, one needs to solve a complex eigenvector equation. A complex eigenvector for $\lambda = \alpha + j\beta$ is a vector $x = u + jv$ such that

$$\begin{aligned}A(u + jv) &= (\alpha + j\beta)(u + jv)\\ &= \alpha u - \beta v + j(\alpha v + \beta u).\end{aligned}$$

Equating the real and imaginary parts from the above equations leads to

$$\begin{aligned}Au &= \alpha u - \beta v,\\ Av &= \alpha v + \beta u.\end{aligned}$$

Solving these equations for u and v gives us the required transformation $T = [u\ v]$ such that

$$T^{-1}AT = \begin{bmatrix} \alpha & \beta \\ -\beta & \alpha \end{bmatrix}.$$

Thus, in the given example, let $x = u + jv$. Then, with $\lambda = -1 + j$, the equation $(A - \lambda I)x = 0$ becomes

$$\begin{bmatrix} 1-j & 1 \\ -2 & -1-j \end{bmatrix}\begin{bmatrix} x_1 \\ x_2 \end{bmatrix} = \begin{bmatrix} 0 \\ 0 \end{bmatrix},$$

which implies that $(1 - j)x_1 + x_2 = 0$. A solution is given by

$$\begin{bmatrix} x_1 \\ x_2 \end{bmatrix} = \begin{bmatrix} 1 \\ -1 \end{bmatrix} + j\begin{bmatrix} 0 \\ 1 \end{bmatrix} = u + jv.$$

Therefore,

$$T = \begin{bmatrix} 1 & 0 \\ -1 & 1 \end{bmatrix}, \quad \text{and} \quad T^{-1}AT = \begin{bmatrix} -1 & 1 \\ -1 & -1 \end{bmatrix}.$$

Remark 2.2 If $\alpha = 0$, so that the eigenvectors are purely imaginary, $\pm j\beta$, then the solutions are periodic and the fixed point is called a *center*, as in the harmonic oscillator example. Note that a center is not a hyperbolic fixed point.

The remaining case occurs when the determinant $\det(A) = 0$, so that $\lambda = 0$ is an eigenvalue of A. In this case, the matrix A is generically in the form

$$A = \begin{bmatrix} a & b \\ 0 & 0 \end{bmatrix}, \quad a, b \in \mathbb{R}.$$

These are degenerate, nonhyperbolic systems that have infinitely many nonisolated equilibrium points. These systems are of less interest in practical applications and so we will omit the sketches of the phase portraits. Table 2.1 summarizes the classification of fixed points according to the location of the eigenvalues of the system matrix in the complex plane.

Numerical Computation with `eig` and `jordan` Commands

The computation of eigenvalues and eigenvectors inherently involves working with singular matrices, which makes the computations extremely sensitive to errors and roundoff. For example, consider the following matrix A, which has repeated eigenvalues and is in Jordan form:

$$A_1 = \begin{bmatrix} 2 & 1 \\ 0 & 2 \end{bmatrix}.$$

Table 2.1 Summary of isolated fixed points for nonzero eigenvalues. The case where λ_1 or $\lambda_2 = 0$ is not shown. These correspond to nonisolated fixed points

Eigenvalues of A	Type of fixed point
$\lambda_1 < 0, \lambda_2 < 0$	stable node
$\lambda_1 > 0, \lambda_2 > 0$	unstable node
$\lambda_1\lambda_2 < 0$	saddle node
λ_1, λ_2 complex, $\mathrm{Re}(\lambda_i) < 0$	stable focus
λ_1, λ_2 complex, $\mathrm{Re}(\lambda_i) > 0$	unstable focus
λ_1, λ_2 complex, $\mathrm{Re}(\lambda_i) = 0$	center (nonhyperbolic)

An arbitrarily small change in one of the eigenvalues results in a matrix with distinct eigenvalues and which is therefore diagonalizable.

For an $n \times n$ matrix A, the Matlab statement `[T,D]=eig(A)` generates an $n \times n$ diagonal matrix `D` and an $n \times n$ matrix `T` such that `A*T=T*D`. The diagonal elements of D are the eigenvalues of A and the columns of T are unit eigenvectors corresponding to the eigenvalues of A. Because of the sensitivity of numerical eigenvalue/eigenvector calculations, symbolic computation in Matlab may be used for these problems in the Matlab script below.

Eigenvalue/Eigenvector Computation

```
A=sym([0 1;-2 -3]) % symbolic representation of A
[T,D]=eig(A)       % eigenvalue/eigenvector calculation
Abar = T\A*T       % check that T\A*T = D
```

Running the script gives

$$A = \begin{pmatrix} 0 & 1 \\ -2 & -3 \end{pmatrix}, \quad T = \begin{pmatrix} -\frac{1}{2} & -1 \\ 1 & 1 \end{pmatrix}, \quad D = \begin{pmatrix} -2 & 0 \\ 0 & -1 \end{pmatrix}.$$

For matrices with repeated eigenvalues Matlab provides the command `jordan` as illustrated in the script below. A more detailed discussion of the Jordan form of a matrix of arbitrary dimension is included in Chapter 3.

Matlab script to compute the Jordan form

```
A=sym([-4 -9; 4 8]) % A has repeated eigenvalues
[T,D]=jordan(A)     % Jordan form calculation
Abar = T\A*T        % check that T\A*T = D
```

Running this script gives

$$A = \begin{pmatrix} -4 & -9 \\ 4 & 8 \end{pmatrix}, \quad T = \begin{pmatrix} -6 & 1 \\ 1 & 0 \end{pmatrix}, \quad D = \begin{pmatrix} 2 & 1 \\ 0 & 2 \end{pmatrix}.$$

For matrices with complex eigenvalues, the command `eig` will produce complex eigenvectors. To convert these to real-valued matrices, one should use the following commands.

Matlab script for complex eigenvalues/eigenvectors

```
A = [ 0 1; -2 -2]; % complex eigenvalues alpha +-j beta
[V,D] = eig(A);
T=[real(V(:,1)) imag(V(:,1))];
```

One can check that

Matlab script for complex eigenvalues/eigenvectors

```
T\A*T = [alpha beta; -beta alpha].
```

2.7 Linearization of Nonlinear Systems

For linear systems we typically do not distinguish between *local behavior*, i.e., properties that hold for some initial conditions, and *global behavior*, properties that hold for all initial conditions. The case of nonlinear systems is much different and the distinction between local and global properties is important. In this section we discuss the linearization of a nonlinear system in the vicinity of a fixed point. In the case of hyperbolic fixed points, the linearization determines the local properties of the nonlinear system. In other words, if the state of the system is "near" the equilibrium, then the behavior of the linear system will be a good description of the behavior of the nonlinear system. This is made precise by the Hartman–Grobman theorem, which we discuss in Chapter 3.

Why is this important? The importance of this result is that, once again, we can know the nonlinear system response without having to solve the nonlinear equations. It is sufficient to study the linear system approximation near the equilibrium state.

Linearization about a Fixed Point: The Jacobian

Consider the second-order nonlinear system

$$\dot{x} = f(x), \quad x \in \mathbb{R}^2.$$

Definition 2.7 ***(Jacobian)*** The *Jacobian* (or Jacobian matrix) of the vector field $f(x)$ is the matrix of partial derivatives

$$J(x) = \begin{bmatrix} \dfrac{\partial f_1}{\partial x_1} & \dfrac{\partial f_1}{\partial x_2} \\ \dfrac{\partial f_2}{\partial x_1} & \dfrac{\partial f_2}{\partial x_2} \end{bmatrix}.$$

Example 2.15 For the nonlinear system

$$\begin{aligned} \dot{x}_1 &= x_2 \\ \dot{x}_2 &= x_1^2 - x_1 - x_2, \end{aligned} \tag{2.18}$$

the vector field $f(x)$ is given by

$$f(x) = \begin{bmatrix} x_2 \\ x_1^2 - x_1 - x_2 \end{bmatrix}$$

and the Jacobian matrix for this vector field is

$$J(x) = \begin{bmatrix} 0 & 1 \\ 2x_1 - 1 & -1 \end{bmatrix}.$$

Now, suppose $\overline{x} \in \mathbb{R}^2$ is an equilibrium point of a vector field $f(x)$ on $\mathbb{R}^2$. Since $f(\overline{x}) = 0$, the Taylor series of $f(x)$ about $x = \overline{x}$ is given by

$$f(x) = J(\overline{x})(x - \overline{x}) + O(||x - \overline{x}||^2).$$

Set $A = J(\overline{x})$ and $w = x - \overline{x}$, the deviation of the state x from the fixed point $\overline{x}$. Then the linear system

$$\dot{w} = Aw$$

is the called the *linearization* or *linear approximation* of the nonlinear system about the equilibrium point $\overline{x}$.

For the system (2.18), we see that there are two equilibrium points, $\overline{x}_1 = (0, 0)^T$ and $\overline{x}_2 = (1, 0)^T$. Therefore, at the two equilibrium points, we have

$$A_1 = J(\overline{x}_1) = \begin{bmatrix} 0 & 1 \\ -1 & -1 \end{bmatrix}, \quad A_2 = J(\overline{x}_2) = \begin{bmatrix} 0 & 1 \\ 1 & -1 \end{bmatrix}.$$

Thus, with $v = x - \overline{x}_1$ and $w = x - \overline{x}_2$, the two linear systems

$$\dot{v} = A_1 v, \quad \dot{w} = A_2 w$$

represent linear approximations of the nonlinear system about the two equilibrium points $\overline{x}_1$ and $\overline{x}_2$, respectively. The phase portrait of the nonlinear system is shown in Figure 2.22. The phase portraits of the two linear approximations are shown in Figure 2.23. Note that the qualitative behaviors of the linear and nonlinear systems are very similar near the equilibrium points.

Based on our definition of a hyperbolic equilibrium point for a linear system, we can define an equilibrium point for a nonlinear system as hyperbolic if it is

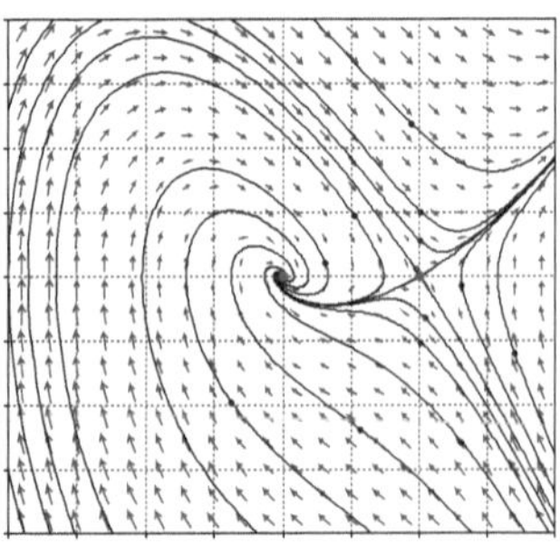

Figure 2.22 Phase portrait of the nonlinear system (2.18).

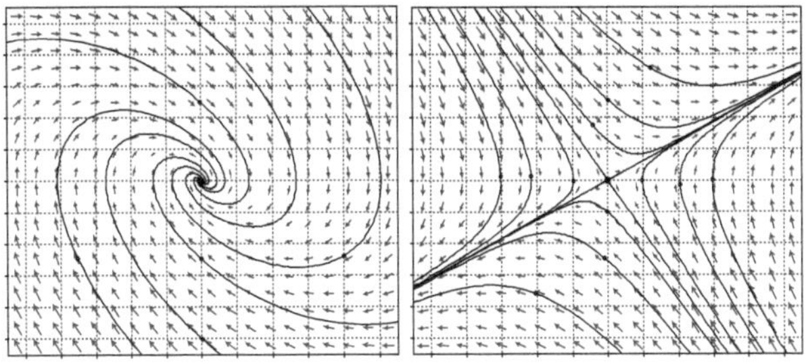

Figure 2.23 Phase portraits of the linear approximations of the nonlinear system (2.18) about the respective equilibrium points.

hyperbolic for the linear approximation of the nonlinear system about the equilibrium point. Specifically, we have the following definition.

Definition 2.8 ***(Hyperbolic Fixed Point for Nonlinear Systems)*** The fixed point $\overline{x}$ of the nonlinear system $\dot{x} = f(x)$ is hyperbolic if and only if the Jacobian $A = \frac{\partial f}{\partial x}(\overline{x})$ has no eigenvalues with zero real part. We say that the equilibrium $\overline{x}$ for the nonlinear system is, respectively, a node, saddle, focus, if the origin $y = 0$ is a node, saddle, focus for the linear approximation.

Example 2.16 Recall the simple pendulum equation

$$\begin{bmatrix} \dot{x}_1 \\ \dot{x}_2 \end{bmatrix} = \begin{bmatrix} x_2 \\ -\frac{g}{\ell}\sin(x_1) \end{bmatrix}.$$

It is straightforward to show that the linear approximations about the two equilibrium points $(0, 0)$ and $(\pi, 0)$ are

$$(1) \quad \dot{v} = \begin{bmatrix} 0 & 1 \\ -\frac{g}{\ell} & 0 \end{bmatrix} v,$$

$$(2) \quad \dot{w} = \begin{bmatrix} 0 & 1 \\ \frac{g}{\ell} & 0 \end{bmatrix} w.$$

The eigenvalues of the linear systems (1) and (2) are, respectively,

$$\pm j\sqrt{\frac{g}{\ell}} \quad \text{and so } (0, 0) \text{ is a center,}$$

$$\pm\sqrt{\frac{g}{\ell}} \quad \text{and so } (\pi, 0) \text{ is a saddle.}$$

The equilibrium $(0, 0)$ for the nonlinear system is also a center, although this is not implied by the linearization since $(0, 0)$ is not a hyperbolic fixed point.

Example 2.17 ***(Damped Simple Pendulum)***
Next, we consider the simple pendulum equation with damping,

$$\ddot{\theta} + B\dot{\theta} + \frac{g}{\ell}\sin(\theta) = 0.$$

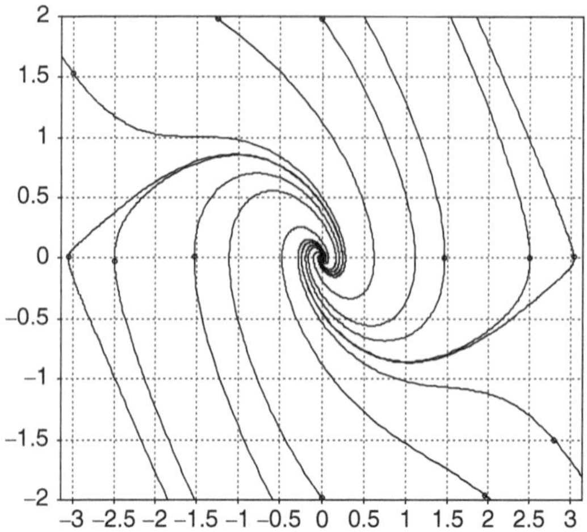

Figure 2.24 Phase portrait of the damped simple pendulum. Oscillatory solutions now will decay to zero, making the fixed point (0, 0) a sink.

It is easy to show that (0, 0) and (π, 0) are still fixed points of this system and that the linearized systems about these fixed points are

$$(1) \quad \dot{y} = \begin{bmatrix} 0 & 1 \\ -\frac{g}{\ell} & -B \end{bmatrix} y,$$

$$(2) \quad \dot{y} = \begin{bmatrix} 0 & 1 \\ \frac{g}{\ell} & -B \end{bmatrix} y.$$

In this case the fixed point (π, 0) is still a saddle but the fixed point (0, 0) becomes a stable focus (Problem 2.18); see Figure 2.24.

Remark 2.3 ***(More on Symbolic Computation)*** We can use Matlab's symbolic computation features to find and classify fixed points for nonlinear systems. For example, given the nonlinear system

$$\dot{x} = y$$
$$\dot{y} = -y + x(x-1)(x+1),$$

consider the following Matlab script.

Matlab script to find and classify equilibrium points

```
syms x y
f=sym([y -y+x*(x-1)*(x+1)]);      % define the vector field
eqn=f==0;                     % set f=0 to find the fixed points
s=solve(eqn);                 % solve for the fixed points
FixedPoints=double([s.x s.y]); % matrix of fixed points
```

```
J=jacobian(f);                 % symbolic Jacobian
Ja=matlabFunction(J);          % function handle for Jacobian
%
% return the eigenvalues at each fixed point
S1=eig(Ja(0));
S2=eig(Ja(1));
S3=eig(Ja(-1));
```

Running this script finds three `FixedPoints` at $(0,0)$, $(-1,0)$, and $(1,0)$ and computes the symbolic Jacobian as

```
J =
[                                          0,  1]
[ (x - 1)*(x + 1) + x*(x - 1) + x*(x + 1), -1]
```

Evaluating the Jacobian function *Ja* at the fixed points gives

$$S1 = -0.5 \pm j0.866 \quad \text{which is a sink}$$

$$S2 = S3 = (-1, 2) \quad \text{which are saddle points.}$$

2.8 Periodic Trajectories and Limit Cycles

We have seen that both the harmonic oscillator and simple pendulum systems exhibit pure oscillatory or periodic motion. Periodic motion in dynamical systems is found in many other examples, such as electronic oscillator circuits, planetary motion, population dynamics, and chemical reactions. For linear systems like the harmonic oscillator, there are infinitely many periodic solutions, which are non-isolated in the sense that there are periodic trajectories arbitrarily close to any given trajectory. A nonlinear system may have *isolated periodic* trajectories, which are known as *limit cycles*. This means that there exists a region containing the periodic trajectory that contains no other periodic trajectory. Nearby trajectories will either converge toward the limit cycle or diverge away from the limit cycle.

2.8.1 Relaxation Oscillators

Examples of systems possessing limit cycles are so-called *relaxation oscillators*. A relaxation oscillation is a periodic system in which a slow change in the state over a finite interval alternates with a rapid change in the state. In electronics, relaxation oscillator circuits can be built from transistors, operational amplifiers, or negative resistance devices like tunnel diodes in a feedback loop. A good example is a circuit that controls blinking lights, such as the turn signal for an automobile.

There are many other examples of physical systems whose dynamics can be modeled as relaxation oscillators. Systems with stick-slip friction, such as earthquakes, a violin bow drawn across a string, and squeaking chalk on a blackboard are examples of relaxation oscillations. Biological examples include firing of nerve cells and a beating heart.

Van der Pol Oscillator

One of the most important examples of a relaxation oscillator from a historical perspective is the *Van der Pol equation*

$$\ddot{y} - \mu(1 - y^2)\dot{y} + y = 0, \tag{2.19}$$

which is a second-order nonlinear system with nonlinear damping. Note that, for $\mu = 0$, Equation (2.19) is the harmonic oscillator. Balthasar Van der Pol discovered this equation while building oscillator circuits using vacuum tubes in the 1920s [68], and coined the term "relaxation oscillator." In state space Equation (2.19) can be written as

$$\begin{aligned} \dot{x}_1 &= x_2 \\ \dot{x}_2 &= -x_1 + \mu x_2(1 - x_1^2). \end{aligned}$$

The Van der Pol system has a unique limit cycle in phase space for all values of $\mu > 0$. All other trajectories converge asymptotically to the limit cycle. The direction field and a portion of the phase portrait showing the limit cycle are shown in Figure 2.25. The significance of the limit cycle is that the long-term behavior of the system is essentially independent of where it starts, i.e., independent of the initial conditions. This is in sharp contrast to the harmonic oscillator and simple pendulum, where the amplitude of oscillation is different for each different initial condition.

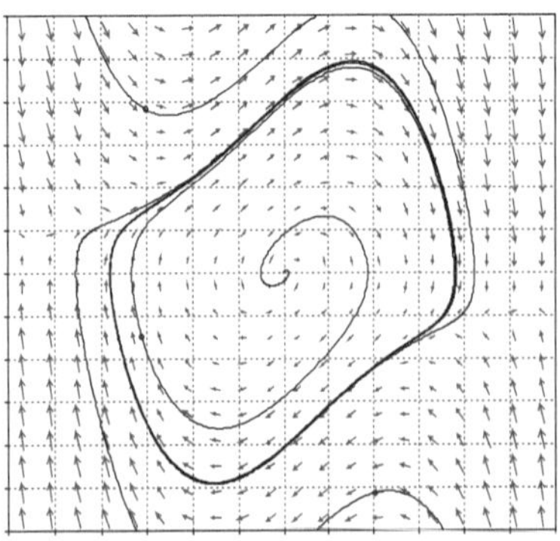

Figure 2.25 Direction field and phase portrait of the Van der Pol oscillator with $\mu = 1.8$ showing the existence of a limit cycle.

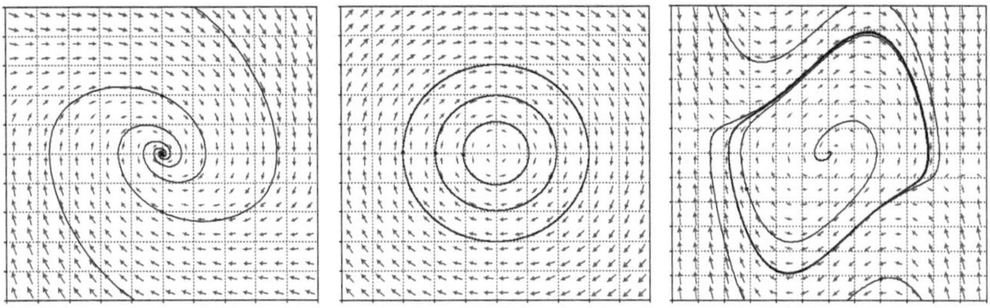

Figure 2.26 Illustrating a Hopf bifurcation in the Van der Pol system as the parameter μ changes from negative to positive.

The Jacobian for the Van der Pol system at (0, 0) is given by

$$\begin{bmatrix} 0 & 1 \\ -1 & \mu \end{bmatrix}.$$

As the parameter μ varies between −1 and +1, the linearized system has the following properties, illustrated in Figure 2.26:

1. $-1 \le \mu < 0$: $\overline{x} = 0$ is a stable focus (left plot);
2. $\mu = 0$: $\overline{x} = 0$ is a center (center plot);
3. $0 < \mu \le 1$: $\overline{x} = 0$ is an unstable focus (right plot).

This is an example of a *Hopf bifurcation*. Hopf proved that if the equilibrium of the Jacobian linearization of a nonlinear system changes from a sink to a source as a parameter μ passes through a value μ_0, then for $\mu > \mu_0$ there must be a periodic orbit of the nonlinear system.

Bifurcation theory deals with qualitative changes in a dynamical system as a function of changes in a parameter as in the above Hopf bifurcation example. The term "bifurcation" was introduced by Henri Poincaré in 1885 [57]. Bifurcations occur in both continuous and discrete systems and have applications in biological systems, quantum systems, electronics, and other systems. We will give a few examples of bifurcations; however, a detailed study of bifurcation theory is outside the scope of this text.

FitzHugh–Nagumo Oscillator

Another important example of a system exhibiting limit cycle behavior is the *FitzHugh–Nagumo model*. Starting with the Van der Pol equation

$$\ddot{y} - \mu(1 - y^2)\dot{y} + y = 0,$$

we apply the so-called *Lienard transformation*,

$$x = y - \frac{1}{3}y^3 - \dot{y}/\mu, \tag{2.20}$$

from which we obtain

$$\dot{y} = \mu \left(y - \frac{1}{3}y^3 - x \right).$$

Differentiating Equation (2.20), and substituting for $\ddot{y}$ from Van der Pol's equation, yields

$$\dot{x} = \dot{y} - y^2\dot{y} - \frac{\ddot{y}}{\mu} = \frac{1}{\mu}y.$$

Defining state variables $x_1 = y, x_2 = x$, we have

$$\dot{x}_1 = \mu \left(x_1 - \frac{1}{3}x_1^3 - x_2 \right) \tag{2.21}$$

$$\mu\dot{x}_2 = x_1. \tag{2.22}$$

Beginning with the above system, FitzHugh and Nagumo modified the equations as

$$\dot{x}_1 = x_1 - \frac{1}{3}x_1^3 - x_2 + I \tag{2.23}$$

$$\mu\dot{x}_2 = x_1 - bx_2 + a, \tag{2.24}$$

where a, b, μ, and I are constants.

The FitzHugh–Nagumo model is a model of a spiking biological neuron and is a simplification of an earlier model by Hodgkin and Huxley, who were studying action potentials in the giant squid axon [38]. The state x_1 is the *membrane potential* and x_2 is a *recovery variable*. The constant I is the magnitude of stimulus current applied to the neuron.

The limit cycle for the FitzHugh–Nagumo model and the spiking membrane potential x_1 are shown in Figure 2.27.

The Brusselator

As a final example, we present the *Brusselator* model [39]. The Brusselator (or *Brussels oscillator*) was proposed by Ilya Prigogine to model a type of autocatalytic chemical reaction. An autocatalytic reaction is one in which one of the reaction products is a catalyst for the reaction. Autocatalysis plays a major role in replication of RNA molecules and has even been postulated as an explanation for abiogenesis, or the origin of life.

The Brusselator equations are

$$\dot{x}_1 = A + x_1^2x_2 - Bx_1 - x_1 \tag{2.25}$$

$$\dot{x}_2 = Bx_1 - x_1^2x_2, \tag{2.26}$$

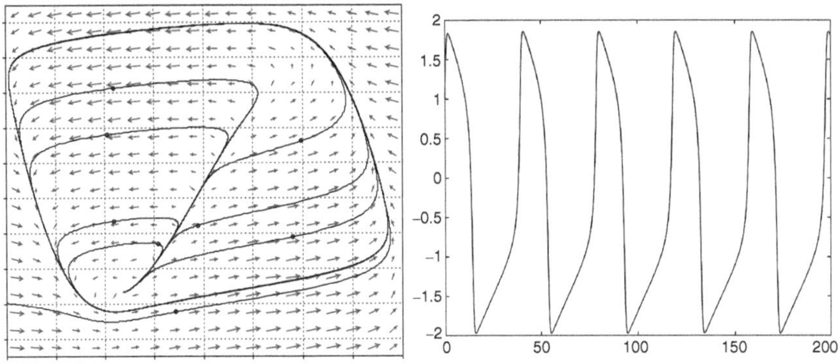

Figure 2.27 Phase portrait of the FitzHugh–Nagumo system with parameters $a = 0.7$, $b = 0.8$, $I = 0.5$, $\mu = 12.5$ and the membrane potential $x_1(t)$, showing the spiking behavior [23].

Figure 2.28 Direction field and a portion of the phase portrait of the Brusselator system with parameters $A = 1$, $B = 3$.

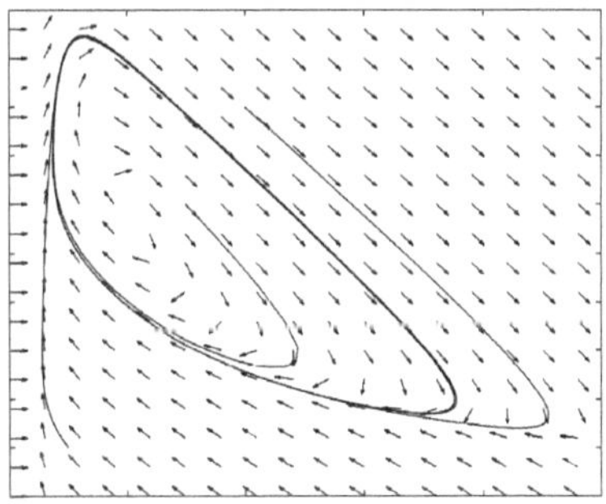

where A and B are constants. The system has an equilibrium point at $x_1 = A$, $x_2 = \frac{B}{A}$. For $B > 1 + A^2$ the equilibrium is unstable (Problem 2.20) and the trajectories approach a limit cycle. The limit cycle is shown in Figure 2.28 for $A = 1$, $B - 3$.

Existence of Limit Cycles

An important question is how to determine the existence or nonexistence of periodic trajectories. We next discuss two results, known as Bendixson's theorem and the Poincaré–Bendixson theorem that can be used to investigate the existence of periodic solutions for second-order systems.

2.8.2 Bendixson's Theorem

We first present a result known as *Bendixson's theorem*, which is a negative result in the sense that it can be used to infer the nonexistence of limit cycles. First we need some additional notation. For a vector field $f = (f_1, f_2)^T$ in $\mathbb{R}^2$, the expression

$$\nabla \cdot f = \partial f_1/\partial x_1 + \partial f_2/\partial x_2$$

is called the *divergence* of $f(x)$, and is a measure of the extent to which the vector field is "outgoing" or "incoming." For example, a vector field near a source has positive divergence and a vector field near a sink has negative divergence. With this notation we can state the following theorem.

Theorem 2.2 ***(Bendixson's Theorem)*** *Consider the second-order nonlinear system*

$$\begin{aligned}\dot{x}_1 &= f_1(x_1, x_2)\\ \dot{x}_2 &= f_2(x_1, x_2).\end{aligned}$$

Suppose $D \subset \mathbb{R}^2$ is a simply connected region such that the divergence

$$\nabla \cdot f = \frac{\partial f_1}{\partial x_1} + \frac{\partial f_2}{\partial x_2}$$

is not identically zero over any subregion of D and does not change sign in D. Then D does not contain any periodic solutions of the nonlinear system.

Proof: See Appendix D.

Example 2.18 Consider the nonlinear system

$$\begin{aligned}\dot{x}_1 &= x_2 + x_1 x_2^2\\ \dot{x}_2 &= -x_1 + x_1^2 x_2.\end{aligned}$$

The equilibrium $(0, 0)$ is a center for the Jacobian linearization and so the linearized system has periodic trajectories. However,

$$\nabla \cdot f = x_1^2 + x_2^2 > 0,$$

which means that there are no periodic solutions for the nonlinear system.

Remark 2.4 It is interesting to see how Bendixson's theorem applies in the case of linear systems. For a second-order linear system

$$\begin{aligned}\dot{x}_1 &= ax_1 + bx_2\\ \dot{x}_2 &= cx_1 + dx_2,\end{aligned}$$

$\partial f_1/\partial x_1 + \partial f_2/\partial x_2 = a + d$ and so Bendixson's theorem implies that a necessary condition for a periodic solution is that the trace $a + d$ of the coefficient matrix

A must be equal to zero. Recall that the eigenvalues of this linear system are roots of the polynomial

$$p(\lambda) = \lambda^2 - \text{tr}(A)\lambda + \det(A) = \lambda^2 - (a + d)\lambda + (ad - bc).$$

Thus, if $a + d = 0$, the eigenvalues are $\lambda = \pm\sqrt{\det(A)}$. It follows that the equilibrium $\bar{x} = 0$ is a center if $\det(A) < 0$ and a saddle point if $\det(A) > 0$.

2.8.3 Poincaré–Bendixson Theorem

Bendixson's theorem can be used to show the nonexistence of periodic solutions. In this section we present the *Poincaré–Bendixson theorem*, which gives conditions for the existence of periodic orbits for two-dimensional systems. Consider the second-order autonomous system

$$\dot{x} = f(x), \quad x \in \mathbb{R}^2, \tag{2.27}$$

and suppose that $\bar{x} = 0$ is an isolated equilibrium point.

Definition 2.9 ***(Positively Invariant Set)*** Let M be a closed, bounded region in $\mathbb{R}^2$ and let $x(t)$ be a solution trajectory of Equation (2.27). The set M is said to be *positively invariant* for the system (2.27) if $x(T) \in M$, for some $T > 0$, implies that $x(t) \in M$ for all $t > T$. In other words, any trajectory entering M or starting in M remains in M.

Theorem 2.3 ***(Poincaré–Bendixson)*** *Let M be a closed, bounded region in* $\mathbb{R}^2$ *such that M is positively invariant for the nonlinear system (2.27). Then M contains either a periodic orbit or an equilibrium point for (2.27).*

Typically the way one applies the Poincaré–Bendixson theorem in practice is to find a positively invariant set M containing a single equilibrium point and show that the equilibrium is unstable. In this case, the Poincaré–Bendixson theorem says that there is a periodic trajectory in M. Figure 2.29 illustrates this concept.

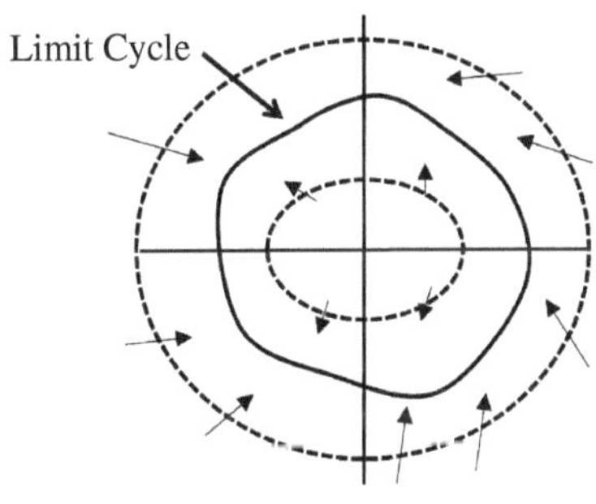

Figure 2.29 Illustrating the Poincaré–Bendixson theorem. The annular region between the two dotted closed curves is positively invariant and does not contain any equilibrium points. Therefore, it must contain a periodic orbit.

Suppose that the region bounded by the larger dotted circle is positively invariant. If the equilibrium at the origin is unstable, then every trajectory starting inside the region bounded by the smaller dotted circle will leave that region but remain inside the region bounded by the larger circle. Thus the annular region M bounded by these two circles is positively invariant and contains no equilibrium points. Therefore, it must contain a periodic trajectory.

In two dimensions, a positively invariant set "traps" the trajectories of the system in its interior. The Poincaré–Bendixson theorem says that the the behavior of a two-dimensional system inside a positively invariant set is limited to only the two possibilities of a fixed point or a limit cycle. In particular, a second-order autonomous system cannot have chaotic solutions. The behavior of systems of order greater than 2 can be much more complex, including the possibility of chaotic solutions. Of course, the theorem does not preclude the possibility that there are more than one fixed point and/or more than one limit cycle inside the set.

Example 2.19 Consider the nonlinear system

$$\begin{aligned}\dot{x}_1 &= 2x_2 + x_1(0.5 - x_1^2 - x_2^2)\\ \dot{x}_2 &= -2x_1 + x_2(0.5 - x_1^2 - x_2^2).\end{aligned} \tag{2.28}$$

The eigenvalues of the Jacobian at the origin are $0.5 \pm 1.41j$ and hence the origin is unstable. If we compute the directional derivative of the system vector field on the unit circle $x_1^2 + x_2^2 = 1$ we get, after some calculation (Problem 2.21),

$$2x_1\dot{x}_1 + 2x_2\dot{x}_2 = -1 < 0. \tag{2.29}$$

Thus, the interior of the unit circle is positively invariant and therefore contains a periodic solution. Figure 2.30 shows the phase portrait for this system

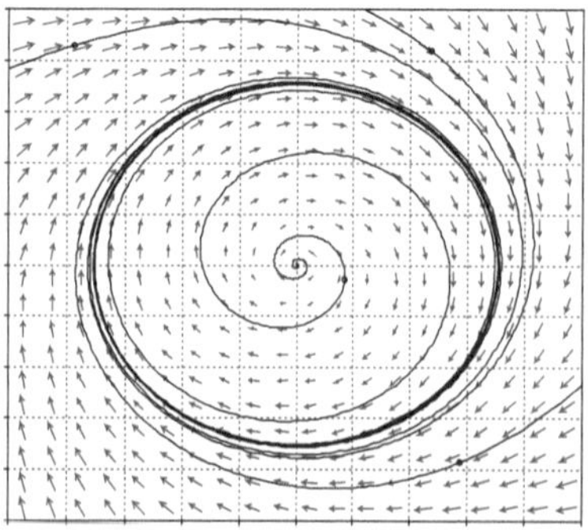

Figure 2.30 Phase portrait for the system (2.28).

2.9 Coupled Second-Order Systems

Many higher-order systems in mechanics, robotics, and other areas are governed by coupled second-order systems, as a consequence of Newton's laws. Although the state space for such higher-order systems is not planar, the structure of the dynamic equations still offers some insight that can often be exploited in analysis and control.

Example 2.20 ***(Coupled Oscillators)***

Consider two identical pendula coupled by a spring as shown in Figure 2.31. If we assume that the spring is linear with stiffness k, then the equations of motion of this system are given by

$$\ddot{\alpha} + mg\ell \sin(\alpha) + k(\alpha - \beta) = 0$$
$$\ddot{\beta} + mg\ell \sin(\beta) + k(\beta - \alpha) = 0.$$

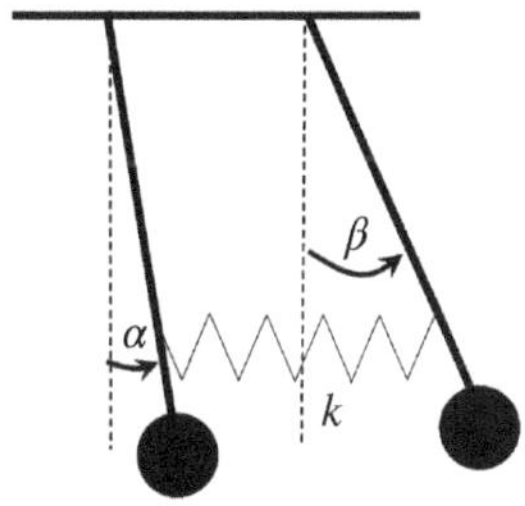

Figure 2.31 Two pendula coupled by a linear spring with stiffness k.

Equations similar to these arise, for example, in the problem of synchronization of generators in a power system. We note that the order of this coupled second-order system, i.e., the dimension of the state space, is twice the number of second-order equations in the model. Thus, the trajectories evolve in the four-dimensional state space $\mathbb{R}^4$.

A typical response of this system is shown in Figure 2.32. Note how the amplitudes of the oscillators alternately increase and decrease out of phase. If the initial conditions of the oscillator equations are identical, the variables α and β will oscillate in phase with one another. If the initial conditions have the same value but opposite signs, the variables will oscillate out of phase. Verifying these statements is left as an exercise (Problem 2.25).

The Matlab script below simulates the coupled pendulum system.

Matlab script for the coupled pendula

```
f=@(t,x) [x(2);-om^2*sin(x(1))-k*(x(1)-x(3));
x(4);-om^2*sin(x(3))+k*(x(1)-x(3))];
tspan = [0 40]; x0 = [1.1;0;0.1;0];
[t,x] = ode45(f,tspan,x0);
alpha = x(:,1); beta = x(:,3);
```

(Continued)

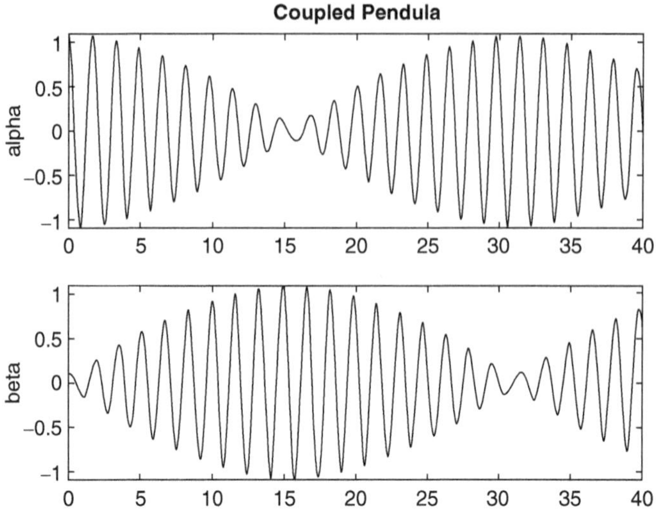

Figure 2.32 Response of the coupled pendula. With the given initial conditions the oscillations alternately increase and decrease out of phase.

(Continued)

```
ax1 = nexttile;
plot(t,alpha,'k')
title(ax1,'Coupled Pendula')
ylabel(ax1,'alpha')
ax2 = nexttile;
plot(t,beta,'k')
ylabel(ax2,'beta')
```

Example 2.21 ***(Double Pendulum)***
Consider the double pendulum shown in Figure 2.33. This can also be viewed as a simple model of a two-link robot. The equations of motion are given by [64]

$$(m_1 + m_2)\ell_1^2\ddot{\theta}_1 + m_2\ell_1\ell_2\cos(\theta_1 - \theta_2)\ddot{\theta}_2 + m_2\ell_1\ell_2\sin(\theta_1 - \theta_2)\dot{\theta}_2^2 + (m_1 + m_2)g\ell_1\sin(\theta_1) = 0$$

$$m_2\ell_2^2\ddot{\theta}_2 + m_2\ell_1\ell_2\cos(\theta_1 - \theta_2)\ddot{\theta}_1 - m_2\ell_1\ell_2\dot{\theta}_1^2\sin(\theta_1 - \theta_2) + m_2g\ell_2\sin(\theta_2) = 0.$$

This is a fourth-order nonlinear system consisting of two coupled second-order equations.

We can write the double pendulum equations in vector form as

$$M(q)\ddot{q} + h(q, \dot{q}) + g(q) = 0,$$

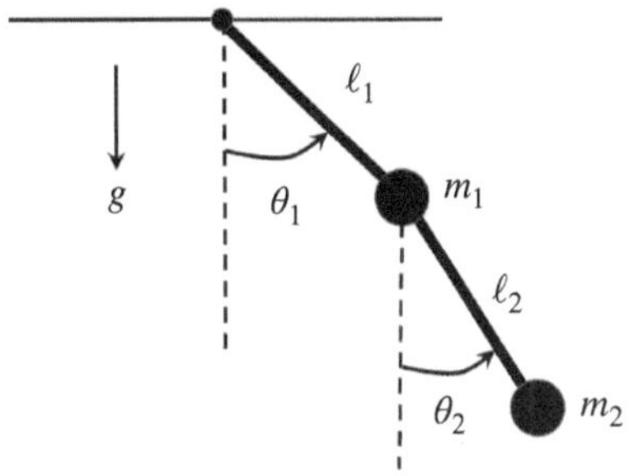

Figure 2.33 Double Pendulum

where $q = (q_1, q_2) = (\theta_1, \theta_2)$. The matrix $M(q)$ is given by

$$M(q) = \begin{bmatrix} (m_1 + m_2)\ell_1^2 & m_2\ell_1\ell_2\cos(q_1 - q_2) \\ m_2\ell_1\ell_2\cos(q_1 - q_2) & m_2\ell_2^2 \end{bmatrix}$$

and is called the *inertia matrix*. $M(q)$ is symmetric and positive-definite for each q. The vectors $h(q, \dot{q})$ and $g(q)$ are given by

$$h(q,\dot{q}) = \begin{bmatrix} m_2\ell_1\ell_2\sin(q_1 - q_2)\dot{q}_2^2 \\ -m_2\ell_1\ell_2\sin(q_1 - q_2)\dot{q}_1^2 \end{bmatrix}, \quad g(q) = \begin{bmatrix} (m_1 + m_2)g\ell_1\sin(q_1) \\ m_2 g\ell_2\sin(q_2) \end{bmatrix}.$$

The vector $h(q, \dot{q})$ contains centrifugal force terms and $g(q)$ is a vector of gravitational forces.

For initial conditions close to the stable equilibrium the solutions of the double pendulum system are sinusoidal. For large amplitude initial conditions, the double pendulum system exhibits chaotic behavior (Problem 2.27).

Problems

2.1 Derive Equation (2.2) for the *RLC* circuit.

2.2 Define state variables x_1 and x_2 and write Equation (2.2) in state space. With input $V = 0$, classify the equilibrium at the origin (node, saddle, etc.) if $L = 1, C = 1/4$, and $R = 2$. Explain the qualitative behavior of the system, starting at any initial conditions.

2.3 Repeat Problem 2.2 if $L = 1, C = 1/4$, and $R = 6$.

2.4 Derive Equation (2.3) for the mass–spring–damper system.

2.5 Define state variables x_1 and x_2 and write Equation (2.3) in state space. With input $F = 0$, classify the equilibrium at the origin (node, saddle, etc.) if $M = 2$, $k = 2$, and $b = 4$. Explain the qualitative behavior of the system starting at any initial conditions.

2.6 Repeat Problem 2.5 if $M = 2$, $k = 10$, and $b = 16$.

2.7 Show that the eigenvalues λ_i of the A matrix in Equation (2.16) are given by

$$\lambda = \frac{1}{2}p \pm \frac{1}{2}\sqrt{p^2 - 4q},$$

where p and q are the trace and determinant, respectively, of the matrix A.

2.8 Add a damping term $-B\dot{x}$ to Equation (2.5) of the harmonic oscillator and modify the symbolic Matlab scripts to generate the solution of the modified equation with $B = 1$. Plot the response.

2.9 Show that the general solution of Equation (2.7) is given by Equation (2.8).

2.10 Consider the linear system $\dot{x} = Ax$. Prove that if the matrix A is nonsingular, then $\overline{x} = 0$ is the only equilibrium point of the system.

2.11 Consider the linear second-order system

$$\ddot{y} - 4\dot{y} + 4y = 0.$$

1. Show by direct calculation that both $y(t) = e^{2t}$ and $y(t) = te^{2t}$ are solutions of this equation.
2. Write the system in state space and find T such that $\bar{A} = T^{-1}AT$ is in Jordan form.

2.12 Consider the linear system $\dot{x} = Ax$, where the coefficient matrix A is

$$A = \begin{bmatrix} 1 & 2 \\ 2 & 4 \end{bmatrix}.$$

Find the set of all equilibrium points of the system. Are they isolated or nonisolated?

2.13 Consider the scalar second-order system

$$\ddot{y} + (1 + \dot{y})y^3 + 3y^2 = 0.$$

1. Define state variables and write the system in state space as a system of two first-order equations.
2. Find the equilibrium points.
3. Compute the Jacobian linearization about each equilibrium point.

2.14 Construct a second-order system that has exactly five isolated equilibrium points.

2.15 Construct a differential equation that has both isolated and nonisolated equilibrium points.

2.16 For each of the linear systems below:

1. Find the linear transformation T so that $T^{-1}AT$ is in normal form.
2. Classify the equilibrium at $(0, 0)$, i.e., stable node, source, etc.
3. Sketch the phase portraits.

(a) $$\dot{x}_1 = 0.5x_1 - 2.5x_2$$
$$\dot{x}_2 = -2.5x_1 + 0.5x_2$$

(b) $$\dot{x}_1 = 2x_1 + x_2$$
$$\dot{x}_2 = -x_1 + 2x_2$$

(c) $$\dot{x}_1 = 2.5x_1 + 0.5x_2$$
$$\dot{x}_2 = -0.5x_1 + 1.5x_2$$

(d) $$\dot{x}_1 = -2.25x_1 - 1.25x_2$$
$$\dot{x}_2 = -1.75x_1 - 2.75x_2$$

2.17 For each of the nonlinear systems below, find and classify all equilibrium points and compute the respective linear approximations about the equilibrium points. Write a Matlab script to draw the direction field and phase portrait for the nonlinear system.

(a) $$\dot{x}_1 = x_2$$
$$\dot{x}_2 = -x_2 + x_1^4 + x_1^3 + 2x_1^2 + 2x_1$$

(b) $$\dot{x}_1 = 2x_1 - x_1x_2$$
$$\dot{x}_2 = -0.5x_2 + x_1x_2$$

(c) $$\dot{x}_1 = 2x_1(1 - x_1) - x_1x_2$$
$$\dot{x}_2 = -0.5x_1 + x_1x_2$$

2.18 For the simple pendulum with damping, verify by calculating the eigenvalues of the Jacobian that the equilibrium $(0, 0)$ is a sink and the equilibrium $(\pi, 0)$ is a saddle.

2.19 For the FitzHugh–Nagumo system (2.23)–(2.24) compute the linear approximation about the equilibrium and show that it is unstable.

2.20 For the Brusselator system compute the linear approximation about the equilibrium and show that it is unstable if $B > 1 + A^2$.

2.21 Verify inequality (2.29) by directly calculating the directional derivative on the unit circle.

2.22 Consider the nonlinear equation

$$\ddot{y} + y - y^3 = 0.$$

Write the system is state space and sketch the phase portrait.

2.23 Does the system

$$\dot{x}_1 = x_2^2 + x_1$$
$$\dot{x}_2 = x_2 + x_1^3$$

have any periodic orbits? Prove your answer.

2.24 Consider the nonlinear system

$$\dot{x}_1 = \beta + x_1^2$$
$$\dot{x}_2 = -x_2.$$

Investigate the behavior of the system for $\beta = -1, 0, 1$. In each case, find the equilibrium points. Are they hyperbolic or nonhyperbolic? Sketch the direction fields and phase portraits. This system is an example of a saddle-node bifurcation as the parameter β changes from negative to positive.

2.25 Run the Matlab script for the coupled pendulum system and verify that, with initial conditions $(a, 0, a, 0)$ the angles α and β oscillate in phase. Repeat the simulation with initial conditions $(a, 0, -a, 0)$ and verify that the angles oscillate out of phase.

2.26 For small angles, the coupled pendulum system can be modeled as coupled harmonic oscillators

$$\ddot{\alpha} + \omega^2\alpha + k(\alpha - \beta) = 0$$
$$\ddot{\beta} + \omega^2\beta + k(\beta - \alpha) = 0,$$

where $\omega^2 = mg\ell$. Modify the Matlab script for the coupled pendula to simulate the couple harmonic oscillators. Experiment with various initial conditions.

2.27 Write a Matlab script to simulate the double pendulum system from Example 2.21. Experiment with different initial conditions and notice the chaotic behavior of the solutions with large-amplitude initial conditions.

2.28 Consider the coupled mass–spring–damper system below, where u is considered as a external force acting on the system.

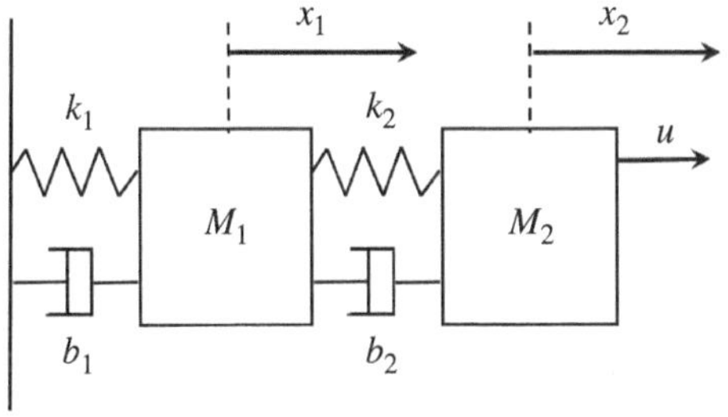

(a) Derive the equations of motion for the system.
(b) Write a Matlab script to simulate the response of the system. With $u = 0$, experiment with various values of the masses, damping constants, and stiffness constants and various initial conditions.
(c) Set the initial conditions to zero and apply a sinusoidal force $u = A\sin(\omega t)$. Simulate the response for various values of A and ω.

3

System Fundamentals

3.1 Introduction

In the previous chapter we discussed continuous-time second-order systems. In this chapter we consider continuous-time systems of any order. We discuss existence and uniqueness of solutions, linearization about hyperbolic equilibrium points, and the matrix exponential, which allows a compact representation of solutions of linear systems of any order. We then present a result, known as the Hartman–Grobman theorem, which relates the response of a continuous-time nonlinear system in a neighborhood of a hyperbolic fixed point to the corresponding linear approximation about the fixed point. We also discuss singular perturbations, which are used to model systems with very large or very small parameters.

3.2 Existence and Uniqueness of Solution

In this section we consider the question of existence and uniqueness of solutions for an nth-order autonomous system of the form

$$\dot{x} = f(x), \; x(0) = x_0, \quad x \in \mathbb{R}^n,$$

where $f(x)$ is a vector field on the state space $\mathbb{R}^n$. For an nth-order autonomous linear system

$$\dot{x} = Ax, \; x(0) = x_0, \quad x \in \mathbb{R}^n,$$

the initial-value problem always has a unique solution defined for all $t \geq 0$. The situation is more complicated for nonlinear systems, which may have no solution, a unique solution, or multiple solutions. Moreover, even when a nonlinear system is known to have a unique solution, it may not be possible to solve the equations without resorting to numerical approximation methods.

Introduction to Modeling and Simulation: A Systems Approach, First Edition. Mark W. Spong.

Companion Website: www.wiley.com/go/spong/engineeringmodeling

Example 3.1 ***(No Solution)***
Consider the nonlinear system

$$\dot{x} = -\text{sgn}(x), \quad x(0) = 0,$$

where

$$\text{sgn}(x) = \begin{cases} +1, & \text{if } x \geq 0 \\ -1, & \text{if } x < 0 \end{cases}$$

is the *signum* function. There is no differentiable function $x(t)$ satisfying this initial-value problem.

Proof: The proof is by contradiction. Suppose such a function $x(t)$ exists. If, for $t > 0$, we have $x(t) > 0$, then $\dot{x}(t) = -\text{sgn}(x(t)) = -1$. Thus, $x(t) = -t < 0$. Likewise if $x(t) < 0$, then $\dot{x}(t) = +1$, which implies that $x(t) = +t > 0$. Therefore, no solution can exist.

Example 3.2 ***(Multiple Solutions)***
A nonlinear system may also have multiple distinct solutions. Consider the nonlinear system

$$\dot{x}(t) = \frac{1}{2x(t)}, \quad x(0) = 0.$$

It is easy to verify by direct calculation that

$$x(t) = \sqrt{t} \quad \text{and} \quad x(t) = -\sqrt{t}$$

are both solutions to this initial-value problem (Problem 3.1).

Nonlinear systems can also exhibit a property known as *finite escape*. By "finite escape" we mean that a solution asymptotically approaches $\pm\infty$ or "blows up" in finite time. Finite escape is strictly a nonlinear phenomenon. Linear systems can increase at most exponentially and so are bounded on any finite time interval.

Example 3.3 ***(Finite Escape)***
Consider the nonlinear system

$$\dot{x} = x^2, \quad x(0) = x_0.$$

This equation has the unique solution (Problem 3.2)

$$x(t) = \frac{x_0}{1 - tx_0},$$

which is defined for $t < \frac{1}{x_0}$. The solution blows up at $t = \frac{1}{x_0}$ (Figure 3.1).

Next, we give a sufficient condition on the vector field defining the initial-value problem for there to be a unique solution. For simplicity, we only state the result for autonomous systems, but a similar result holds for nonautonomous systems.

Figure 3.1 Trajectories for the system $\dot{x} = x^2$ exhibiting finite escape. The amount of time that the trajectory remains finite depends on the initial condition x_0.

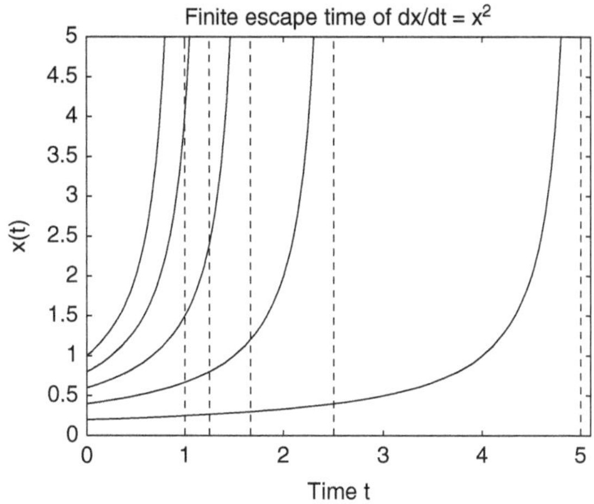

Theorem 3.1 ***(Existence and Uniqueness of Solutions)*** *Consider the initial-value problem*

$$\dot{x} = f(x), \ x(0) = x_0, \quad x \in \mathbb{R}^n, \tag{3.1}$$

and suppose that the vector field $f(x)$ *is* continuously differentiable *in* $\mathbb{R}^n$. *Then there is a unique solution* $x(t) \in \mathbb{R}^n$ *defined for* $0 \leq t \leq \delta$ *for some* $\delta > 0$.

The possibility of finite escape is the reason that a solution may only be guaranteed to exist on an interval $[0, \delta)$. The proof of this theorem relies on some concepts in functional analysis that are beyond the scope of this text. The proof uses a method known as *Picard iteration*, which is of independent interest in its own right. The method of Picard iteration begins by writing the differential equation (3.1) as an equivalent *integral equation*

$$x(t) = x(0) + \int_0^t f(u)du.$$

With $x(0) = x_0$, we then define a sequence of functions $x_1, x_2, \dots$ as

$$x_1(t) = x_0 + \int_0^t f(x_0)du,$$

$$x_2(t) = x_0 + \int_0^t f(x_1(u))du,$$

$$\vdots$$

$$x_k(t) = x_0 + \int_0^t f(x_{k-1}(u))du.$$

Then, under the conditions of the theorem, the sequence $\{x_k(t)\}$ converges to a unique solution $x(t)$ of the initial-value problem.

Example 3.4 Consider the initial-value problem

$$\dot{x} = ax, \quad x(0) = x_0,$$

which has solution $x(t) = e^{at}x_0$. Then the method of Picard iteration generates the sequence

$$\begin{aligned} x_1(t) &= x_0 + \int_0^t ax_0 du = x_0(1 + at), \\ x_2(t) &= x_0 + \int_0^t ax_0(1 + au)du = x_0\left(1 + at + \frac{1}{2}a^2t^2\right), \\ &\vdots \\ x_k(t) &= x_0\left(1 + at + \frac{1}{2}a^2t^2 + \cdots + \frac{1}{k!}a^kt^k\right), \end{aligned}$$

which converges to x_0e^{at} as $k \to \infty$.

Remark 3.1 Existence and uniqueness of solutions can be shown under weaker conditions than continuous differentiability of the vector field $f(x)$. However, the vector fields defining most of the systems that we present in this text are at least continuously differentiable and so the additional effort needed to relax this assumption in Theorem 3.1 does not seem justified here. The reader can refer to [44] for a more detailed treatment of existence and uniqueness of solutions.

It is interesting to note that, if a solution trajectory is known to be contained in a bounded set M, i.e., if a bounded set M is positively invariant, then finite escape cannot occur and, hence, the solution will exist for all time.

Example 3.5 Consider the initial-value problem

$$\dot{x} = -x^3, \quad x(0) = x_0 \in \mathbb{R}.$$

Then it is easy to see that the interval $[-x_0, x_0]$ is positively invariant for this system and therefore there is a unique solution that exists for all $t \geq 0$.

3.3 The Matrix Exponential

In this section we introduce the matrix exponential, which characterizes the solutions of n-dimensional linear systems, in direct analogy to the scalar exponential function. Consider an n-dimensional autonomous linear system

$$\dot{x} = Ax, \quad \text{with } x \in \mathbb{R}^n,\ A \in \mathbb{R}^{n\times n},$$

with initial condition $x(0) = x_0$. By analogy to the scalar case, $\dot{x} = ax$, which has solution $x(t) = e^{at}x_0$, we look for a solution of the form

$$x(t) = e^{At}x_0,$$

where e^{At} is an $n \times n$ matrix called the *matrix exponential*, which we define next.

We know that the series expansion for the exponential function e^a, for a real number a, is given by

$$e^a = 1 + a + \frac{1}{2}a^2 + \frac{1}{3!}a^3 + \cdots .$$

We therefore define the exponential of an $n \times n$ matrix A as

$$e^A = I + A + \frac{1}{2}A^2 + \frac{1}{3!}A^3 + \cdots .$$

The series expansion for e^A is well defined and converges for any $n \times n$ matrix A (see, for example, [37]).

Some Properties of the Matrix Exponential

The following properties of the matrix exponential are not too difficult to show using the series expansion for e^A.

1. If matrices A and B commute, i.e., if $AB = BA$, then the product $e^A e^B = e^{A+B}$.
2. The determinant of e^A satisfies $\det(e^A) = e^{\mathrm{tr}(A)}$, where $\mathrm{tr}(A)$ is the trace of the matrix A.
3. As a consequence of the above, the matrix exponential e^A is always nonsingular and its inverse is given by $(e^A)^{-1} = e^{-A}$.
4. If T is a nonsingular $n \times n$ matrix then $T^{-1}e^A T = e^{T^{-1}AT}$.
5. If $A = \mathrm{diag}(\lambda_1, \cdots, \lambda_n)$ is a diagonal matrix, then e^A is the diagonal matrix

$$e^A = \mathrm{diag}(e^{\lambda_1}, \ldots, e^{\lambda_n}).$$

Next, using the series expansion for e^{At}, we have

$$\begin{aligned}\frac{d}{dt}e^{At} &= \frac{d}{dt}\left(I + At + \frac{1}{2}(At)^2 + \frac{1}{3!}(At)^3 + \cdots\right)\\ &= A + A^2t + \frac{1}{2}A^3t^2 + \cdots\\ &= A\left(I + At + \frac{1}{2}A^2t^2 + \cdots\right) = Ae^{At}.\end{aligned}$$

Therefore, $x(t) = e^{At}x_0$ satisfies $\dot{x}(t) = Ax(t)$ and, moreover, $x(0) = Ix_0 = x_0$. The matrix exponential thus gives us a compact way to express the solution of any system of linear differential equations.

Example 3.6 Consider the linear system

$$\dot{x}_1 = -x_1 - 2x_2$$
$$\dot{x}_2 = -2x_2,$$

which we write as $\dot{x} = Ax$ with

$$A = \begin{bmatrix} -1 & -2 \\ 0 & -2 \end{bmatrix}.$$

We note that $T = \begin{bmatrix} 1 & 2 \\ 0 & 1 \end{bmatrix}$ gives us

$$\bar{A} = T^{-1}AT = \begin{bmatrix} 1 & -2 \\ 0 & 1 \end{bmatrix} \begin{bmatrix} -1 & -2 \\ 0 & -2 \end{bmatrix} \begin{bmatrix} 1 & 2 \\ 0 & 1 \end{bmatrix} = \begin{bmatrix} -1 & 0 \\ 0 & -2 \end{bmatrix}.$$

Therefore, we have

$$T^{-1}e^{At}T = e^{T^{-1}ATt} = \begin{bmatrix} e^{-t} & 0 \\ 0 & e^{-2t} \end{bmatrix}$$

and so

$$x(t) = T\begin{bmatrix} e^{-t} & 0 \\ 0 & e^{-2t} \end{bmatrix} T^{-1}x_0 = \begin{bmatrix} e^{-t} & -2e^{-t} + 2e^{-2t} \\ 0 & e^{-2t} \end{bmatrix} x_0.$$

The complete solution is therefore

$$x_1(t) = e^{-t}x_1(0) + 2(e^{-2t} - e^{-t})x_2(0)$$
$$x_2(t) = e^{-2t}x_2(0).$$

Example 3.7 Consider the initial-value problem $\dot{x} = Ax, x(0) = x_0$, where

$$A = \begin{bmatrix} 0 & 1 & 0 \\ -3 & -2 & 0 \\ -2 & -3 & -2 \end{bmatrix}, \quad x_0 = \begin{bmatrix} 1 \\ 0 \\ 0 \end{bmatrix}.$$

We can use Matlab's matrix exponential function to solve this problem as follows.

Using Matlab `expm()` function

```
clear
A = [0 1 0; -3 -2 0; -2 -3 -2];
x0 = [1 0 0]';                % initial condition
t= sym('t','real');           % define symbolic t
x=latex(expm(A*t)*x0);        % format the matrix exponential
```

Running this script returns the solution

$$x(t) = \begin{pmatrix} e^{-t}\cos\left(\sqrt{2}\,t\right) + \left(\sqrt{2}\,e^{-t}\sin\left(\sqrt{2}\,t\right)\right)/2 \\ -\left(3\sqrt{2}\,e^{-t}\sin\left(\sqrt{2}\,t\right)\right)/2 \\ 3\,e^{-2\,t} - 3\,e^{-t}\cos\left(\sqrt{2}\,t\right) + \left(\sqrt{2}\,e^{-t}\sin\left(\sqrt{2}\,t\right)\right)/2 \end{pmatrix}.$$

3.4 The Jordan Canonical Form

We saw in the previous chapter that matrices with repeated eigenvalues may not be diagonalizable by a similarity transformation but may be transformed to a nondiagonal Jordan form. In the multi-dimensional case, the Jordan canonical form is

$$J = \begin{bmatrix} J_1 & & \\ & \ddots & \\ & & J_p \end{bmatrix},$$

in which each submatrix J_i has the form

$$J_i = \begin{bmatrix} \lambda_i & 1 & & \\ & \lambda_i & \ddots & \\ & & \ddots & 1 \\ & & & \lambda_i \end{bmatrix},$$

where J_i has dimensions $\ell_i \times \ell_i$ if the eigenvalue λ_i has multiplicity ℓ_i. Each J_i is called a *Jordan block*.

Example 3.8 Consider the matrix

$$A = \begin{bmatrix} -1 & 1 & 0 & -1 \\ -6 & -5 & -2 & 0 \\ 0.5 & 0 & -3 & 0.5 \\ -4 & -1 & -2 & -4 \end{bmatrix}.$$

The eigenvalues of A are $-3, -2, -4, -4$, so the Jordan form should have two blocks of size 1×1 and one block of size 2×2. Using the command `jordan`, we find that the matrix

$$T = \begin{bmatrix} -1 & 2 & 0 & 0 \\ 3 & -4 & -1 & 1 \\ 0 & 0 & 0.5 & 0 \\ 1 & -2 & -1 & 1 \end{bmatrix}$$

gives the Jordan form

$$J = T^{-1}AT = \begin{bmatrix} -3 & 0 & 0 & 0 \\ 0 & -2 & 0 & 0 \\ 0 & 0 & -4 & 1 \\ 0 & 0 & 0 & -4 \end{bmatrix}.$$

As in the second-order case, the columns of the coordinate transformation matrix T are eigenvectors and generalized eigenvectors corresponding to the eigenvalues λ_i. If λ_i is complex, then the entries of both T and J_i will be complex. In the case of (repeated) complex eigenvalues $\alpha \pm j\beta$, real-valued Jordan blocks are of the form

$$\begin{bmatrix} D & I & & \\ 0 & D & I & \\ 0 & 0 & D & \\ & & & \ddots \end{bmatrix},$$

where each block D is

$$\begin{bmatrix} \alpha & -\beta \\ \beta & \alpha \end{bmatrix}.$$

The number of such D-blocks is equal to the multiplicity of the complex conjugate pairs of eigenvalues.

Example 3.9 Consider the 4×4 matrix

$$A = \begin{bmatrix} -0.5 & 2 & 1 & -0.5 \\ -3 & -4 & 0 & 1 \\ 1 & 0 & -2 & 1 \\ -1.5 & -2 & -5 & -1.5 \end{bmatrix}. \tag{3.2}$$

The matrix A has repeated complex eigenvalues at $-2 \pm 2j$, $-2 \pm 2j$. It is left as an exercise (Problem 3.9) to show, using the Matlab command `jordan`, that the matrix

$$T = \begin{bmatrix} -0.25 & 0.25 & 0.5 & 0.5 \\ 0.5 & 0 & 0 & -1 \\ 0 & 0 & 0 & 0.25 \\ 0.25 & -0.25 & 0 & -0.5 \end{bmatrix}$$

transforms A into the Jordan form

$$T^{-1}AT = \begin{bmatrix} -2 & -2 & 1 & 0 \\ 2 & -2 & 0 & 1 \\ 0 & 0 & -2 & -2 \\ 0 & 0 & 2 & -2 \end{bmatrix}. \tag{3.3}$$

The Jordan form is useful to compute the matrix exponential solution of a linear ordinary differential equation. If λ_i is an eigenvalue of multiplicity n, then the exponential e^{At} will have the block submatrix

$$e^{J_i t} = \begin{bmatrix} 1 & t & \frac{t^2}{2!} & \cdots & \frac{t^{n-1}}{(n-1)!} \\ 0 & 1 & t & \cdots & \frac{t^{n-2}}{(n-2)!} \\ \vdots & & 1 & \ddots & \vdots \\ \vdots & & & \ddots & t \\ 0 & & \cdots & \cdots & 1 \end{bmatrix} e^{\lambda_i t}.$$

Example 3.10 Consider the system in Jordan form

$$\dot{x} = \begin{bmatrix} 0 & 1 \\ 0 & 0 \end{bmatrix} x,$$

which is called the *double integrator* system. The exponential of the coefficient matrix from the above formula is

$$e^{At} = \begin{bmatrix} 1 & t \\ 0 & 1 \end{bmatrix},$$

which leads to the solution of the initial-value problem

$$\begin{aligned} x_1(t) &= x_1(0) + tx_2(0) \\ x_2(t) &= x_2(0). \end{aligned}$$

Stability of Linear Systems

It is evident from the Jordan form that, if the eigenvalues of the system coefficient matrix A are negative or have negative real part, then the exponential terms will decay to zero and all solutions of the linear system will converge to the equilibrium $\bar{x} = 0$, just as we saw in the case of second-order systems. We refer to this property as *exponential stability*. We introduce the following terminology to describe these systems.

Definition 3.1 ***(Hurwitz matrix)*** An $n \times n$ matrix A is *Hurwitz* if and only if all n eigenvalues of A have negative real part.

As a consequence, we note that if A is Hurwitz then $\bar{x} = 0$ is a hyperbolic equilibrium point for the linear system $\dot{x} = Ax$ and the equilibrium is exponentially stable. It further follows that if any eigenvalue of A has positive real part then the origin is an unstable equilibrium. The case where one or more eigenvalues of A has zero real part is more subtle and we will say more about this and the stability of equilibrium points in Chapter 5.

Other Matrix Functions

The matrix exponential is the most important matrix function for our purposes. Here we will briefly mention a few other matrix functions and how to compute them in Matlab. First note that if a function f has a convergent Taylor series expansion,

$$f(a) = f(0) + f'(0)a + \frac{f''(0)}{2!}a^2 + \frac{f'''(0)}{3!}a^3 + \dots$$

then we can define $f(A)$, for a matrix A, as

$$f(A) = f(0)I + f'(0)A + \frac{f''(0)}{2!}A^2 + \frac{f'''(0)}{3!}A^3 + \dots,$$

for example,

$$\sin(A) = A - \frac{1}{3!}A^3 + \frac{1}{5!}A^5 - \dots,$$
$$\cos(A) = I - \frac{1}{2!}A^2 + \frac{1}{4!}A^4 - \dots.$$

Note also that our familiar trigonometric identities hold for these matrix functions, such as

$$\sin^2(A) + \cos^2(A) = I,$$
$$\sin(A \pm B) = \sin(A)\cos(B) \pm \cos(A)\sin(B),$$
$$\cos(A \pm B) = \cos(A)\cos(B) \mp \sin(A)\sin(B)$$

although the latter two identities hold only if $AB = BA$. We also note that

$$e^{jA} = \cos(A) + j\sin(A), \quad j = \sqrt{-1},$$

which can also be used to define the sine and cosine of a matrix in the usual way from the matrix exponential.

Matlab has the built-in functions `expm(A)`, `logm(A)`, and `sqrtm(A)`, which respectively return the exponential, natural logarithm, and square root of the matrix A. In addition, the Matlab command `funm(A, fun)` computes `fun(A)` where `fun` is a handle to one of `@exp`, `@sin`, `@cos`, `@log`, `@sinh`, or `@cosh`.

Example 3.11 Consider the matrix differential equation

$$\ddot{y} + Ay = 0, \quad y(0) = y_0, \quad \dot{y}(0) = y_1,$$

where A is an $n \times n$ matrix and y is an n-dimensional vector. The solution of this equation is

$$y = \cos\left(\sqrt{A}\right)y_0 + \left(\sqrt{A}\right)^{-1}\sin\left(\sqrt{A}\right)y_1.$$

Remark 3.2 There are several subtleties associated with computing matrix functions, such as `logm(A)` and `sqrtm(A)`, that are left out of the above discussion. The interested reader should consult the Matlab documentation of these functions for details.

Note also that the Matlab command `exp(A)` does not compute the matrix exponential. Rather it computes the exponential of each element of A separately, and similarly for other Matlab functions, `sin`, `cos`, and so on.

3.5 Linearization

For an n-dimensional nonlinear system $\dot{x} = f(x)$, $x \in \mathbb{R}^n$, the linear approximation about an equilibrium point is defined in the same way as the second-order case discussed in the previous chapter. In the n-dimensional case, the Jacobian matrix is an $n \times n$ matrix of partial derivatives

$$J(x) = \begin{bmatrix} \frac{\partial f_1}{\partial x_1} & \cdots & \frac{\partial f_1}{\partial x_n} \\ \frac{\partial f_2}{\partial x_1} & \cdots & \frac{\partial f_2}{\partial x_n} \\ \vdots & \ddots & \vdots \\ \frac{\partial f_n}{\partial x_1} & \cdots & \frac{\partial f_n}{\partial x_n} \end{bmatrix}.$$

The linear approximation of the nonlinear system about an equilibrium point $\overline{x}$ is then

$$\dot{z} = Az,$$

where $A = J(\overline{x})$ is the Jacobian evaluated at $\overline{x}$, and $z = x - \overline{x}$.

Example 3.12 Consider the third-order nonlinear system

$$\begin{aligned} \dot{x}_1 &= x_2 + x_1^3 \\ \dot{x}_2 &= -2x_3 \\ \dot{x}_3 &= -x_3^3 + x_2 - 2x_1. \end{aligned} \tag{3.4}$$

It is straightforward to show (Problem 3.8) that the origin $(0,0,0)^T$ is the unique equilibrium point for this system and the linear approximation about the origin is $\dot{x} = Ax$, where

$$A = \begin{bmatrix} 0 & 1 & 0 \\ 0 & 0 & -2 \\ -2 & 1 & 0 \end{bmatrix}.$$

3.6 The Hartman–Grobman Theorem

In this section we present a result known as the *Hartman–Grobman theorem*, which says that the behavior of a nonlinear system near a hyperbolic equilibrium point is qualitatively the same as its linear approximation about the equilibrium point. The definition of a hyperbolic fixed point in higher dimensions is the same as for second-order systems.

Definition 3.2 ***(Hyperbolic Fixed Point)*** The fixed point $\overline{x}$ of an nth-order nonlinear system is hyperbolic if and only if the $n \times n$ Jacobian $A = \frac{\partial f}{\partial x}(\overline{x})$ has no eigenvalues with zero real part. Equivalently, $x = \overline{x}$ is a hyperbolic fixed point for the nonlinear system if and only if $\overline{z} = x - \overline{x} = 0$ is a hyperbolic fixed point for the linear approximation of the nonlinear system about $x = \overline{x}$.

Theorem 3.2 ***(Hartman–Grobman)*** *Let $\Omega \subset \mathbb{R}^n$ be an open subset of $\mathbb{R}^n$ and let $f : \Omega \to \mathbb{R}^n$ be a continuously differentiable vector field. Suppose $\overline{x} \subset \Omega$ is a hyperbolic fixed point of the autonomous nonlinear system*

$$\dot{x} = f(x).$$

Let A be the Jacobian of $f(x)$ at $x = \overline{x}$ and $x_\ell(t,x) = \overline{x} + e^{At}(x - \overline{x})$ be the phase portrait of the linearized system about $\overline{x}$. (These are trajectories starting at x near $\overline{x}$.) Let $x_{n\ell}(t,x)$ be the local phase portrait generated by the nonlinear system. Then there are neighborhoods $\mathcal{U}$ and $\mathcal{V}$ and a continuous (and invertible) transformation $h : \mathcal{U} \to \mathcal{V}$ such that

$$x_{n\ell}(t, h(x)) = h(x_\ell(t,x))$$

whenever x and $\overline{x} + e^{At}(x - \overline{x})$ are in $\mathcal{U}$.

Figure 3.2 illustrates the Hartman–Grobman theorem for a two- dimensional system. The phase portrait of the nonlinear system (left) near the equilibrium can be thought of as a distorted version of the phase portrait of the linear system (right). Roughly speaking, the Hartman–Grobman theorem says that the phase portrait

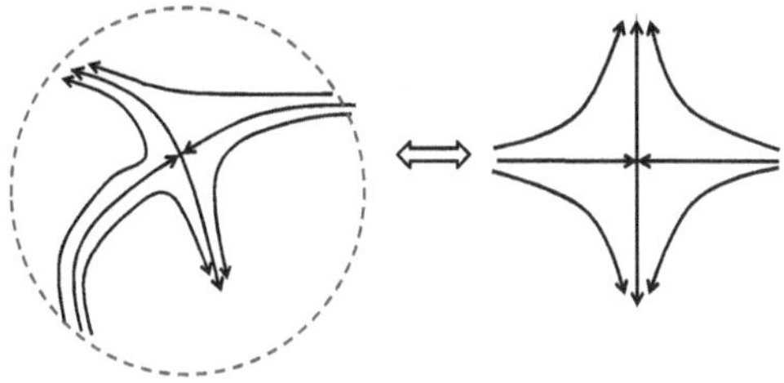

Figure 3.2 Illustrating the Hartman–Grobman theorem.

of the nonlinear system can be mapped or deformed to the phase portrait of the linearization by a suitable function $h(x)$ in a sufficiently small neighborhood of hyperbolic fixed points. The theorem does not tell us how to determine the function $h(x)$.

Thus, for hyperbolic fixed points, the qualitative local behavior of the nonlinear system can be determined from the linear approximation. Global behavior and the behavior near nonhyperbolic fixed points are more difficult to determine.

Note that the above property does not, in general, hold for nonhyperbolic equilibrium points. For example, consider the first-order system

$$\dot{x} = \alpha x^3.$$

The equilibrium $\overline{x} = 0$ is not hyperbolic and linearization about the origin is $\dot{x} = 0$. The origin of the nonlinear system is asymptotically stable for $\alpha < 0$ and unstable for $\alpha > 0$. Thus the linearization about $\overline{x} = 0$ does not give us sufficient information to determine the behavior of the nonlinear system in this case.

3.7 Singular Perturbations

All models of physical systems are approximations of reality. A good model should capture the essential features of the system and, at the same time, it should be as simple as possible. For example, an electrical circuit will generally have stray capacitances. A mechanical or hydraulic system will have elasticity in gears and hoses. The decision whether or not to include such effects in the model affects both the accuracy of the model and its mathematical tractability. The notion of singular perturbations helps to make this precise.

Example 3.13 Consider the third-order linear system

$$\begin{aligned}\dot{x}_1 &= x_2\\ \dot{x}_2 &= -x_1 + x_3\\ \dot{x}_3 &= -10x_3.\end{aligned} \tag{3.5}$$

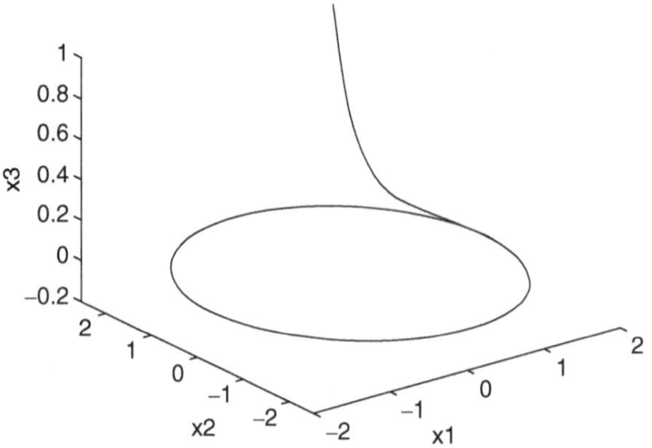

Figure 3.3 Solution trajectory of the system (3.5) in $\mathbb{R}^3$. The trajectory rapidly converges to the (x_1, x_2) plane and asymptotically tracks the solution of the harmonic oscillator.

We note that the third equation has solution $x_3(t) = e^{-10t}x_3(0)$, which rapidly decays to zero for any initial value of x_3. Note also that with $x_3 = 0$, the first two equations are the harmonic oscillator equations. Figure 3.3 shows the trajectory of the system in $\mathbb{R}^3$ with initial conditions (2,2,1).

Thus, the long-term behavior of this system is accurately modeled by the harmonic oscillator. The advantage of neglecting the third component x_3 of the system is that we are able to use a second-order model, which is simpler to analyze than the full-order system.

Standard Singular Perturbation Form

If we set $z = x_3$ and $\epsilon = \frac{1}{10}$, we can write the system (3.5) as

$$\begin{aligned} \dot{x}_1 &= x_2 \\ \dot{x}_2 &= -x_1 + z \\ \epsilon \dot{z} &= -z. \end{aligned} \tag{3.6}$$

The system (3.6) is said to be in the *standard form* for singular perturbations. The term *singular perturbation* refers to the fact that, for all $\epsilon > 0$, the system (3.6) is third order but when $\epsilon = 0$ the system becomes

$$\dot{x}_1 = x_2 \tag{3.7}$$

$$\dot{x}_2 = -x_1 + z \tag{3.8}$$

$$0 = -z, \tag{3.9}$$

which is a second-order system (3.7)–(3.8), together with an algebraic equation (3.9).

The states x_1 and x_2 are called *slow* variables and z is called a *fast* variable. The idea is that, provided the fast variable decays sufficiently fast to zero, the second-order model

$$\dot{x}_1 = x_2$$
$$\dot{x}_2 = -x_1$$

will be a sufficient model for the system.

Example 3.14 This example shows the importance of choosing the right state variables for a given model. Consider the second-order linear system

$$\ddot{y} + 11\dot{y} + 10y = 0.$$

With $x_1 = y$ and $x_2 = \dot{y}$ we have the state-space model

$$\dot{x}_1 = x_2$$
$$\dot{x}_2 = -10x_1 - 11x_2.$$

With these state variables it is not evident that there may be a reduced-order model that is a good approximation of the full-order model. However, if we change variables by defining

$$z = x_1 + x_2,$$

then a straightforward calculation (Problem 3.12) gives

$$\dot{x}_1 = -x_1 + z \tag{3.10}$$
$$\dot{z} = -10z. \tag{3.11}$$

Therefore, we see that, in these coordinates, the system can be put into standard singular perturbation form with $\epsilon = 1/10$ and so the long-term behavior of the system is well described by the first-order model $\dot{x} = -x$.

Example 3.15 ***(DC Motor)***
Let us consider a simple model of a permanent-magnet DC motor connected to a load (Figure 3.4, adapted from [44, 46, 64]).

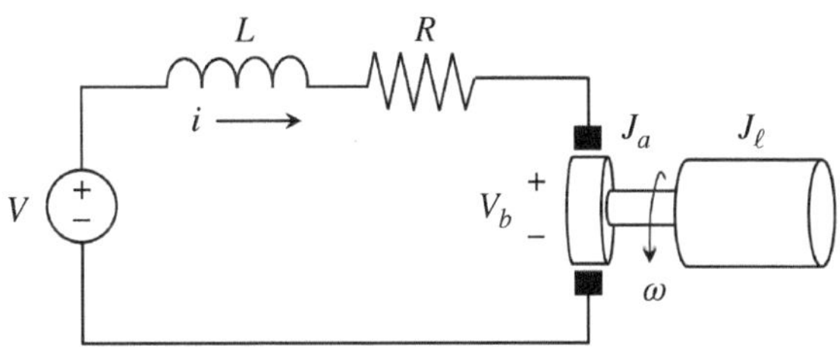

Figure 3.4 Schematic of a permanent-magnet DC motor connected to a load.

The dynamics of the system consist of the electrical dynamics

$$L\frac{di}{dt} + Ri = V - V_b,$$

where

$$\begin{aligned} V &= \text{the armature voltage,} \\ L &= \text{the armature inductance,} \\ R &= \text{the armature resistance,} \\ V_b &= \text{the back electromotive force (emf),} \\ i &= \text{the armature current,} \end{aligned}$$

together with the mechanical dynamics

$$J\frac{d\omega}{dt} = \tau_m,$$

where $J = J_a + J_\ell$ is the sum of the rotor and load inertias, and

$$\begin{aligned} \omega &= \text{the rotor speed,} \\ \tau_m &= \text{the generated torque.} \end{aligned}$$

For a permanent-magnet DC motor, these equations are coupled according to

$$\begin{aligned} \tau_m &= K_m i, \\ V_b &= K_b \omega, \end{aligned}$$

where K_m is the motor torque constant and K_b is the back-emf constant. We can take $K_m = K_b = K$, which holds if the torque and back-emf constants are expressed in meter-kilogram-second units. This leads to the model

$$J\frac{d\omega}{dt} = Ki \tag{3.12}$$

$$L\frac{di}{dt} + Ri = V - K\omega. \tag{3.13}$$

Frequently it is assumed that the inductance L is small enough to neglect in these equations. Setting $\epsilon = L$, we have

$$\begin{aligned} J\frac{d\omega}{dt} &= Ki \\ \epsilon\frac{di}{dt} + Ri &= V - K\omega. \end{aligned}$$

At $\epsilon = 0$ we have

$$i = \frac{V}{R} - \frac{K}{R}\omega.$$

Substituting i into the first equation leads to the reduced-order model

$$J\frac{d\omega}{dt} = -\frac{K^2}{R}\omega + \frac{V}{R}.$$

Scaling

The assumption that ϵ is "small" can introduce modeling ambiguities because of dimensional scaling. For example, $\epsilon = 0.1$ meters/second may be considered a small parameter. But in terms of meters per hour, $\epsilon = 360$, which we may not recognize as a small parameter. For this reason, it is useful to scale time and space variables in such a way that ϵ becomes a dimensionless parameter. We will illustrate this for the above DC-motor example.

First, define the dimensionless variables

$$\omega_r = \frac{\omega}{\Omega}, \quad i_r = \frac{iR}{K\Omega}, \quad V_r = \frac{V}{K\Omega},$$

and rewrite the dynamic equations (3.12)–(3.13) as

$$T_m \frac{d\omega_r}{dt} = i_r$$
$$T_e \frac{di_r}{dt} = -\omega_r - i_r + V_r,$$

where $T_m = JR/K^2$ is the mechanical time constant and $T_e = L/R$ is the electrical time constant. Next we introduce a dimensionless time variable $t_r = t/T_m$, so that $dt_r = dt/T_m$. With these variables the state equations become

$$\frac{d\omega_r}{dt_r} = i_r \tag{3.14}$$

$$\frac{T_e}{T_m}\frac{di_r}{dt_r} = -\omega_r - i_r + V_r. \tag{3.15}$$

Since we assume that the mechanical time constant T_m is much larger than the electrical time constant T_e, we define ϵ as the dimensionless parameter

$$\epsilon = \frac{T_e}{T_m} = \frac{LK^2}{JR^2}$$

and note that the singularly perturbed system (3.14)–(3.15) depends only on the single parameter ϵ and so can be used as a generic model for a DC motor.

The General Case for Linear Systems

Consider the singularly perturbed linear system in the (standard) form

$$\dot{x} = A_{11}x + A_{12}z \tag{3.16}$$

$$\epsilon\dot{z} = A_{21}x + A_{22}z. \tag{3.17}$$

Here we assume that $x \in \mathbb{R}^n$ and $z \in \mathbb{R}^m$ so the order of the system is $n + m$. The dimensions of the coefficient matrices are as follows: A_{11} is $n \times n$, A_{12} is $n \times m$, A_{21} is $m \times n$, and A_{22} is $m \times m$. When $\epsilon = 0$, equation (3.17) becomes

$$0 = A_{21}x + A_{22}z.$$

Assuming that A_{22} is nonsingular, we can solve for z as

$$z = -A_{22}^{-1}A_{21}x := h(x)$$

and substitute this expression into Equation (3.16), which yields

$$\dot{x} = (A_{11} - A_{12}A_{22}^{-1}A_{21}) = \bar{A}x. \tag{3.18}$$

Equation (3.18) is called the *quasi-steady state*, also called the *slow subsystem*.

The equation $z = h(x)$ defines an $(n - m)$-dimensional subspace in $\mathbb{R}^n$ called the *slow subspace*. Next, define

$$y = z - h(x) = z + A_{22}^{-1}A_{21}x.$$

The variable y represents the deviation of z from the slow subspace defined by $h(x)$ and satisfies

$$\begin{aligned}\epsilon\dot{y} &= \epsilon\dot{z} + \epsilon(A_{22}^{-1}A_{21}\dot{x})\\ &= A_{21}x + A_{22}(y - A_{22}^{-1}A_{21}x) + \epsilon(A_{22}^{-1}A_{21}\dot{x})\\ &= A_{22}y + \epsilon(A_{22}^{-1}A_{21}\dot{x}).\end{aligned}$$

Next, we define a *stretched time scale* $\tau = t\epsilon$ so that $\epsilon\frac{dy}{dt} = \frac{dy}{d\tau}$. At $\epsilon = 0$ we consider $x(t) = x(\epsilon\tau) = x(0)$ as a constant and we define the *boundary-layer system* as

$$\frac{dy}{d\tau} = A_{22}y.$$

In summary, if the parameter ϵ is small, then we can approximate the original system by two reduced-order models, the quasi-steady state and the boundary-layer system

$$\dot{x} = (A_{11} - A_{12}A_{22}^{-1}A_{21})x = \bar{A}x \tag{3.19}$$

$$\frac{dy}{d\tau} = A_{22}y. \tag{3.20}$$

Thus, if the matrix A_{22} is Hurwitz, the quasi-steady-state model (3.19) can be taken as a reduced-order model for the full-order system provided that ϵ is suitably small.

Example 3.16 Consider the linear system

$$\begin{aligned}\dot{x} &= A_{11}x + A_{12}z\\ \epsilon\dot{z} &= A_{21}x + A_{22}z,\end{aligned}$$

where

$$A_{11} = \begin{bmatrix} 0 & 1\\ -2 & -3\end{bmatrix},\ A_{12} = \begin{bmatrix} 0 & 1\\ 1 & 1\end{bmatrix},$$

$$A_{21} = \begin{bmatrix} 1 & 0\\ -1 & -1\end{bmatrix},\ A_{22} = \begin{bmatrix} 0 & 1\\ -1 & -2\end{bmatrix}.$$

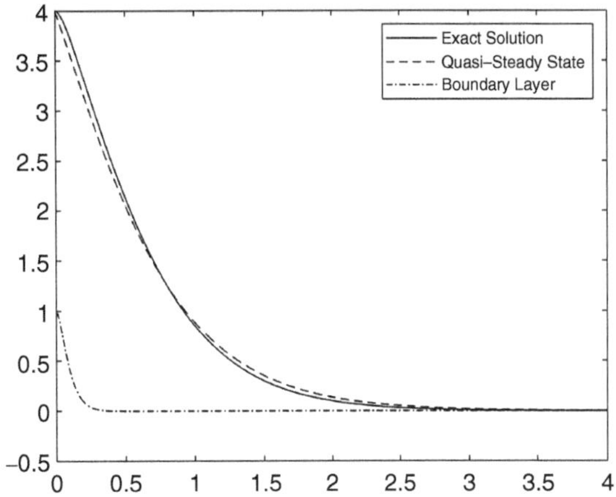

Figure 3.5 Full-order and reduced-order responses for x_1. The other state variables show similar responses. The boundary layer y decays rapidly to zero.

The quasi-steady-state system is then

$$\dot{\bar{x}} = (A_{11} - A_{12}A_{22}^{-1}A_{12})\bar{x} = \begin{bmatrix} -1 & 1 \\ -2 & -4 \end{bmatrix} \bar{x}.$$

The boundary-layer system in the τ time scale is

$$\frac{dy}{d\tau} = A_{22}y.$$

The eigenvalues of A_{22} are at $-1, -1$ and so the boundary-layer system is exponentially stable. Figure 3.5 shows a plot of $x_1(t)$, $\bar{x}_1(t)$, and the boundary layer y, with $\epsilon = 0.05$. Thus we are able to reduce the original fourth-order system to a second-order system without sacrificing accuracy of the model.

Problems

3.1 Consider the initial-value problem in Example 3.2,

$$\dot{x} = \frac{1}{2x}, \quad x(0) = 0.$$

Show by direct calculation that $x(t) = \sqrt{t}$ and $x(t) = -\sqrt{t}$ are both solutions to this initial-value problem.

3.2 Consider the initial-value problem

$$x = x^2, \quad x(0) = x_0.$$

Using the method of separation of variables, show that the solution of this equation is

$$x(t) = \frac{x_0}{1 - tx_0}.$$

3.3 Consider the second-order, nonautonomous equation

$$\ddot{y} - ty = 0. \tag{3.21}$$

Equation (3.21) is called *Airy's equation* and arises in problems in optics [4] and also in solutions of Schrödinger's equation in quantum mechanics. Airy's equation is an example of a system with a so-called *turning point*, where the solution changes from oscillatory to exponential.

1. Define state variables and write the system in state space as $\dot{x} = A(t)x$.
2. With initial conditions $y(0) = 0, \dot{y}(0) = 1$, use the method of Picard iteration to compute the first three terms in the infinite series solution of this equation. Plot the approximate solution for $y(t)$ from time $t = 0$ to $t = 3$.
3. Write a Matlab script to simulation this system using `ode45`. Compare the solution obtained with the approximate solution computed above. (Plot them both on the same graph.)
4. Now simulate the system from time $t = -10$ to $t = 3$ and plot $y(t)$. Notice the turning point at approximately $t = 1$.

3.4 Using Matlab's symbolic integration function, write a script to compute the first few terms of the Picard iteration method for a one-dimensional initial-value problem. Test your program with the following systems:

(a) $\dot{x} = x + 1,\ x(0) = 2$

(b) $\dot{x} = \sin(x),\ x(0) = 0.5$

(c) $\dot{x} = 1/(2x),\ x(0) = 1$

3.5 Prove that, if T is a nonsingular matrix, then

$$T^{-1}e^{At}T = e^{T^{-1}ATt}.$$

3.6 Compute e^{At} for each matrix A below.

$$\text{(a)}\ \begin{bmatrix} 2 & -1 \\ 1 & 2 \end{bmatrix} \quad \text{(b)}\ \begin{bmatrix} 2 & -1 \\ 0 & 2 \end{bmatrix} \quad \text{(c)}\ \begin{bmatrix} 0 & 1 \\ 1 & 0 \end{bmatrix}$$

$$\text{(d)}\ \begin{bmatrix} 0 & 1 & 2 \\ 0 & 0 & 3 \\ 0 & 0 & 0 \end{bmatrix} \quad \text{(e)}\ \begin{bmatrix} 1 & 0 & 0 \\ 0 & 3 & 1 \\ 0 & 0 & 3 \end{bmatrix} \quad \text{(f)}\ \begin{bmatrix} a & 0 & 0 \\ 1 & a & 0 \\ 0 & 1 & a \end{bmatrix}.$$

3.7 Write a Matlab script to compute the first k terms in the Taylor series expansion of $\sin(A)$, where A is an $n \times n$ matrix. With

$$A = \begin{bmatrix} -1 & -2 \\ 0 & -2 \end{bmatrix}$$

compute `funm(A,@sin)` and determine how many terms k in your series expansion for $\sin(A)$ are needed to approximate `funm(A,@sin)` to four decimal places.

3.8 Consider the nonlinear system (3.4). Show that the origin in $\mathbb{R}^3$ is the unique equilibrium point and compute the linear approximation of the nonlinear system about the origin.

3.9 Carry out the details using the Matlab command `jordan` to show that the matrix in Equation (3.3) is the Jordan canonical form of the matrix A in Equation (3.2).

3.10 Consider the nonlinear system

$$\dot{x}_1 = -x_1 + \frac{2x_2}{1+x_2^2}$$
$$\dot{x}_2 = -x_2 + \frac{2x_1}{1+x_1^2}.$$

(a) Sketch the phase portrait in the region $-2 \le x_1 \le 2, -2 \le x_2 \le 2$.
(b) Find the equilibrium points and compute the linear approximation about each equilibrium point.
(c) Sketch the phase portraits about each equilibrium point.
(d) Interpret the results from item (c) in light of the Hartman–Grobman theorem.

3.11 Consider the nonlinear system

$$\dot{x}_1 = x_2$$
$$\dot{x}_2 = x_3$$
$$\dot{x}_3 = -2x_1 - 4x_2 - 2x_3 - x_1^2 - x_2^2.$$

(a) Write a Matlab script to simulate this system for 10 seconds starting at the initial conditions $(1, 0, 0)$.
(b) Find the equilibrium points and computer the linear approximation about each equilibrium point.

3.12 Carry out the calculations in Example 3.14 to derive equations (3.10) and (3.11).

3.13 Consider the singularly perturbed system

$$\dot{x} = -x + z$$
$$\epsilon \dot{z} = -2x^3 - z.$$

(a) Compute the quasi-steady-state system.
(b) Compute the boundary-layer system.

3.14 Consider the DC-motor example

$$\frac{d\omega}{dt} = i$$
$$\epsilon \frac{di}{dt} = -\omega - i + V.$$

Write a Matlab script to simulate this system.

(a) Set the input voltage $V = 1$ and test the response of the system starting at initial conditions $i(0) = \omega(0) = 0$ for various values of $\epsilon < 1$.
(b) Compute the reduced-order system with $\epsilon = 0$. Compare the response of the reduced-order and full-order systems.

4

Compartmental Models

4.1 Introduction

In this chapter we introduce the notion of compartmental modeling, which is widely used in the fields of biology, epidemiology, pharmacokinetics, pharmacodynamics, physiology, and chemical or nuclear reactions [13, 14, 24, 36]. Compartmental modeling is a modeling method that decomposes a system into *homogeneous compartments*, where each compartment may represent a population of like individuals, material, chemicals, or other quantities [28].

Figure 4.1 is a representation of a simple two-compartment model. The arrows represent *flows* or *transfers* between compartments. We assume that the rate of change of the state $x(t)$ in each compartment is equal to the input flow minus the output flow. The governing equations of this model are therefore

$$\begin{aligned}\dot{x}_1 &= -k_1x_1 + k_2x_2\\ \dot{x}_2 &= k_1x_1 - k_2x_2.\end{aligned}$$

The coefficients k_1 and k_2 may be constant, time varying, or functions of the states x_1 and x_2. Therefore, this model is general enough to encompass both autonomous and nonautonomous systems as well as linear and nonlinear systems.

A more complex example is shown in Figure 4.2. In this model u represents an external input. It is easy to write the governing equations by inspection as

$$\begin{aligned}\frac{dx_1}{dt} &= -k_1x_1 + k_2x_2 + bu\\ \frac{dx_2}{dt} &= -(k_2 + k_3 + k_6)x_2 + k_1x_1 + k_4x_3 + k_5x_4\\ \frac{dx_3}{dt} &= -k_4x_3 + k_3x_2\\ \frac{dx_4}{dt} &= -k_5x_4 + k_6x_2.\end{aligned}$$

Introduction to Modeling and Simulation: A Systems Approach, First Edition. Mark W. Spong.

Companion Website: www.wiley.com/go/spong/engineeringmodeling

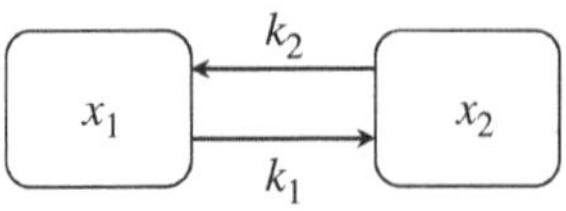

Figure 4.1 Example of a compartmental model. The arrows represent flows or transfers between compartments.

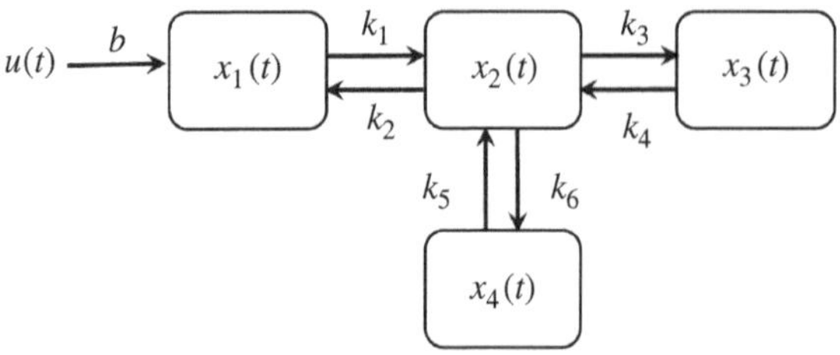

Figure 4.2 Example of a multi-compartmental model.

4.2 Exponential Growth and Decay

Exponential growth or exponential decay occurs when the rate of change of a certain quantity is proportional to its current value. Bacterial growth, compound interest, Ponzi schemes, computer processing power, and nuclear chain reactions are just a few examples where exponential growth can occur.

Drug Distribution Models

An example of exponential decay, and an early use of compartmental modeling, is in the modeling of alcohol distribution in the body [75]. If we let $x(t)$ represent the concentration of alcohol in the blood stream, then we have a one-compartment model as shown in Figure 4.3, where $u(t)$ and $y(t)$ represent the rate of alcohol intake and rate of absorption by the body, respectively.

Let us assume that a fixed amount of alcohol is ingested, leading to an initial concentration x_0 at time t_0, and we take the input $u(t) = 0$ for $t > t_0$. If the rate of absorption is proportional to the current concentration, we have the linear model

$$\dot{x}(t) = -cx(t), \quad x(t_0) = x_0,$$

which has solution

$$x(t) = e^{-c(t-t_0)}x_0.$$

For $c > 0$, this shows that the concentration is exponentially decreasing over time.

A more sophisticated model representing drug distribution in the body is shown in Figure 4.4, consisting of two compartments. The compartment labeled *blood*

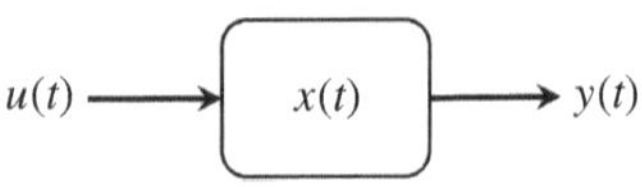

Figure 4.3 One-compartment model.

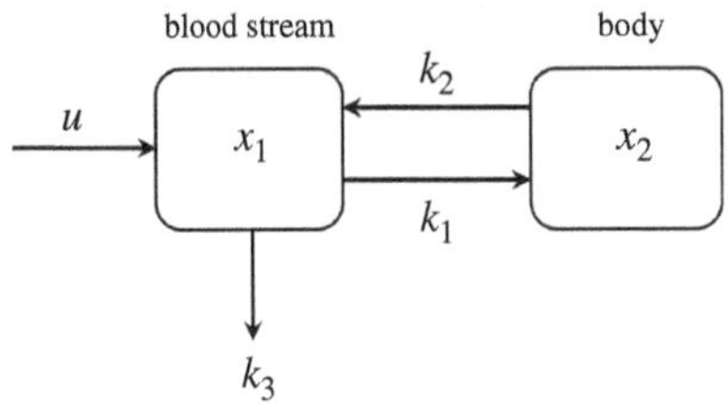

Figure 4.4 Two-compartment drug distribution model.

stream represents the circulatory system and the compartment labeled *body* represents organs, tissue, or other parts of the body.

We suppose that a drug is introduced into the blood stream at a rate $u \geq 0$. The parameters k_1 and k_2 represent rates of drug exchange between the two compartments. Also, the drug is eliminated from the blood stream by the liver at a rate k_3. The system equations are therefore given by

$$\dot{x}_1 = -k_1 x_1 - k_3 x_1 + k_2 x_2 + u$$
$$\dot{x}_2 = k_1 x_1 - k_2 x_2.$$

With the parameters shown below in the simulation, the response is given in Figure 4.5, showing that the respective drug concentrations increase exponentially to a steady-state value.

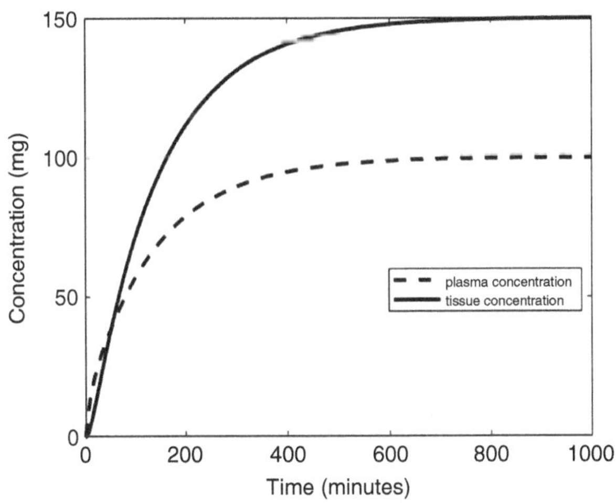

Figure 4.5 Drug concentrations in the blood and tissue with a constant input rate $u = 2$ mg/min.

Drug Distribution Model

```
u = 2;        % input rate of the drug
k1 = 0.04;    % fractional exchange rates in 1/min
k2 = 0.06;
k3 = 0.02;
tfinal = 1000; % simulation time
X0= [0;0];      % initial concentration
f=@(t,x) [u-k1*x(1)-k3*x(1)+k2*x(2); k1*x(1)-k2*x(2)];
[t,x]= ode45(f,[0,tfinal],X0);
```

We may also model the case where we assume that a single dose of a drug is administered all at once and is perfectly mixed in the blood at time t_0. In this case we take $u = 0$ and initial conditions $x_0 = [x_1(0); 0]$. Figure 4.6 shows the response with $x_1(0) = 60$ mg.

The concentration in the blood stream decreases exponentially to zero. The concentration in the tissue reaches a maximum value and then decreases exponentially to zero.

Population Dynamics

In his 1798 book, *An Essay on the Principle of Population*, Thomas Malthus argued that with unlimited food supply a population will grow exponentially. A simple

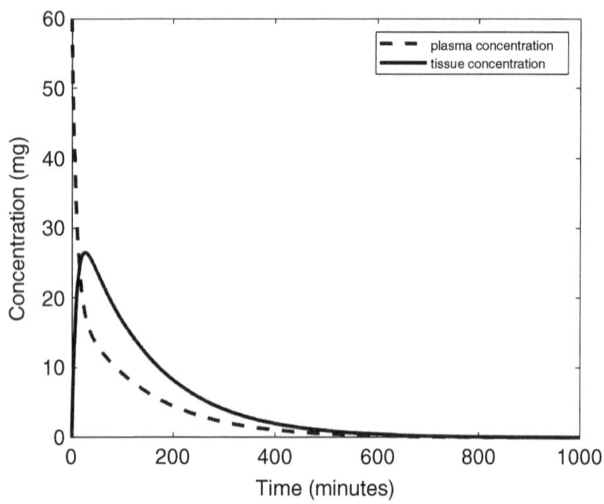

Figure 4.6 Drug concentrations in the blood and tissue with initial concentration $x_1(0) = 60$ and $x_2(0) = 0$.

model of the population N of a given species that captures the exponential growth phenomenon is

$$\frac{dN}{dt} = (b - d)N, \tag{4.1}$$

where b is the birth rate and d is the death rate. Then, if the birth rate exceeds the death rate, $b > d$, the solution represents exponential growth. Malthus also argued that in human society, the food supply increases only arithmetically and hence exponential growth cannot continue forever. His ideas greatly influenced Charles Darwin in his development of the theory of natural selection based on competition among species for limited resources. Other factors that may limit population growth include disease, wars, and overcrowding. In economic systems, limited resources, limited money supply, and so on will limit economic growth. Thus the exponential growth model (4.1) does not adequately describe what happens in many real-world situations. A more realistic model is the so-called *logistic model*, described next.

4.3 The Logistic Equation

The *logistic equation*, which is a model that describes limited growth. is given by

$$\frac{dN}{dt} = rN(t) - cN^2(t), \quad r > 0, \ c > 0. \tag{4.2}$$

For small values of N the linear term dominates, leading to exponential growth For large values of N the quadratic term dominates, leading to a negative rate of change and saturation of N. The logistic equation is an important model and has been discovered and rediscovered several times, first by Pierre-François Verhulst in 1838 [70], who was motivated by the work of Malthus, and later by A. G. McKendrick [43] to model the growth of bacteria in broth. Later, R. Pearl and L. Reed rediscovered the logistic equation in their work on predicting population growth in the United States [55]. For this reason, the logistic equation is sometimes called the *Verhulst–Pearl equation*.

The logistic equation has two equilibrium points, namely $N = 0$ and $N = K = r/c$. The parameter K is called the *carrying rate* and determines the steady-state value of N. With this value of K we can rewrite the logistic equation as

$$\frac{dN}{dt} = rN\left(1 - \frac{N}{K}\right).$$

The logistic equation can be solved by separation of variables. It is left as an exercise (Problem 4.2) to show that the solution starting at $N(0) = N_0$ is

$$N(t) = \frac{K}{1 + Ae^{-rt}}, \quad \text{where } A = \frac{K - N_0}{N_0}.$$

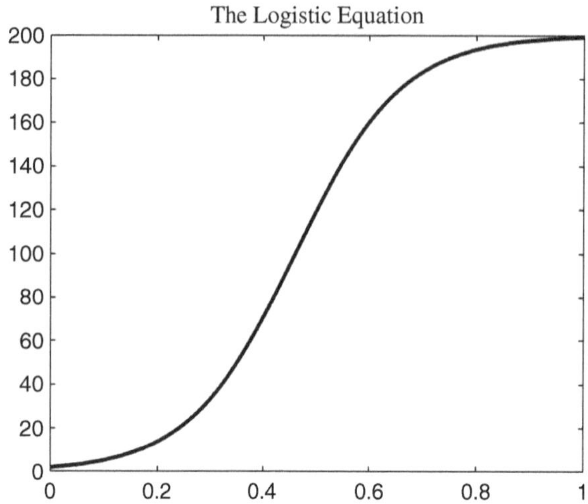

Figure 4.7 Solution of the logistic equation.

The solution $N(t)$ of the logistic equation is the familiar S-curve shown in Figure 4.7.

The S-curve also arises in models of product adoption, spread of infections, project completion, logistic regression, autocatalytic chemical reactions, and others.

For example, in the so-called *diffusion of innovation*, a new product or new innovation initially receives enormous interest and sales or adoptions rapidly increase. Later, as the idea or interest matures, the market saturates, resulting in the classic S-curve.

4.4 Models of Epidemics

In this section we discuss compartmental models of disease transfer and epidemics. These models not only describe diseases such as colds and flu, AIDS, SARS, and coronavirus, but also can be used to model computer viruses, the spread of fads, rumors, political opinions and other phenomena.

We start with a fixed-size population of K individuals and partition the population into compartments, for example those who are infected (I), those who are susceptible to infection (S), those who have recovered from infection (R), those who have been exposed (E) but are not yet symptomatic, and so on. Transfers between compartments occur when, for example, a susceptible individual becomes infected or an infected individual recovers.

The SI Model

We start with the simplest epidemic model, which is the so-called SI model.

Figure 4.8 SI epidemic model consisting of two compartments, susceptible (S) and infected (I) individuals.

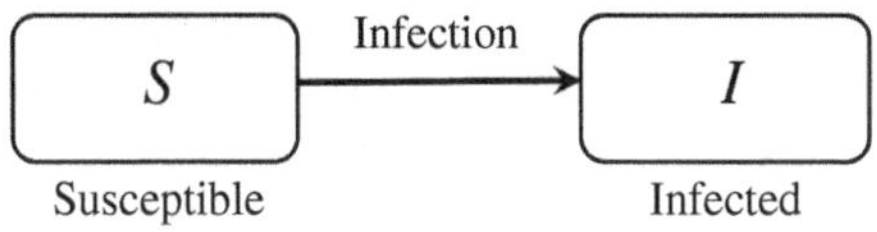

In this model there are two compartments, susceptible individuals (S) and infected individuals (I). The transfer between compartments occurs based not simply on the number of susceptible and infected individuals, but rather on the rate at which susceptible and infected individuals come in contact with one another, which is proportional to the product SI. Thus we can write the model as

$$\dot{S} = -\frac{\beta}{K}SI \tag{4.3}$$

$$\dot{I} = +\frac{\beta}{K}SI. \tag{4.4}$$

The parameter β is the *contact rate*, which means that an infected individual comes in contact with βK other individuals per unit of time. The fraction of the population susceptible to infection is S/K. We can think of β as the probability of transmitting the disease between an infectious individual and a susceptible individual.

In the SI model, once infected, individuals stay infected and infectious throughout their life. This model matches the behavior of diseases like cytomegalovirus, which is related to the virus that causes chickenpox or herpes simplex. With a constant population size $K = S + I$, then $S = K - I$ and we can substitute this expression for S in equation (4.4) to obtain

$$\dot{I} = \frac{\beta}{K}I(K - I) = \beta I\left(1 - \frac{I}{K}\right). \tag{4.5}$$

This is again the logistic equation. As the number of infected individuals rises, there are fewer and fewer remaining individuals that can become infected. We see from equation (4.5) that there are two equilibrium points, $I = 0$ and $I = K$. Figure 4.9 shows the response of the SI model for a population size of 1000 starting with a single infected individual. In this model, all individuals eventually become infected.

SI Epidemic Model: $x(1) = S, x(2) = I$

```
K= 1000;            % population size
beta = 0.2;         % contact rate
tfinal = 100;       % simulation time
X0= [K-1; 1];       % initial conditions
f=@(t,x)[-beta*x(1)*x(2)/K; beta*x(1)*x(2)/K];
[t,x] = ode45(f,[0,tfinal],X0);
```

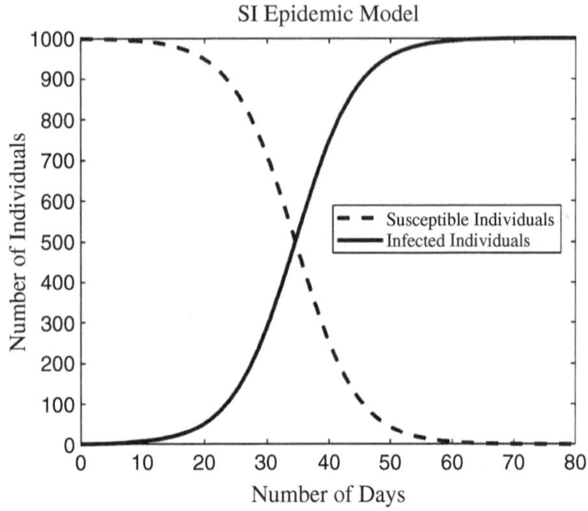

Figure 4.9 Response of the SI model with $\beta = 0.2$.

The SIS Model

A related model is the SIS model, which is useful for diseases that have repeat infections, like the common cold. In this model, infected individuals recover (without immunity) and return to the susceptible state (Figure 4.10). The SIS model is described by the equations

$$\dot{S} = -\frac{\beta}{K}SI + \gamma I$$
$$\dot{I} = \frac{\beta}{K}SI - \gamma I,$$

where β is the contact rate and γ is the recovery rate. The quantity $\frac{1}{\gamma}$ is the average time that an infected individual remains infectious. As before, with a constant population size $K = S + I$, we write the second equation for I as

$$\dot{I} = \beta I \left(1 - \frac{I}{K}\right) - \gamma I.$$

We see that there are two equilibria, $I = 0$ and $I = (1 - \frac{\gamma}{\beta})K$.

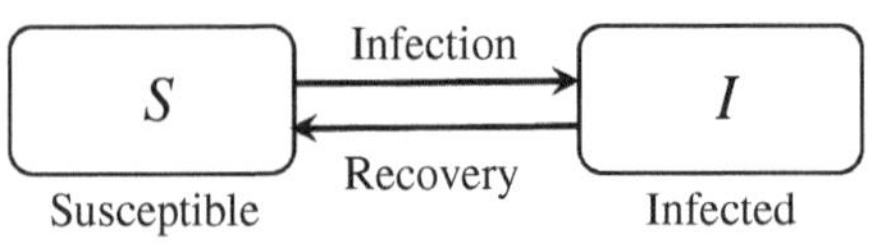

Figure 4.10 The SIS epidemic model. Infected individuals recover at a certain rate and then become susceptible to reinfection.

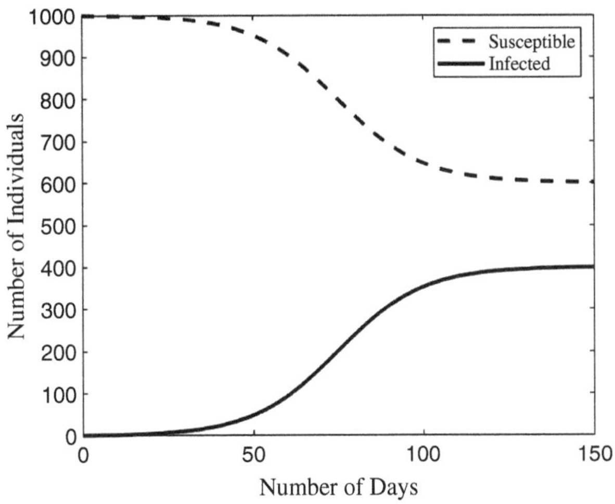

Figure 4.11 Response of the SIS model with $\beta = 0.2$, $\gamma = 0.12$, and $K = 1000$. The basic reproduction number is $\beta/\gamma = 1.667$. The number of infected individuals reaches the equilibrium value of $I = 400$.

Figure 4.11 shows a typical response of the SIS model.

Basic Reproduction Number

The ratio $R_0 = \beta/\gamma$ is called the *basic reproduction number* and is an extremely important quantity to consider as it is a determining factor in whether or not an epidemic persists or dies out. Roughly speaking, since R_0 is the product of the contact rate, β, and the average time, $1/\gamma$, spent in the infectious compartment, it indicates how many people a single infectious person is likely to infect. If $R_0 < 1$ then the disease will die out, while if $R_0 > 1$ the disease will persist and the number of infected individuals will reach the equilibrium $I = (1 - \frac{\gamma}{\beta})K$ in the steady state.

SIS Epidemic Model: $x(1) = S, x(2) = I$

```
K = 1000;         % population size
beta =   0.2;     % contact rate
gamma =  0.1;     % recovery rate
tfinal  = 160;    % simulation time
X0= [K-1; 1];     % initial conditions
f=@(t,x)[-beta*x(1)*x(2)/K+gamma*x(2);
          beta*x(1)*x(2)/K-gamma*x(2)];
[t,x]= ode45(f,[0, tfinal],X0);
```

The SIR Model

A further refinement to the SI model is the SIR model (Figure 4.12) in which we also compute the number of recovered individuals R. Recovered individuals may either be immune from further infection or they may be infected again depending on the nature of the particular disease being modeled. The SIR model is the starting point for modeling the recent Covid-19 pandemic.

In the case of permanent immunity, we can write

$$\begin{aligned}\dot{S} &= -\frac{\beta}{K}SI \\ \dot{I} &= \frac{\beta}{K}SI - \gamma I \\ \dot{R} &= \gamma I.\end{aligned}$$

We can see from these equations that any point of the form $(S, I, R) = (S, 0, R)$ is an equilibrium for the SIR model. In this case if the basic reproduction number $R_0 > 1$ then the number of infected individuals will initially rise before declining to zero.

Remark 4.1 Note that, under the assumption of a constant population size $K = S + I + R$, only two of the above three equations are needed. We could,

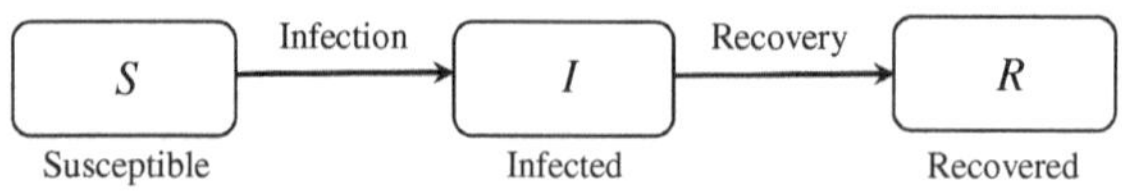

Figure 4.12 The SIR epidemic model.

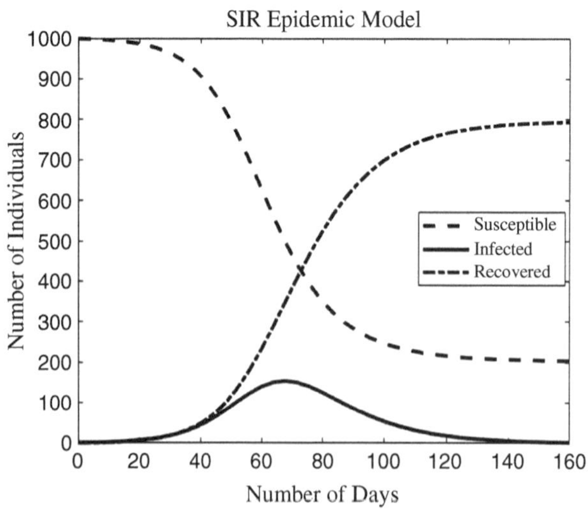

Figure 4.13 Response of the SIR model with $\beta = 0.2$ and $\gamma = 0.1$.

for example, solve the first two equations for S and I and simply compute R as $R = K - S - I$, in other words

$$\dot{S} = -\frac{\beta}{K}SI$$
$$\dot{I} = \frac{\beta}{K}SI - \gamma I$$
$$R = K - S - I.$$

Figure 4.13 shows the response of the SIR system with $\beta = 0.2$ and $\gamma = 0.1$, so that $R_0 = 2$.

SIR Epidemic Model: $x(1) = S, x(2) = I, x(3) = R$

```
K = 1000;          % population size
beta = 0.2;        % contact rate
gamma=0.1;         % recovery rate
tfinal = 160;      % simulation time
X0= [N-1; 1;0]; % initial conditions
f=@(t,x) [-beta*x(1)*x(2)/N;
          beta*x(1)*x(2)/N-gamma*x(2);gamma*x(2)];
[t,x]= ode45(f,[0,tfinal],X0);
```

Flattening the Curve

One frequently hears the expression *flatten the curve* with respect to epidemics like the Covid-19 pandemic. This refers to lowering the peak value of infected individuals as a function of time so that the healthcare systems do not become overwhelmed. This can be accomplished by reducing the contact among susceptible and infected individuals using social distancing, quarantines, wearing masks, and so on, which affects the value of the parameter β and hence the value of R_0. Figure 4.14 shows the number of infected individuals over time with three different values of the basic reproduction number R_0.

Further Extensions of the SIR Model

There are several increasingly sophisticated models that are used for epidemic modeling. The SEIR model adds the compartment of *exposed* individuals E. These are individuals who become exposed at the rate β but are not yet infectious.

The governing equations for the SEIR model are

$$\dot{S} = -\frac{\beta}{K}SI$$
$$\dot{E} = \frac{\beta}{K}SI - \delta E$$
$$\dot{I} = \delta E - \gamma I$$
$$\dot{R} = \gamma I.$$

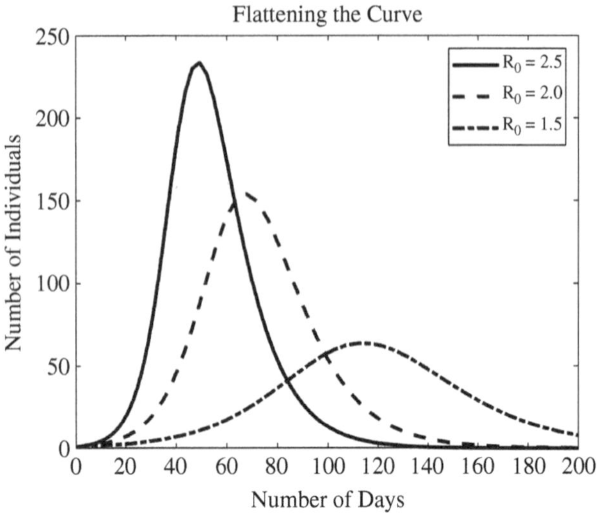

Figure 4.14 Flattening the curve with three different values for R_0, namely, $R_0 = 2.5, 2.0, 1.5$.

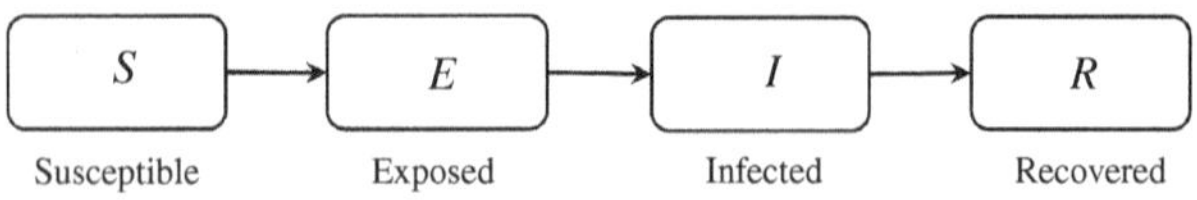

Figure 4.15 The SEIR compartmental model. In this model individuals transfer from susceptible to exposed rather than directly to infected.

We note that, in this model, individuals transfer from S to E and then from E to I rather than directly from S to I as in the SIR model.

Another modification of either the SIR or SEIR model is to include so-called *vital dynamics*, which means including births and deaths in the population. If we assume a birth rate α and an identical death rate so that the total population size remains constant, then the SIR model with vital dynamics becomes

$$\begin{aligned}\dot{S} &= -\beta SI - \alpha S + \alpha\\ \dot{I} &= \beta SI - \gamma I - \alpha I\\ \dot{R} &= \gamma I - \alpha R.\end{aligned}$$

The compartmental model of the SIR model with vital dynamics is shown in Figure 4.16.

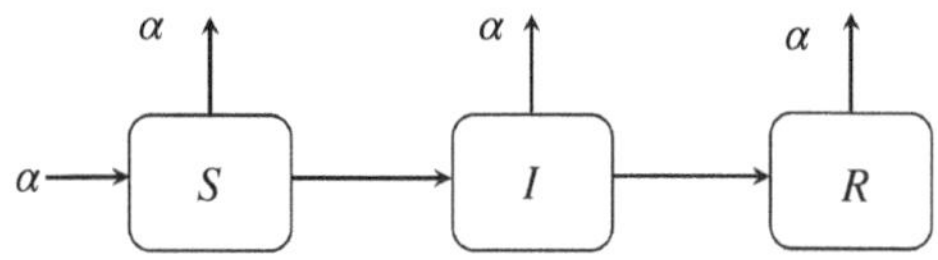

Figure 4.16 SIR model with vital dynamics. The birth and death rates are both equal to α in this example.

4.5 Predator–Prey System

In this section we discuss the *predator–prey* system. The predator–prey system, also called the *Volterra–Lotka* system, was originally motivated by a study of populations of fish in the Adriatic Sea [7]. This model is an extension of the logistic model to include the interaction of two species: a predator and a prey. It was motivated by an observation that the percentage of predator fish caught increased during the First World War, during a time that the fishing effort had greatly declined. One might speculate that reduced fishing operations removed fewer prey fish from the ecosystem, which provided more food for the predators, allowing the predator population to increase. Another well-known study was of populations of lynx and hares in Canada, where it was observed that the populations of the two species oscillated over time [65].

To model such a system, let $x(t)$ be the number of prey individuals and $y(t)$ the number of predators at time t. Then the predator–prey system is given by

$$\dot{x} = ax - bxy \tag{4.6}$$

$$\dot{y} = -cy + dxy, \tag{4.7}$$

where a, b, c, d are nonnegative constants. The simplifying assumptions behind this model are that:

1. the prey has unlimited food supply,
2. the predators eat only the single prey species.

In this case, with no interaction among predator and prey, $b = d = 0$, the prey population grows exponentially and the predator population dies out exponentially. With b and d nonzero, the model reflects that contact between the predator and prey tends to increase the predator population and decrease the prey population.

It is straightforward to show that the predator–prey equations have two equilibrium points, (0,0) and $(c/d, a/b)$, which are a saddle point and a center, respectively, for the linear approximation (Problem 4.7). Since the populations of each species cannot be negative, we are interested only in the portion of the phase portrait in the first quadrant, $x \geq 0, y \geq 0$. Figure 4.17 shows the phase portrait of

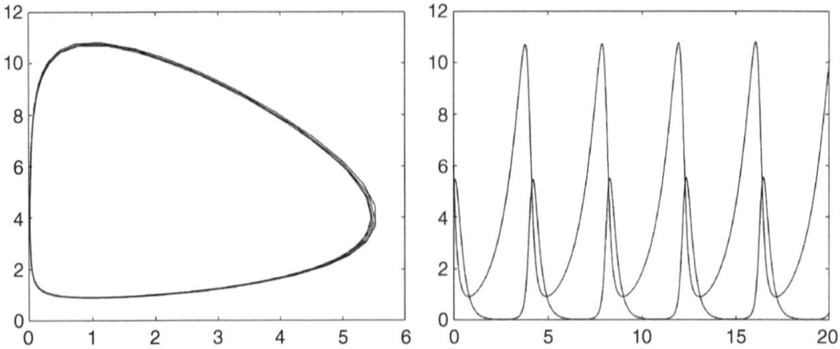

Figure 4.17 Phase portrait of the predator–prey system in the first quadrant (left) and the time responses of predator and prey (right). The populations oscillate around the equilibrium at $(c/d, a/b)$.

the system and the time response of $x(t)$ and $y(t)$, showing the phase difference in the oscillation of the two populations.

The Volterra–Lotka model has also been used to model business cycles in economics, the so-called *Goodwin model* [31], which is based on the theory that the driving forces behind fluctuations in economic activity are class struggles between capitalists and workers over income distribution.

Multiple Predator and Prey Species

In a more complex food network, there will be multiple predator and prey species. Here we give an example of a system with two predator and two prey species. The diagram in Figure 4.18 indicates that predator 1 feeds on both prey species and predator 2 feeds on prey 1 only.

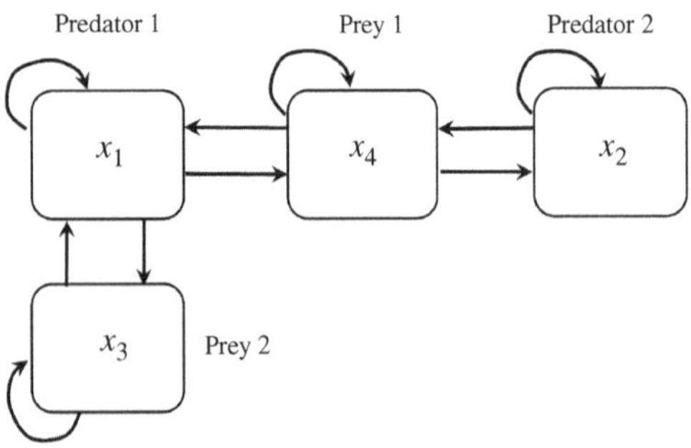

Figure 4.18 A two-predator, two-prey model. Predator 1 feeds on both prey species while predator 2 feeds only on prey species 1.

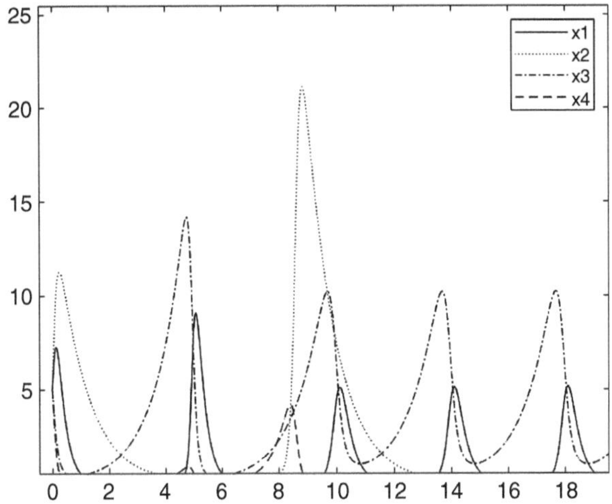

Figure 4.19 Response of the two-predator, two-prey system with $a_1 = 4, a_2 = 1, a_3 = 1, b_1 - 1, b_2 = 2, c_1 = 1, c_2 = 1, d_1 = 2, d_2 = 0.5, d_3 = 1.$

We model the dynamics of this system as a simple extension of the Volterra–Lotka two-species system as follows:

$$\begin{aligned}\dot{x}_1 &= -a_1x_1 + a_2x_1x_3 + a_3x_1x_4\\ \dot{x}_2 &= -b_1x_2 + b_2x_2x_4\\ \dot{x}_3 &= c_1x_3 - c_2x_3x_1\\ \dot{x}_4 &= d_1x_4 - d_2x_2x_4 - d_3x_1x_4.\end{aligned} \tag{4.8}$$

Figure 4.19 shows the response with the indicated parameter values. Systems such as (4.8) are extremely sensitive to the parameters defining the interactions and can exhibit very complex behaviors, including chaotic behavior. We will be able to say more about such systems in Chapter 13 when we discuss complex networks.

Problems

4.1 Consider a supply chain with suppliers (S), producers (P), and retailers (R). The supplier takes in raw material at input rate $u(t)$. Raw material that does not meet standards is discarded at rate k_1 and raw material meeting standards is shipped to the producer at rate k_2. The producer manufactures various products, which are sent to retailers at rate k_3. The retailers sell products at output rate $y(t)$ and return products to the producer at rate k_4.

1. Draw a three-compartment model to represent this supply chain.
2. Write the differential equations for this model in terms of states x_s, x_p, and x_r.

4.2 Using the method of separation of variables derive the solution of the logistic equation

$$\frac{dN}{dt} = rN\left(1 - \frac{N}{K}\right)$$

with $N(0) = N_0$ as

$$N(t) = \frac{K}{1 + Ae^{-rt}}, \quad \text{where } A = \frac{K - N_0}{N_0}.$$

4.3 Consider the SEIR model with a population of $K = 1000$ individuals and parameters $\beta = 0.2, \gamma = 0.1, \delta = 0.05$. Find the equilibrium points and compute the Jacobian linearization about each equilibrium.

4.4 Suppose in the SEIR model with $K = 1000$ that there is initially one person exposed, none infected and none recovered. Simulate the SEIR model with these initial conditions and with the above parameters. Interpret your results.

4.5 Consider the SIR model with vital dynamics with a population of K individuals. Find the equilibrium points and compute the Jacobian linearization about each equilibrium. Determine the stability of each equilibrium.

4.6 Simulate the SIR model with vital dynamics with parameters $\beta = 0.2$, $\gamma = 0.1$, and $\alpha = 0.01$, assuming a population size of $K = 1000$ with initially one infected individual.

4.7 Consider the predator–prey model

$$\dot{x} = ax - bxy$$
$$\dot{y} = -cy + dxy.$$

Show that the equilibrium points are (0,0) and $(c/d, a/b)$ and verify that (0,0) is a saddle point and $(c/d, a/b)$ is a center for the linear approximations about these equilibrium points.

4.8 With $a = 2$, $b = d = 1$, $c = 0.5$ plot the direction field in the first quadrant for the predator–prey system.

4.9 Suppose that in the prey population x is governed by the logistic model when $y = 0$, reflecting the fact that the food supply for the prey is limited. The modified predatory-prey model is now

$$\dot{x} = ax(1 - x) - bxy$$
$$\dot{y} = -cy + dxy.$$

Setting $a = 2, b = d = 1, c = 0.5$, find the nonzero equilibrium point in the first quadrant and show that it is asymptotically stable, i.e., no solutions of this system are periodic. What does this mean physically for the two populations?

4.10 With $a = 2,\ b = d = 1,\ c = 0.5$, plot the direction field for the modified predator–prey model above.

5
Stability

5.1 Introduction

In this chapter we consider the problem of stability of equilibrium states of nonlinear systems. In this context, stability refers to the response of a system if it is perturbed away from an equilibrium state. Perturbations, or disturbances, are always present in physical systems. Gusts of wind, ocean currents, tripped circuit breakers, temperature fluctuations, and so on cause a system state to deviate from an equilibrium and it is important to understand the effect of such deviations on the behavior of the system.

The intuitive notions of a stable and an unstable equilibrium are represented in Figure 5.1 showing a ball rolling in a bowl under the influence of gravity. If the ball on the left is perturbed slightly, it will stay close to the equilibrium position, whereas any perturbation of the ball on the right will cause it to roll away. We say that the equilibrium on the left is *stable* and the equilibrium on the right is *unstable*. In addition, if there is friction present, the ball on the left will settle back to the equilibrium position, which we call *asymptotic stability*.

In this chapter we present a rigorous treatment of stability of equilibria for systems that matches this intuitive notion of stability. We introduce the notion of stability in the sense of Lyapunov or Lyapunov stability. We also discuss an important extension of Lyapunov stability, known as LaSalle's invariance principle.

Previously, we defined *exponential stability* for a linear system in terms of the location in the complex plane of the eigenvalues of the system coefficient matrix. In this chapter we relate this notion of stability to nonlinear systems and show how the local stability properties of a nonlinear system near a hyperbolic equilibrium point are determined by the stability properties of the corresponding linear approximation about the equilibrium.

Introduction to Modeling and Simulation: A Systems Approach, First Edition. Mark W. Spong.

Companion Website: www.wiley.com/go/spong/engineeringmodeling

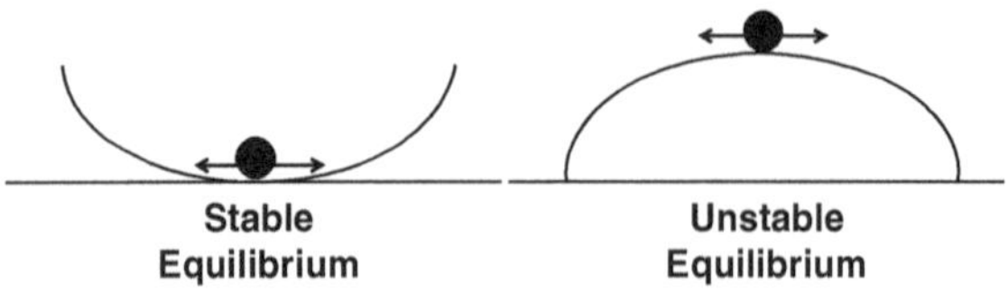

Figure 5.1 The intuitive notion of stable and unstable equilibria. The equilibrium on the left is stable. The equilibrium on the right is unstable.

5.2 Lyapunov Stability

Consider an autonomous system of the form

$$\dot{x} = f(x), \quad x \in \mathbb{R}^n, \tag{5.1}$$

where the vector field $f : \mathbb{R}^n \to \mathbb{R}^n$ is continuously differentiable. Suppose $\overline{x} = 0 \in \mathbb{R}^n$ is an equilibrium point of $f(x)$. The mathematical definition of stability of an equilibrium that matches our intuitive notion described above is the following.

Definition 5.1 ***(Stability)*** The equilibrium point $\overline{x} = 0$ of the system (5.1) is said to be:

1. stable, if for any $\epsilon > 0$, there exists a $\delta > 0$ such that
$$||x(0)|| < \delta \Longrightarrow ||x(t)|| < \epsilon, \quad \text{for all } t > 0;$$
2. asymptotically stable, if it is stable and, in addition,
$$\lim_{t\to\infty} x(t) = 0;$$
3. exponentially stable, if there exist constants $\alpha, \lambda, \delta > 0$ such that
$$||x(t)|| \leq \alpha e^{-\lambda t}||x(0)||, \quad \text{for} \quad ||x(0)|| < \delta;$$
4. unstable, if it is not stable.

The above stability properties are said to be *global* if the respective property holds for every initial condition $x(0) \in \mathbb{R}^n$. Otherwise, the stability is said to be *local*. Note that we define stability only of equilibria rather than stability of systems, since nonlinear systems may have multiple isolated equilibrium points, some stable and some unstable, as in the simple pendulum example.

Figure 5.2 shows the conceptual idea behind the ϵ–δ definition of stability of an equilibrium point. In order to bound the solution of the equation within a ball of radius ϵ, one must find a (possibly smaller) $\delta > 0$ that defines the allowable set of the initial conditions. Stability of the equilibrium requires one to find such a δ for any $\epsilon > 0$.

In order to state conditions for stability of an equilibrium of a nonlinear system, we need to define the notion of a positive-definite scalar function and the derivative of such a function along trajectories of the system.

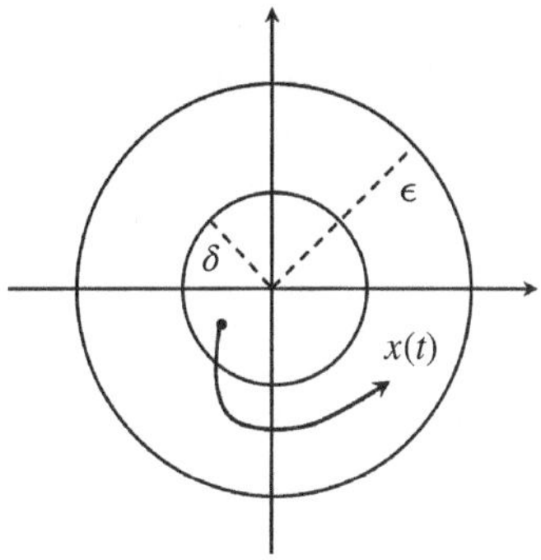

Figure 5.2 Illustrating the ϵ–δ definition of stability. Solutions starting at any point within a distance δ of the equilibrium will remain within ϵ of the equilibrium for all time $t > 0$.

Definition 5.2 ***(Positive-Definite Function)*** A continuously differentiable scalar function $V(x) : \mathbb{R}^n \to \mathbb{R}$ satisfying

(a) $V(0) = 0$
(b) $V(x) > 0$ for $x \neq 0$

is called *positive definite*.

If $V(x) \geq 0$ it is called *positive semidefinite* or *nonnegative definite*. $V(x)$ is *negative (semi)definite* if $-V(x)$ is positive (semi)definite.

Example 5.1 A simple example of a positive-definite function in two dimensions is

$$V(x_1, x_2) = x_1^2 + x_2^2.$$

Note that $V(x) = ||x||^2$ is just the square of the length (norm) of the vector x. On the other hand, the function

$$V(x_1, x_2) = x_1^2$$

is only positive semidefinite because $V = 0$ when $x_1 = 0$ for any nonzero value of x_2, and so is not strictly positive for all (x_1, x_2).

Example 5.2 A more general example of a positive-definite function is a quadratic function of the form

$$V(x) = \sum_{i=1}^{n} \sum_{j=1}^{n} p_{ij} x_i x_j = x^T P x,$$

where $x = (x_1, \ldots, x_n)^T \in \mathbb{R}^n$ and $P > 0$ is a real, symmetric, positive-definite matrix. In fact the condition that $x^T P x > 0$ for all $x \neq 0$ is the definition of a positive-definite (symmetric) matrix.

Recall from Appendix A that a positive-definite matrix P is one that has all leading principle minors positive, or equivalently, real and positive eigenvalues.

Intuitively, a positive-definite function is like a norm, measuring the distance away from the origin, which is why it is useful for investigating stability of equilibrium points. Moreover, the level sets of the function $x^T Px$ are *ellipsoids* (or *ellipses* in two dimensions).

The Lyapunov Derivative

Let $V : D \to \mathbb{R}$ be a continuously differentiable scalar function defined on a region $D \subset \mathbb{R}^n$. The derivative of V along trajectories of the system (5.1), denoted $\dot{V}(x)$, is defined as

$$\dot{V}(x) = \sum_{i=1}^{n} \frac{\partial V}{\partial x_i} \dot{x}_i = \sum_{i=1}^{n} \frac{\partial V}{\partial x_i} f_i(x).$$

For simplicity we will often write the above expression as

$$\dot{V} = \frac{\partial V}{\partial x} f(x) \quad \text{or} \quad \dot{V} = \nabla V(x)^T f(x).$$

We see from the above definition that $\dot{V}$ is the directional derivative of V in the direction of $f(x)$, in other words the inner product of the gradient of V and the vector field $f(x)$.

Example 5.3 Suppose that $V(x) : \mathbb{R}^2 \to \mathbb{R}$ is given by $V(x) = x_1^2 + x_2^2$ and suppose the vector field $f(x)$ is

$$f(x) = \begin{bmatrix} -x_1 - x_1^3 \\ -x_2 \end{bmatrix}.$$

Then the derivative of V in the direction of the vector field f is

$$\dot{V} = [2x_1, 2x_2] \begin{bmatrix} -x_1 - x_1^3 \\ -x_2 \end{bmatrix} = -2x_1^2 - 2x_1^4 - 2x_2^2.$$

The next theorem gives sufficient conditions for stability of an equilibrium and is the most fundamental result in this chapter. These conditions are referred to as *stability in the sense of Lyapunov.*

Theorem 5.1 **(Lyapunov)** *Let $\bar{x} = 0$ be an equilibrium for the system (5.1). Let $V : D \to \mathbb{R}$ be continuously differentiable scalar function in a neighborhood D of $\bar{x} = 0$, such that*

(a) $V(0) = 0$ and $V(x) > 0$ for $x \neq 0$ (positive definite),
(b) $\dot{V}(x) \leq 0$ in $D \backslash \{0\}$ (negative semidefinite).

Then the equilibrium $\overline{x} = 0$ *is stable. Moreover, if*

(c) $\dot{V}(x) < 0$ *in* $D\backslash\{0\}$ *(negative definite)*

then $\overline{x} = 0$ *is asymptotically stable.*

Proof: See Appendix D.2. Note that embedded in the proof is the guarantee that the solutions of the system exist for all time $t > 0$, which is not guaranteed just by the assumption that $f(x)$ is continuously differentiable.

Remark 5.1 We call a positive-definite scalar function $V(x)$ a *Lyapunov function candidate* or a *candidate Lyapunov function*. If $V(x)$ satisfies the conditions of Theorem 5.1 it is then called a *Lyapunov function* for the system (5.1). Figure 5.3 illustrates the intuitive idea behind this result in two dimensions.

We know that the inner product of the gradient of V and the vector field $f(x)$ in $\mathbb{R}^2$ satisfies

$$\nabla V(x)^T f(x) = ||\nabla V(x)|| \cdot ||f(x)|| \cos(\theta),$$

where θ is the angle between the vectors $\nabla V(x)$ and $f(x)$. The condition $\dot{V} < 0$ means that $\frac{\pi}{2} < \theta < \frac{3\pi}{2}$ and hence $f(x)$ must point to the interior of the region defined by the level set $V^{-1}(c)$ of V. Thus this region is positively invariant for the system, and so trajectories are trapped inside the level sets.

Another way to understand this result is to note that the condition $\dot{V} < 0$ means that the Lyapunov function $V(x)$ is decreasing along trajectories of the system. Therefore, since $V(x)$ is like a norm, the distance as defined by this norm between $x(t)$ and the equilibrium must be decreasing also.

Example 5.4 Consider the two-dimensional nonlinear system

$$\dot{x}_1 = -x_1 + x_2 \tag{5.2}$$

$$\dot{x}_2 = -x_1 - x_2 - x_2^3 \tag{5.3}$$

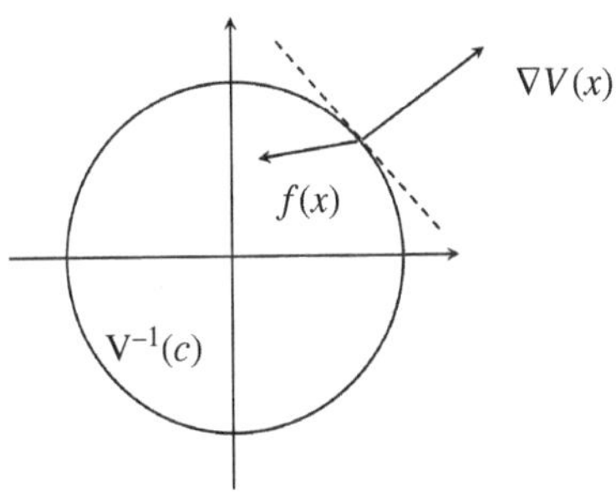

Figure 5.3 The intuitive idea of Lyapunov stability. The condition $\dot{V}(x) < 0$ means that the vector field $f(x)$ points toward the interior of the region bounded by level sets of $V(x)$.

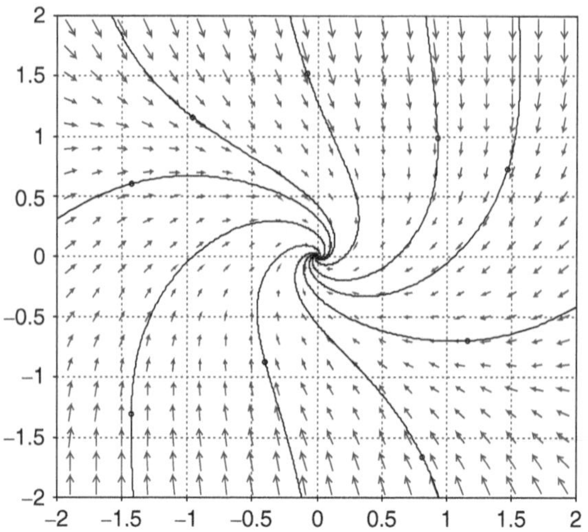

Figure 5.4 Phase portrait for the system (5.2)–(5.3), showing that the origin is asymptotically stable.

and let $V = \frac{1}{2}x_1^2 + \frac{1}{2}x_2^2$ be a Lyapunov function candidate. Then computing $\dot{V}$ along trajectories of the system gives us

$$\begin{aligned}\dot{V} &= x_1\dot{x}_1 + x_2\dot{x}_2 \\ &= x_1(-x_1 + x_2) + x_2(-x_1 - x_2 - x_2^3) = -x_1^2 - x_2^2 - x_2^4 < 0.\end{aligned}$$

Therefore, by Theorem 5.1 the equilibrium $\overline{x} = 0$ is asymptotically stable. The phase portrait for this system is shown in Figure 5.4.

The significance of Lyapunov's second method for nonlinear systems like equations (5.2)–(5.3) is that it does not rely on linear methods, such as computation of eigenvalues.

Example 5.5 As another example, consider the dynamics of a simple pendulum of length ℓ:

$$\begin{aligned}\dot{x}_1 &= x_2 \\ \dot{x}_2 &= -\frac{g}{\ell}\sin(x_1).\end{aligned}$$

A natural Lyapunov function candidate for mechanical systems is the total energy (kinetic plus potential),

$$V(x) = \frac{1}{2}m\ell^2x_2^2 + mg\ell(1 - \cos(x_1)).$$

The energy $V(x)$ is zero at the equilibrium (0,0) and strictly positive away from the equilibrium and is, therefore, a valid Lyapunov function candidate. Computing the derivative of V along trajectories of the pendulum gives

$$\dot{V} = mg\ell\sin(x_1)x_2 + m\ell^2x_2\left(-\frac{g}{\ell}\sin(x_1)\right) = 0.$$

Therefore, $\dot{V}$ is semidefinite, implying that the equilibrium $\overline{x} = 0$ is stable.

Remark 5.2 The condition that $\dot{V} = 0$ along trajectories of the simple pendulum means that the energy V is constant along trajectories. Thus, the trajectories lie on curves of constant energy, which are called *first integrals of the motion* in classical mechanics.

Remark 5.3 A. M. Lyapunov was a Russian mathematician mainly interested in the stability of motion of dynamical systems [50]. His work extended the notion of energy dissipation in mechanical systems to more general systems in which the function $V(x)$ need not have a physical interpretation like energy. This opened up numerous applications in such wide-ranging fields as economics, control theory, power systems, robotics, chemical reactions, and biological systems. Today Lyapunov stability theory is the primary tool in the analysis of stability of nonlinear systems.

Remark 5.4 It is important to emphasize that the conditions in Theorem 5.1 are only *sufficient conditions* for stability. This means that, if $\dot{V}$ fails to be negative along trajectories of the system, it does not necessarily imply that the equilibrium is unstable. For example, consider the nonlinear system

$$\begin{aligned} \dot{x}_1 &= x_2 \\ \dot{x}_2 &= -x_2 - x_1^3 \end{aligned}$$

and suppose a Lyapunov function candidate is chosen as

$$V(x_1, x_2) = \frac{1}{2}x_1^2 + \frac{1}{2}x_2^2.$$

Then the derivative of V along trajectories of the system is

$$\begin{aligned} \dot{V} &= x_1\dot{x}_1 + x_2\dot{x}_2 \\ &= x_1x_2 - x_2^2 - x_2x_1^3 \end{aligned}$$

which has both positive and negative values, for example, $\dot{V}(1,1) = -1$ and $\dot{V}(2,-1) = +5$. Thus, no conclusion can be drawn about stability of the origin. However, if we take

$$V = \frac{1}{4}x_1^4 + \frac{1}{2}x_2^2,$$

then we obtain

$$\begin{aligned} \dot{V} &= x_1^3x_2 - x_2^2 - x_2x_1^3 \\ &= -x_2^2 \leq 0, \end{aligned}$$

which shows that the origin is stable. (Later we will be able to show that the origin is, in fact, asymptotically stable.) The key is therefore to find the right Lyapunov function candidate for the given system. Unfortunately, there are few available methods for finding Lyapunov functions.

Global Stability

The condition $\dot{V} < 0$ is a sufficient condition for local asymptotic stability, but it is not sufficient to guarantee global asymptotic stability of the equilibrium even if $\dot{V}(x) < 0$ for all $x \in \mathbb{R}^n$. The problem is that V can decrease along solution trajectories of the system without the norm of the trajectory decreasing.

As an example, consider the Lyapunov function candidate

$$V(x) = \frac{x_1^2}{1 + x_1^2} + x_2^2. \tag{5.4}$$

$V(x)$ is positive definite and the region $V(x) \leq c$ is closed for $c < 1$. However, for $c > 1$ the region is open as shown in Figure 5.5, and hence, for initial conditions far from the equilibrium, solution trajectories can diverge even though V is decreasing along these trajectories. We need an additional condition on the Lyapunov function candidate V in order to prove global stability for the equilibrium $\overline{x} = 0$. That condition turns out to be *radial unboundedness* of V.

Definition 5.3 ***(Radial Unboundedness)*** A scalar function $V : \mathbb{R}^n \to \mathbb{R}$ is radially unbounded if and only if

$$V(x) \to \infty \quad \text{as } ||x|| \to \infty.$$

Theorem 5.2 *Let $\overline{x} = 0$ be an equilibrium point for (5.1). Let $V : \mathbb{R}^n \to \mathbb{R}$ be a continuously differentiable function such that*

(1) $V(0) = 0, \ V(x) > 0, \quad \text{for all } x \neq 0;$

(2) $V(x) \to \infty \quad \text{as } ||x|| \to \infty;$

(3) $\dot{V}(x) < 0, \quad \text{for all } x \neq 0.$

Then $\overline{x} = 0$ is globally asymptotically stable.

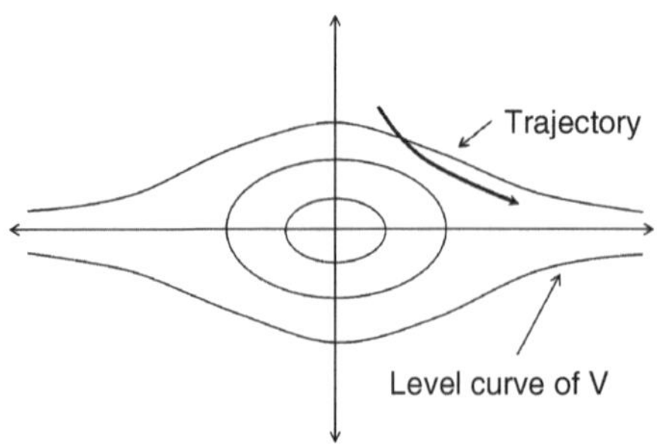

Figure 5.5 Level sets for Lyapunov function candidate in Equation (5.4).

Proof: The radially unbounded condition for V means that the set

$$\Omega_c = \{x \in \mathbb{R}^n \mid V(x) \leq c\}$$

is bounded. For any initial condition $x \in \Omega_c$, the condition $\dot{V} < 0$ implies the trajectory remains in Ω_c for all t. An argument similar to the proof of the first theorem now shows that $V(x(t)) \to 0$ and $x(t) \to 0$, as $t \to \infty$.

Example 5.6 Consider the nonlinear system

$$\dot{x}_1 = -x_1 - x_1 x_2^2$$
$$\dot{x}_2 = -x_2 - x_1^2 x_2.$$

It is straightforward to show that (0,0) is the unique equilibrium point. If we take as a Lyapunov function candidate

$$V = \frac{1}{2}x_1^2 + \frac{1}{2}x_2^2,$$

then $\dot{V}$ satisfies

$$\dot{V} = x_1(-x_1 - x_1 x_2^2) + x_2(-x_2 - x_1^2 x_2) = -x_1^2 - x_2^2 < 0.$$

Therefore, $\bar{x} = 0$ is asymptotically stable and, because V is radially unbounded, the asymptotic stability is global.

Remark 5.5 It should be obvious that for an equilibrium $\bar{x}$ to be globally asymptotically stable it must be the only equilibrium since, if $\bar{x}_0$ is another equilibrium, then the solution starting at $\bar{x}_0$ remains at $\bar{x}_0$ and therefore cannot converge asymptotically to $\bar{x}$.

5.3 Basin of Attraction

In the case where the asymptotic stability of an equilibrium for a nonlinear system is not global, an interesting problem is to determine the set of points such that solutions starting at these points converge to the equilibrium. The set of all such points is called the basin of attraction of the equilibrium. Formally, we have the following definition.

Definition 5.4 ***(Basin of Attraction)*** Let $x(t, x_0)$ be the solution of $\dot{x} = f(x)$ starting at x_0 at time $t = 0$. Then the *basin of attraction* of the equilibrium $\bar{x}$ is the set

$$\mathcal{B} = \left\{x_0 \mid x(t, x_0) \text{ is defined for all } t > 0 \text{ and } \lim_{t\to\infty} x(t, x_0) = \bar{x}\right\}.$$

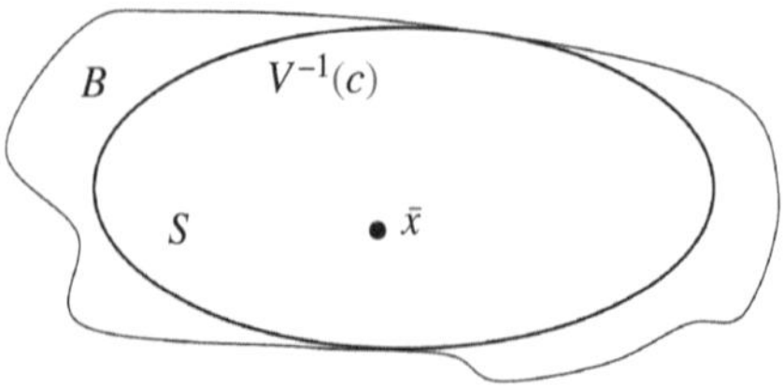

Figure 5.6 Illustrating the basin of attraction and an estimate using the level sets of a Lyapunov function. The set B is the basin of attraction, S is the set with boundary $V^{-1}(c)$, where V is a Lyapunov function for the system. The set S is then an approximation of the true basin of attraction.

Example 5.7 Consider the Van der Pol equation with $\mu = -1.8$, which changes the equilibrium $\bar{x} = 0$ from unstable to asymptotically stable. All trajectories starting inside the limit cycle will converge to the origin. Thus the basin of attraction of the origin is the set of points inside the limit cycle.

In general, the basin of attraction is difficult to determine exactly. However, a Lyapunov function can be used to estimate the basin of attraction. If $\dot{V}$ is negative (away from the equilibrium) in a set $S = \{V(x) \leq c\}$, then S is positively invariant, which means that the solutions starting in S exist for all time $t > 0$ and S is contained in the basin of attraction. This is illustrated in Figure 5.6.

Example 5.8 Consider the one-dimensional nonlinear system

$$\dot{x} = x(x^4 - 1)$$

and set $V = \frac{1}{2}x^2$. Then we have

$$\dot{V} = -x^2(1 - x^4) < 0 \quad \text{if } x^4 < 1.$$

Therefore, the equilibrium $\bar{x} = 0$ is locally asymptotically stable and the set $S = \{x \mid -1 < x < 1\}$ is contained in the basin of attraction. Note that, in this case, the set S is the exact basin of attraction of the equilibrium $\bar{x} = 0$ since $x = \pm 1$ are equilibrium points.

5.4 The Invariance Principle

We next discuss an important extension to the basic Lyapunov stability theorem for autonomous systems due to LaSalle [47]. To motivate the need for such an extension, let us consider again the simple pendulum equation but suppose now that there is friction at the pendulum pivot so that the dynamic equations become

$$\begin{aligned}\dot{x}_1 &= x_2\\ \dot{x}_2 &= -\frac{g}{\ell}\sin(x_1) - \frac{B}{m\ell^2}x_2,\end{aligned}$$

where B is the coefficient of friction. From the physics of this problem we know that the friction will dissipate energy and the pendulum will eventually come

Figure 5.7 Pendulum with friction example. The trajectories point toward the interior of the level sets of the Lyapunov function $V(x)$ at each point except when the velocity $x_2 = 0$. But, this is enough so that we cannot conclude asymptotic stability of the equilibrium.

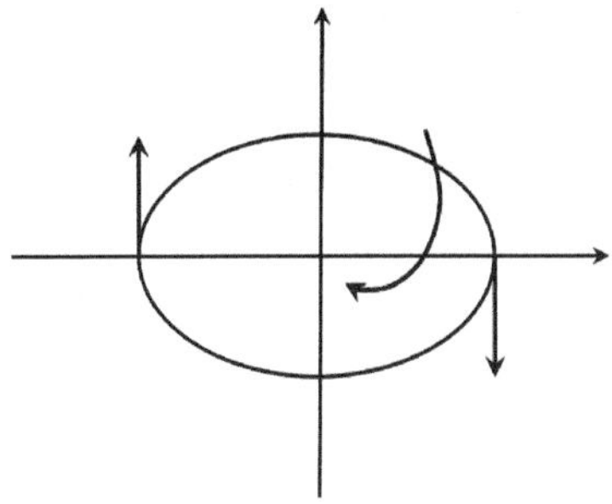

to rest at the equilibrium (0,0). In other words, the equilibrium (0,0) is now asymptotically stable. If we use the total energy as a Lyapunov function candidate as before, a straightforward calculation gives (Problem 5.1)

$$\dot{V} = -Bx_2^2 \leq 0. \tag{5.5}$$

But, from Equation (5.5), we can conclude only that the equilibrium $x = 0$ is stable since $\dot{V}$ is only negative semidefinite along solution trajectories.

The problem with the above analysis arises from the fact that, when $x_2 = 0$, the trajectory is tangent to the level curves of the total energy function (Figure 5.7), which means that no energy is dissipated at those points. As a result, V is not strictly decreasing at those points. The basic Lyapunov stability theorem requires that V be decreasing at all points in order to conclude asymptotic stability.

LaSalle's theorem, which addresses issues like this problem with the damped pendulum, gives an additional condition on the derivative $\dot{V}$ of the Lyapunov function candidate that can show asymptotic stability in cases where $\dot{V}$ is only negative semidefinite. A weak form of LaSalle's theorem can be stated as follows.

Theorem 5.3 *Suppose $V : D \to \mathbb{R}$ is positive definite and $\dot{V}(x(t)) \leq 0$ along trajectories of the system. If $\dot{V}$ does not vanish identically along any solution other than the equilibrium solution, then $\bar{x} = 0$ is asymptotically stable.*

Example 5.9 Let us see how this result applies to the simple pendulum with friction. Recall that choosing V as the total energy gave us

$$\dot{V} = -Bx_2^2 \leq 0.$$

Suppose now that $\dot{V} \equiv 0$, i.e., $\dot{V}$ is identically zero. Then $x_2 \equiv 0 \Rightarrow \dot{x}_2 \equiv 0$. Therefore, from the pendulum dynamic equations, we must have

$$\begin{aligned} \dot{x}_1 &\equiv 0 \\ 0 &\equiv -\frac{g}{\ell}\sin(x_1), \end{aligned}$$

which implies that $x_1 = n\pi$. Therefore, $\dot{V} \equiv 0$ only at equilibrium points and so $\bar{x} = 0$ is asymptotically stable.

We next show a more general result, known as LaSalle's theorem, or LaSalle's invariance principle. We first need some additional background and notation.

Definition 5.5 Let M be a positively invariant set for $\dot{x} = f(x)$. We say that $x(t)$ approaches M as $t \to \infty$ if, for each $\epsilon > 0$, there exists a time $T > 0$ such that $\text{dist}(x(t), M) < \epsilon$ for $t > T$, where $\text{dist}(p, M)$ means the distance from a point to a set and is defined as

$$\text{dist}(p, M) = \inf_{x \in M} ||p - x||.$$

Theorem 5.4 **(LaSalle)** *Let Ω be a compact, positively invariant set for $\dot{x} = f(x)$. Let $V :\to \mathbb{R}$ be continuously differentiable and*

$$\dot{V}(x(t)) \leq 0 \quad \text{in } \Omega.$$

Let $E = \{x \in \Omega \mid \dot{V}(x(t)) = 0\}$. Let M be the largest invariant set in E. Then every solution starting in Ω approaches M as $t \to \infty$ [44].

Remark 5.6 For the damped pendulum example, the set E is the entire x_1-axis and $M \subset E$ is the set of isolated equilibrium points.

Example 5.10 Consider the nonlinear system

$$\begin{aligned} \dot{x}_1 &= x_1 - x_2 x_1 \\ \dot{x}_2 &= x_1^2. \end{aligned}$$

Note that every point on the x_2-axis, where $x_1 = 0$, is an equilibrium point. Consider the Lyapunov function candidate

$$V = \frac{1}{2}x_1^2 + \frac{1}{2}(x_2 - b)^2, \quad \text{with } b > 1.$$

Then

$$\begin{aligned} \dot{V} &= x_1 \dot{x}_1 + (x_2 - b)\dot{x}_2 \\ &= x_1^2(x_2 - 1) + x_1^2(x_2 - b) \\ &= -x_1^2(b - 1) \leq 0. \end{aligned}$$

Hence, we have that $\dot{V} \leq 0$. Now, $\dot{V} = 0$ implies that $x_1 = 0$ and so the set E equals the set $\{(x_1, x_2) \mid x_1 = 0\}$. Since E consists entirely of equilibrium points, it is an invariant set, and it follows that $M = E$. Thus any trajectory approaches E. Since V is radially unbounded the result is global. Figure 5.8 shows the phase portrait for this system.

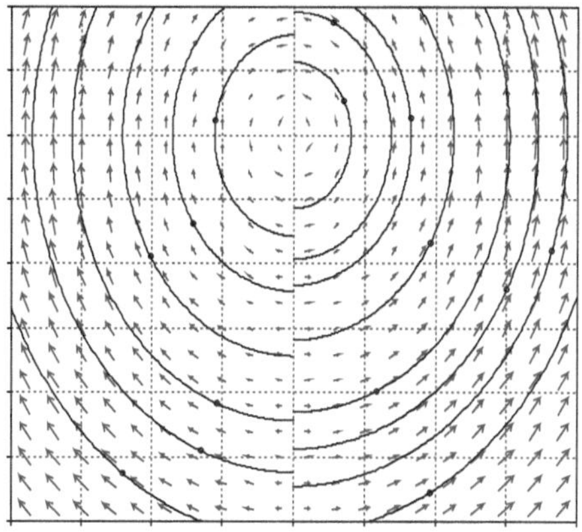

Figure 5.8 Phase portrait for the system. For any initial conditions, the value of x_1 converges to zero and x_2 converges to a constant value on the vertical axis. Every point on the vertical axis is an equilibrium point. The steady-state value of x_2 depends on the initial conditions.

5.5 Linear Systems and Linearization

Here we review Lyapunov stability theory for autonomous linear systems $\dot{x} = Ax,\ x \in \mathbb{R}^n$. Let V be defined as the quadratic form

$$V(x) = x^T Px,$$

where P is a symmetric, positive-definite matrix (so V is a Lyapunov function candidate). Computing $\dot{V}$ along trajectories of the linear system yields

$$\dot{V} = \dot{x}^T Px + x^T P\dot{x} = x^T(A^T P + PA)x = -x^T Qx,$$

where we define $-Q = A^T P + PA$. If the matrix Q is positive definite it follows that the equilibrium $\bar{x} = 0$ for the linear system is globally asymptotically stable. The difficulty lies in the fact that we do not know a priori how to pick the matrix P for a given system matrix A. But we can turn the problem around and choose any symmetric, positive-definite matrix Q and solve the equation

$$A^T P + PA = -Q \tag{5.6}$$

for P. This equation is called the *Lyapunov equation*. If the resulting solution P is positive definite we can conclude that the equilibrium $\bar{x} = 0$ is globally asymptotically stable for the linear system $\dot{x} = Ax$.

Recall our definition of a Hurwitz matrix as one whose eigenvalues have negative real part. Then we can state the following theorem.

Theorem 5.5 *The $n \times n$ matrix A is Hurwitz if and only if, for any symmetric positive-definite $n \times n$ matrix Q, there is a unique $n \times n$ symmetric positive-definite matrix P satisfying the Lyapunov equation*

$$A^T P + PA = -Q. \tag{5.7}$$

Proof: See Appendix D.3. We note that, if A is Hurwitz, then the matrix P satisfies

$$P = \int_0^\infty e^{A^T t} Q e^{At} dt.$$

Example 5.11 Consider the linear system

$$\dot{x}_1 = x_2$$
$$\dot{x}_2 = -x_1 - x_2,$$

and set

$$P = \begin{bmatrix} P_1 & P_2 \\ P_2 & P_3 \end{bmatrix}, \quad Q = \begin{bmatrix} 1 & 0 \\ 0 & 1 \end{bmatrix}.$$

Note that, since P is a symmetric matrix, there are only three independent entries. Substituting these matrices into the Lyapunov equation $A^T P + PA = -Q$ yields

$$\begin{bmatrix} 0 & -1 \\ 1 & -1 \end{bmatrix} \begin{bmatrix} P_1 & P_2 \\ P_2 & P_3 \end{bmatrix} + \begin{bmatrix} P_1 & P_2 \\ P_2 & P_3 \end{bmatrix} \begin{bmatrix} 0 & 1 \\ -1 & -1 \end{bmatrix} = \begin{bmatrix} -1 & 0 \\ 0 & -1 \end{bmatrix},$$

which leads to the three simultaneous linear equations

$$-2P_2 = -1$$
$$P_1 - P_2 - P_3 = 0$$
$$2P_2 - 2P_3 = -1,$$

with solution $P_1 = 1.5, P_2 = 0.5, P_3 = 1$. Now, we can easily check that the matrix

$$P = \begin{bmatrix} P_1 & P_2 \\ P_2 & P_3 \end{bmatrix} = \begin{bmatrix} 1.5 & 0.5 \\ 0.5 & 1 \end{bmatrix}$$

is positive definite, which implies that A is a Hurwitz matrix.

Remark 5.7 Fortunately, we do not have to solve the Lyapunov equation by hand because Matlab has a built-in function `lyap` to solve the Lyapunov Equation (5.6). For example,

Solving a Lyapunov Equation in Matlab

```
A = [-4 -3 2;1 -2 -2; 0 -1 -2];
Q = eye(3);
P = lyap(A',Q)
```

returns the symmetric matrix

$$P = \begin{bmatrix} 0.1266 & 0.0066 & -0.0148 \\ 0.0066 & 0.4046 & -0.3289 \\ -0.0148 & -0.3289 & 0.5641 \end{bmatrix}$$

One can check that P is positive definite and therefore A is a Hurwitz matrix. Note the caveat that one should use the transpose A^T and not A when calling the function `lyap`. This is because the Matlab code for the function `lyap` is based on the so-called *Sylvester equation*

$$AX + XB + C = 0.$$

In other words, `lyap(A,Q)` solves

$$AP + PA^T + Q = 0,$$

which is different from Equation (5.6).

Linearization of Nonlinear Systems

Based on the conclusions of the Hartman–Grobman theorem that we presented in Chapter 3, we expect the stability properties of the Jacobian linearization of a nonlinear system about a hyperbolic equilibrium to determine the local stability properties of the nonlinear system. This is indeed the case and can be investigated using Lyapunov stability theory. Let

$$\dot{x} = f(x), \quad x \in \mathbb{R}^n, \tag{5.8}$$

where $f(x)$ is continuously differentiable, and suppose that $\bar{x} = 0$ is a hyperbolic equilibrium. We can then state the following theorem.

Theorem 5.6 *Let $A = J(0)$, where $J(x)$ is the Jacobian of the vector field $f(x)$.*

1. *The origin $x = 0$ is locally exponentially stable for the nonlinear system (5.8) if and only if A is a Hurwitz matrix.*
2. *The origin $\bar{x}$ is unstable for the nonlinear system (5.8) if* $\text{Re}\lambda_i > 0$ *for one or more eigenvalues of A.*

The proof can be found in [44]. If the equilibrium $\bar{x} = 0$ is not hyperbolic, then we cannot necessarily conclude stability or instability by examining the Jacobian matrix A. However, one important conclusion that we can draw is in the following corollary.

Corollary 5.1 *If the equilibrium $\bar{x} = 0$ for Equation (5.8) is not hyperbolic, it may be stable, asymptotically stable, or unstable, but it cannot be exponentially stable.*

The importance of this corollary is, among other things, that exponential stability of hyperbolic equilibria persists under small perturbations or disturbances of the vector field $f(x)$ in Equation (5.8), which is not true in the case of nonhyperbolic equilibria. We can illustrate this property with the following simple example.

Example 5.12 Consider the nonlinear system

$$\dot{x} = -x + ax^3$$

and note that $\bar{x} = 0$ is a hyperbolic equilibrium. Therefore, $\bar{x} = 0$ is locally exponentially stable for any value of a. Now, consider the system

$$\dot{x} = ax^3.$$

In this case, $\bar{x} = 0$ is not hyperbolic and the equilibrium is asymptotically stable for $a < 0$ and unstable for $a > 0$. Thus the linear approximation about $\bar{x} = 0$ does not provide any information to determine stability of the nonlinear system.

Problems

5.1 Verify Equation (5.5) by direct calculation.

5.2 Consider the first-order system

$$\dot{x} = -g(x), \quad x \in \mathbb{R},$$

where $g(x)$ satisfies

$$g(0) = 0, \quad xg(x) > 0.$$

Show that the function

$$V(x) = \int_0^x g(u)du$$

is a valid Lyapunov function candidate. Using this Lyapunov function candidate, show that the equilibrium $x = 0$ is asymptotically stable.

5.3 Repeat the above calculation using the Lyapunov function candidate

$$V(x) = \frac{1}{2}x^2.$$

5.4 Show that the function

$$V(x) = \frac{x_1^2}{1 + x_1^2} + x_2^2$$

is not radially unbounded.

5.5 Consider the second-order system

$$\dot{x}_1 = x_2$$
$$\dot{x}_2 = -x_1^3 - x_2.$$

(a) Is the equilibrium at (0,0) exponentially stable? Explain why or why not.

(b) Use the Lyapunov function $V = \frac{1}{4}x_1^4 + \frac{1}{2}x_2^2$ and LaSalle's theorem to show that the equilibrium at (0,0) is globally asymptotically stable.

5.6 Show that the matrix

$$A = \begin{bmatrix} -12 & -20 & -21 \\ 2 & 1 & 3 \\ 5 & 10 & 9 \end{bmatrix}$$

is Hurwitz without computing its eigenvalues.

5.7 For the matrix A defined in the previous problem, set $Q = I$, the 3×3 identity matrix, and verify that the Matlab commands

$$P = \texttt{lyap}(A', Q) \quad \text{and} \quad P = \texttt{sylvester}(A', A, -Q)$$

return the same matrix P.

5.8 Write a Matlab script to compute the leading principal minors of an $n \times n$ matrix. Test your script on the matrices below.

$$\text{(a)} \begin{bmatrix} 2 & -1 \\ 1 & 2 \end{bmatrix} \quad \text{(b)} \begin{bmatrix} 2 & -1 \\ 0 & 2 \end{bmatrix} \quad \text{(c)} \begin{bmatrix} 0 & 1 \\ 1 & 0 \end{bmatrix}$$

$$\text{(d)} \begin{bmatrix} 0 & 1 & 2 \\ 0 & 0 & 3 \\ 0 & 0 & 0 \end{bmatrix} \quad \text{(e)} \begin{bmatrix} 1 & 0 & 0 \\ 0 & 3 & 1 \\ 0 & 0 & 3 \end{bmatrix}.$$

5.9 Consider the nonlinear system

$$\dot{x} = f(x) + g(x), \quad x \in R^n.$$

Suppose $x^T f(x) < -||x||^2$ and $||g(x)|| < k||x||$. Show that the origin is asymptotically stable if $k < 1$.

5.10 For each of the nonlinear systems below, investigate the stability of the equilibrium point $(0,0)^T$ using the Lyapunov function candidate $V(x_1, x_2) = \frac{1}{2}x_1^2 + \frac{1}{2}x_2^2$. In each case, you should specify whether the equilibrium is stable, asymptotically stable, globally asymptotically stable, or unstable. State also whether or not the equilibrium is locally exponentially stable and justify your answer.

(a) $$\begin{aligned}\dot{x}_1 &= -x_1^3 + x_1x_2^2\\ \dot{x}_2 &= -x_1^2x_2 - x_2^3\end{aligned}$$

(b) $$\begin{aligned}\dot{x}_1 &= x_1^2x_2 + 2x_1x_2^2 + x_1^3\\ \dot{x}_2 &= -x_1^3 + x_2^3\end{aligned}$$

5.11 Use LaSalle's theorem to investigate stability of the origin for each of the systems below.

(a) $$\begin{aligned}\dot{x}_1 &= -x_1^3 + 2x_2^3\\ \dot{x}_2 &= -2x_1x_2^2\end{aligned}$$

(b) $$\begin{aligned}\dot{x}_1 &= x_2\\ \dot{x}_2 &= -x_1 - x_2 - (x_1 + x_2)^2x_2\end{aligned}$$

5.12 Suppose the Jacobian $A = \frac{\partial f}{\partial x}$ for a second-order nonlinear system $\dot{x} = f(x)$ is

$$A = \begin{bmatrix} -x_1 & 1 - x_2 \\ x_1x_2^3 & -x_2 \end{bmatrix}$$

and suppose that (0,0) and (1,1) are both equilibrium points. What can you say about the solution of the nonlinear system near these equilibrium points?

5.13 Consider the first-order nonlinear system

$$\dot{x} = f(x), \quad \text{with } f(0) = 0,\ x \in \mathbb{R},$$

and suppose there is a Lyapunov function $V(x)$ so that

$$\frac{\partial V}{\partial x}f(x) < 0.$$

Explain whether or not $\overline{x} = 0$ is an asymptotically stable equilibrium point for the system

$$\dot{x} = f(x)(1 + x).$$

5.14 Consider the nonlinear system

$$\begin{aligned}\dot{x}_1 &= -\sin(x_1)\cos(x_2)\\ \dot{x}_2 &= -\cos(x_1)\sin(x_2).\end{aligned}$$

Show that the origin is locally exponentially stable.

5.15 Investigate the stability of the FitzHugh–Nagumo model in equations (2.23)–(2.24) for the given parameters.

5.16 Repeat Problem 5.15 for the Brusselator model in equations (2.25)–(2.26).

6

Discrete-Time Systems

6.1 Introduction

In this chapter we discuss discrete-time systems, which are described by difference equations as opposed to the differential equations that we used to model continuous-time systems. A general discrete-time system is one of the form

$$x(k+1) = f(x(k), k), \quad \text{where } x \in \mathbb{R}^n \text{ and } k = 0,1,2,\ldots. \tag{6.1}$$

The equation is autonomous if f does not depend explicitly on k. Discrete-time systems arise in sampling of continuous-time systems, in numerical integration and numerical optimization, signal processing, image processing, digital control, digital filtering, and other applications.

Examples

As a first example, consider the differential equation

$$\dot{x} = f(x), \quad x(0) = x_0. \tag{6.2}$$

Using the difference approximation for the derivative gives

$$\dot{x}(t) \approx \frac{x(t+h) - x(t)}{h} = f(x(t)).$$

Then, with $t = k$ and $t + h = k + 1$, we can approximate the differential equation (6.2) by the discrete-time equation

$$x(k+1) = x(k) + hf(x(k)). \tag{6.3}$$

Equation (6.3) is called the *Euler approximation* of Equation (6.2). We will study this and other approximation methods for solutions of ordinary differential equations in Chapter 7.

As another example, consider the discrete-time equation

$$y(k+2) = y(k+1) + y(k). \tag{6.4}$$

Introduction to Modeling and Simulation: A Systems Approach, First Edition. Mark W. Spong.

Companion Website: www.wiley.com/go/spong/engineeringmodeling

Equation (6.4) is a second-order difference equation because $y(k)$ and $y(k+2)$ are separated by two steps. As in the case of second-order differential equations, in order to solve this equation we must specify two initial conditions, $y(0)$ and $y(1)$. With initial values $y(0) = y(1) = 1$, this equation generates the *Fibonacci sequence* $1, 1, 2, 3, 5, 8, 13, \ldots$. We can write Equation (6.4) in the form of Equation (6.1) by defining

$$x(k) = \begin{bmatrix} x_1(k) \\ x_2(k) \end{bmatrix} = \begin{bmatrix} y(k) \\ y(k+1) \end{bmatrix},$$

which results in the system

$$\begin{aligned} x_1(k+1) &= x_2(k) \\ x_2(k+1) &= x_1(k) + x_2(k), \end{aligned}$$

which is in the form of Equation (6.1) with

$$f(x(k)) = \begin{bmatrix} x_2(k) \\ x_1(k) + x_2(k) \end{bmatrix} = \begin{bmatrix} 0 & 1 \\ 1 & 1 \end{bmatrix} \begin{bmatrix} x_1(k) \\ x_2(k) \end{bmatrix}. \tag{6.5}$$

Existence and Uniqueness of Solutions

The questions of existence and uniqueness of solutions of the discrete-time system (6.1) are easier than for differential equations. Given initial values $x(0) = x_0$, we define the iterated values

$$\begin{aligned} x(1) &= f(x(0)) \\ x(2) &= f(x(1)) = f(f(x(0))) = f^2(x(0)) \\ &\vdots \\ x(k) &= f^k(x(0)). \end{aligned}$$

Then, as long as each x_k lies in the domain of the function f, the solution of the initial-value problem exists and is given by the sequence $x(k)$, $k = 1,2,\ldots$.

Equilibrium Points

The definition of equilibria for discrete-time systems differs from the definition for continuous-time system as follows.

Definition 6.1 ***(Equilibrium Point)*** An equilibrium (or fixed point) of an autonomous discrete-time system $x(k+1) = f(x(k))$ is a vector $\bar{x}$ such that $f(\bar{x}) = \bar{x}$.

We see that if $x(0) = \bar{x}$, then the solution of the equation $x(k+1) = f(x(k))$ is the constant sequence $x(k) = \bar{x},\ k = 1,2,\ldots.$

Example 6.1 Consider the one-dimensional system

$$x(k+1) = ax(k). \tag{6.6}$$

Then $\bar{x} = 0$ is the unique equilibrium point if $a \neq 1$. For $a = 1$ every point is an equilibrium point. The solution of this equation is

$$x(k) = a^k x(0), \quad k = 1,2,\ldots.$$

Figure 6.1 shows the responses starting at $x(0) = 1$ for $a = 1$, $a = 1.1$, and $a = 0.8$. For $a > 1$ the solution diverges. For $0 < a < 1$ the solution converges to zero, and for $a = 1$ the solution is constant at $x(0)$.

We also note in the above system that if $a = 0$, then $x(k) = 0$, $k = 1,\ldots,$ for any initial condition. In other words, the solution of a discrete-time system can reach an equilibrium in a finite number of steps. This is in contrast to continuous-time systems, which can only converge to an equilibrium asymptotically as $t \to \infty$. If $a < 0$, the solution oscillates as shown in Figure 6.2 for $a = -0.75$. If $a = -1$, then the solution is periodic, $x(k) = (-1)^k$.

Remark 6.1 Discrete values can be displayed using several Matlab functions in addition to `plot`, as shown in Figure 6.3.

Next, let us consider the case of an nth-order linear difference equation of the form

$$y(k+n) + a_n y(k+n-1) + \cdots + a_1 y(k) = 0.$$

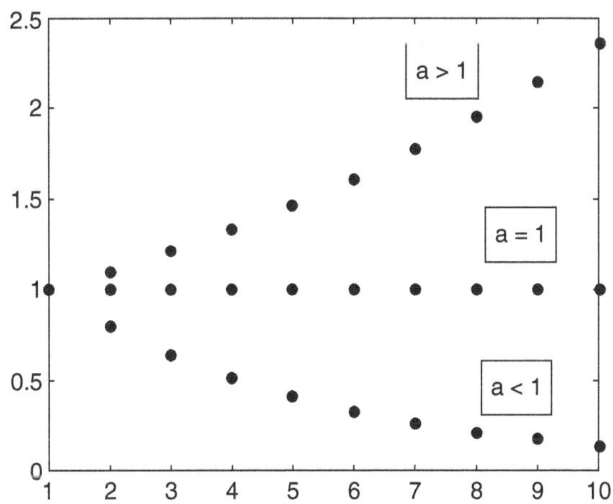

Figure 6.1 Response of the system (6.6) for different values of the coefficient a with initial condition $x(0) = 1$.

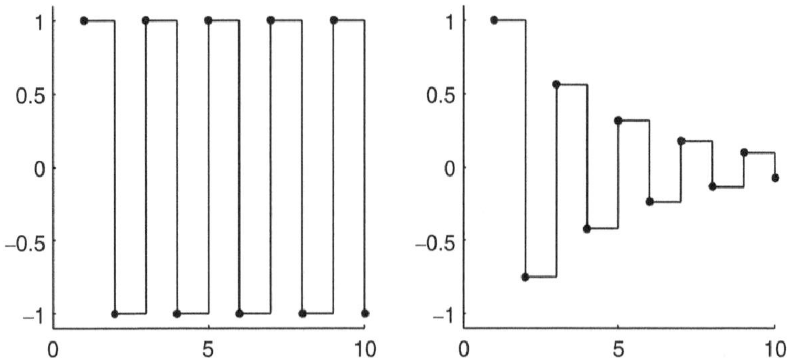

Figure 6.2 Response of the system (6.6) with $a = -0.75$ (right) and with $a = -1$ (left).

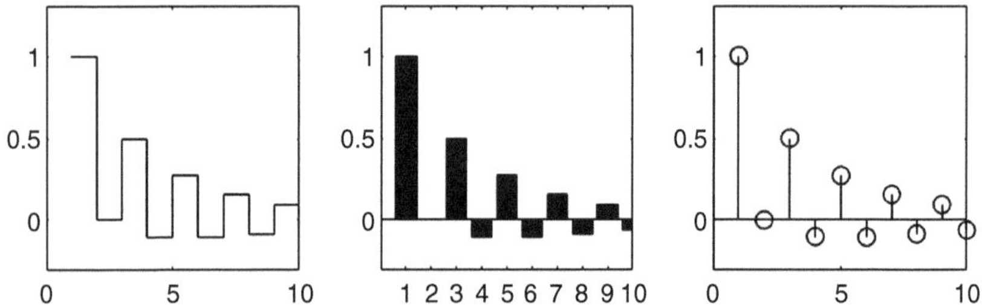

Figure 6.3 `stairs(y)` (left), `bars(y)` (middle), `stem(y)` (right) for a sequence of discrete points.

As in the case of continuous-time ordinary differential equations, we can express this equation as a system of n first-order equations by defining

$$\begin{aligned} x_1(k) &= y(k) \\ x_2(k) &= y(k+1) \\ &\vdots \\ x_n(k) &= y(k+n-1). \end{aligned}$$

Then we have

$$\begin{aligned} x_1(k+1) &= x_2(k) \\ x_2(k+1) &= x_3(k) \\ &\vdots \\ x_n(k+1) &= -a_n x_n(k) - \cdots - a_1 x_1(k), \end{aligned}$$

which we can then write in vector form as

$$x(k+1) = Ax(k)$$

by setting $x(k) = [x_1(k), \dots, x_n(k)]^T$ and

$$A = \begin{bmatrix} 0 & 1 & 0 & \cdots & 0 \\ 0 & 0 & 1 & \cdots & 0 \\ \vdots & \vdots & & \ddots & 0 \\ 0 & 0 & 0 & \cdots & 1 \\ -a_1 & -a_2 & -a_3 & \cdots & -a_n \end{bmatrix}.$$

Given an n-dimensional discrete-time system

$$x(k+1) = Ax(k), \quad x \in \mathbb{R}^n,$$

the solution of the equation, starting at the initial vector $x(0) = x_0$ is easily seen to be

$$x(k) = A^k x_0, \quad k = 1,2, \dots .$$

6.2 Stability of Discrete-Time Systems

The stability theory for discrete-time systems follows rather closely the theory for continuous-time systems. The formal ϵ–δ definition of stability, asymptotic stability, and so on, is the same in discrete time and continuous time and the basic Lyapunov stability results hold with minor modifications, as we show next.

Consider a discrete-time nonlinear system of the form

$$x(k+1) = f(x(k)), \quad x \in \mathbb{R}^n,$$

with $\bar{x} = 0$ an equilibrium point. The following theorem, which we present without proof, gives sufficient conditions for stability analogous to the conditions presented previously for continuous-time systems.

Theorem 6.1 *Let $V : \mathbb{R}^n \to \mathbb{R}$ be a positive-definite scalar function with $V(0) = 0$. Then we compute $\Delta V(k)$, analogous to $\dot{V}(t)$ in continuous time, as*

$$\begin{aligned} \Delta V(k) &= V(x(k+1)) - V(x(k)) \\ &= V(f(x(k))) - V(x(k)), \quad k = 0,1, \dots . \end{aligned}$$

1. *If $\Delta V(k) \leq 0$, then the equilibrium $\bar{x} = 0$ is stable.*
2. *If $\Delta V(k) < 0$, then the equilibrium $\bar{x} = 0$ is asymptotically stable.*
3. *If V is radially unbounded, then the stability or asymptotic stability is global.*

Example 6.2 Consider the discrete-time nonlinear system

$$x_{k+1} = ax_k^2. \tag{6.7}$$

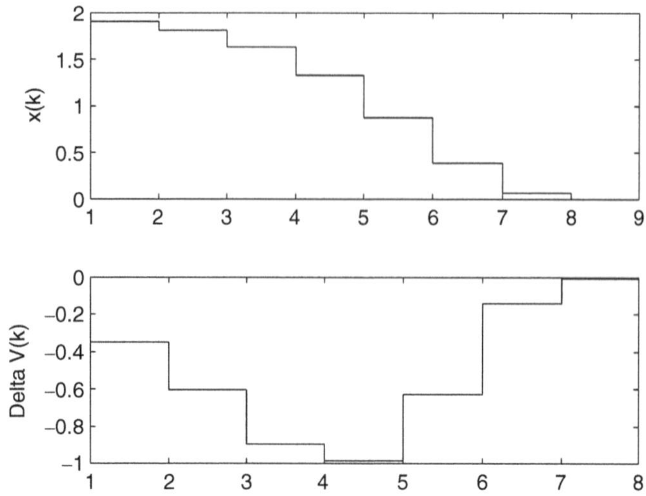

Figure 6.4 Plot of $x(k)$ (top) and $\Delta V(k)$ (bottom) for the discrete-time nonlinear system (6.7).

The fixed points for Equation (6.7) are $\overline{x} = 0, \frac{1}{a}$. Let us choose a Lyapunov function candidate

$$V = x^2.$$

Then we compute the difference

$$\begin{aligned}\Delta V(x_k) &= V(f(x_k)) - V(x_k)\\ &= a^2x_k^4 - x_k^2\\ &= x_k^2(a^2x_k^2 - 1) < 0, \quad \text{if } x_k^2 < 1/a^2.\end{aligned}$$

Thus, the equilibrium point $\overline{x} = 0$ is locally asymptotically stable for the system (6.7) in the region $-1/a < x < 1/a$. Figure 6.4 shows the response with $a = 0.5$ and initial condition $x(0) = 1.9$ as well as a plot of the Lyapunov function $\Delta V(x_k)$. Note that the initial value, $x(0)$, satisfies the condition for local asymptotic stability, namely, that $x(0) < 1/a$.

6.3 Stability of Discrete-Time Linear Systems

Consider an n-dimensional discrete-time linear system

$$x(k+1) = Ax(k), \quad x \in \mathbb{R}^n, \ A \in \mathbb{R}^{n\times n}. \tag{6.8}$$

Theorem 6.2 *The equilibrium $\bar{x} = 0$ of the n-dimensional linear system (6.8) is asymptotically stable if and only if*

$$|\lambda_i| < 1, \quad \text{for all eigenvalues } \lambda_1, \ldots, \lambda_n \text{ of } A.$$

Since the eigenvalues are generally complex, the above inequality means that the eigenvalues of A lie inside the unit circle in the complex plane. Such a matrix A is called a *Schur* matrix. Thus the unit disk in the complex plane plays the same role for discrete-time systems as the left half-plane, $\mathrm{Re}(\lambda) < 0$, does for continuous-time systems.

The Discrete Lyapunov Equation

For a linear system

$$x(k+1) = Ax(k),$$

we let V be a positive-definite quadratic form

$$V = x^T Px,$$

and then a direct computation shows that

$$\begin{aligned} V(x(k+1)) - V(x(k)) &= x(k)^T A^T PAx(k) - x(k)^T Px(k) \\ &= x(k)^T (A^T PA - P)x(k) = -x(k)^T Qx(k), \end{aligned}$$

where

$$A^T PA - P + Q = 0 \tag{6.9}$$

is called the *discrete Lyapunov equation.* The next theorem gives a condition for stability of a discrete-time system analogous to that for a continuous-time system.

Theorem 6.3 *An $n \times n$ matrix A is* Schur *if and only if, for any symmetric positive-definite $n \times n$ matrix Q, there exists a symmetric positive-definite matrix P that satisfies the discrete Lyapunov equation (6.9). Moreover, if A is Schur, then P is the unique solution.*

If A is a Schur matrix, then the matrix P is given by the infinite series

$$P = \sum_{k=0}^{\infty} (A^T)^k QA^k.$$

Matlab function `dlyap`

The counterpart of the Matlab command `lyap` for discrete systems is the command `dlyap`. For discrete-time linear systems, the Matlab command

$$P = \texttt{dlyap}(A', Q)$$

returns the solution of

$$A^T PA - P + Q = 0.$$

Example 6.3 Suppose that

$$A = \begin{bmatrix} 0 & 1 \\ -0.15 & -0.8 \end{bmatrix}, \quad Q = \begin{bmatrix} 1 & 0 \\ 0 & 1 \end{bmatrix}.$$

Then the command

$$P = \texttt{dlyap}(A', Q)$$

returns

$$\begin{bmatrix} 1.0892 & 0.4137 \\ 0.4137 & 3.9647 \end{bmatrix}.$$

One can now check that P is positive definite and therefore A is a Schur matrix. Hence, the equilibrium $\overline{x} = 0$ is globally asymptotically stable.

6.4 Moving-Average Filter

Discrete-time system models are the mainstays of digital signal processing (DSP), digital image processing, and digital filtering, made more important by the fact that most data, such as with audio, video, and telecommunications, are now represented and stored digitally. Consider the discrete-time system

$$y(k) = \frac{x(k) + x(k-1) + x(k-2) + \cdots + x(k-N)}{N+1}, \tag{6.10}$$

where x is vector of input data and y is a vector of output data. We see that Equation (6.10) computes $y(k)$ as the average of $x(k-N), \ldots, x(k)$. Equation (6.10) is called a *moving-average filter*. The number N is called the *window size*. The moving-average filter is the most common filter in DSP because of its simplicity and ease of implementation. The moving-average filter is a special case of the finite impulse response (FIR) filter,

$$y(k) = b_0 x(k) + b_1 x(k-1) + \cdots + b_N x(k-N),$$

so in the moving-average filter the coefficients b_i are each equal to $\frac{1}{N+1}$. A detailed treatment of DSP, including FIR filters, would take us too far afield for the present

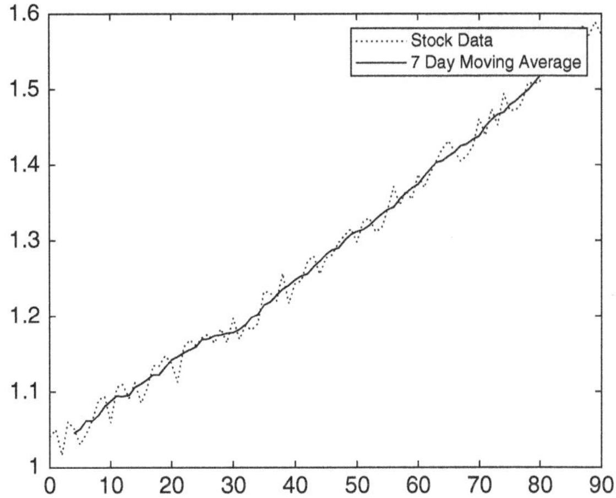

Figure 6.5 Seven-day moving-average filter for a set of stock price data plotted over 90 days.

text. The interested reader is referred to the text [54]. Here we will give just a couple of examples of applications of the moving-average filter.

Example 6.4 Suppose a stockbroker wants to track the value of a certain stock over time but wants to ignore the hour-by-hour or day-by-day fluctuations in the stock price. Figure 6.5 shows a seven-day moving average of the price of a stock, i.e., the moving average with a seven-day window, $N = 7$. Such moving averages are common in economics, not only for stock prices, but also for unemployment data, commodity prices, economic growth, and other quantities. During the Covid-19 pandemic, moving averages of infections, deaths, and number of vaccines given were commonly reported by the media.

Example 6.5 As a second example, we present a moving-average filter of a signal corrupted by noise. All physical measurements are subject to random noise to some degree. Since a random noise signal tends to have zero mean, it is reasonable to assume that averaging the samples will reduce the effect of the noise.

Moving-Average Filter

```
WindowSize = 50;
W = WindowSize;  % number of samples to average
t = -10:.01:10;  % generate 2000 samples
x = sin(t) + 0.15*rand(size(t)); % noisy signal
y=zeros(1,2000-W);
```

(Continued)

(Continued)

```
for i = 1:2000-W
    for j= 1:W
    y(i) = y(i) + x(i+j);
    end
y(i) = y(i)/W;
end
s=zeros(1,2000-W);
for i=1:2000-W
    s(i)=t(i+W);
end
```

Figure 6.6 shows the output of a moving-average filter of a sinusoidal signal with random noise superimposed on it. A window size of 50 is used. Note that there is a delay (or phase shift) in the output response. Such delay is unavoidable and larger for larger window size. More advanced signal processing algorithms are designed to reduce such phase shift.

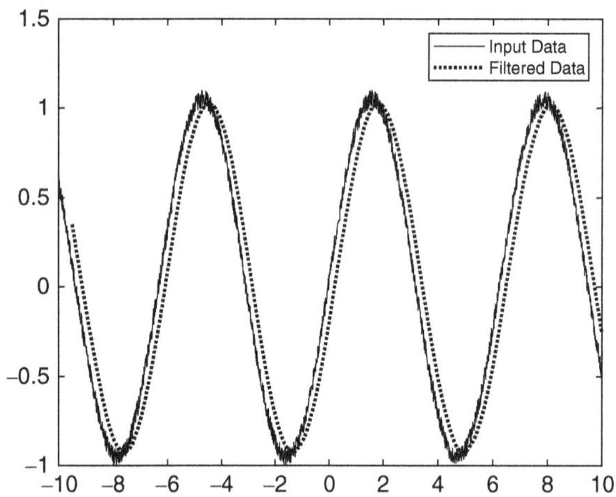

Figure 6.6 Moving-average filter response to noisy input data.

6.5 Cobweb Diagrams

In this section we discuss cobweb diagrams and two important applications, namely, supply and demand in economic systems, and the discrete-time logistic equation. A cobweb diagram is a graphical method to visualize the behavior of one-dimensional discrete-time systems, which are also called *maps*. Beginning at an initial condition x_0, a cobweb diagram shows the response from repeated iterations of a discrete-time map of the form

$$x_{k+1} = f(x_k). \tag{6.11}$$

The procedure to create a cobweb diagram is as follows.

1. Given a map $f(x)$ and an initial condition $x(0)$, draw a vertical line between the points $(x(0), 0)$ and $(x(0), f(x(0)))$.
2. Then draw a horizontal line from $(x(0), f(x(0)))$ to $(f(x(0)), f(x(0)))$. Note that $(f(x(0)), f(x(0))) = (x(1), x(1))$.
3. Next, draw a vertical line from $(x(1), x(1))$ to $(x(1), f(x(1)))$ and continue in this fashion.

Note that the point where $y = f(y)$ is an equilibrium for the discrete-time system (6.11).

Example 6.6 Consider the discrete-time system

$$x_{k+1} = -x_k^2 + 0.5. \tag{6.12}$$

The equilibrium points occur when the line $y = x$ and the curve $y = -x^2 + 0.5$ intersect. The equilibrium point in the first quadrant is $y = 0.3660$.

The Matlab script below draws the cobweb diagram for this equation, which is shown in Figure 6.7.

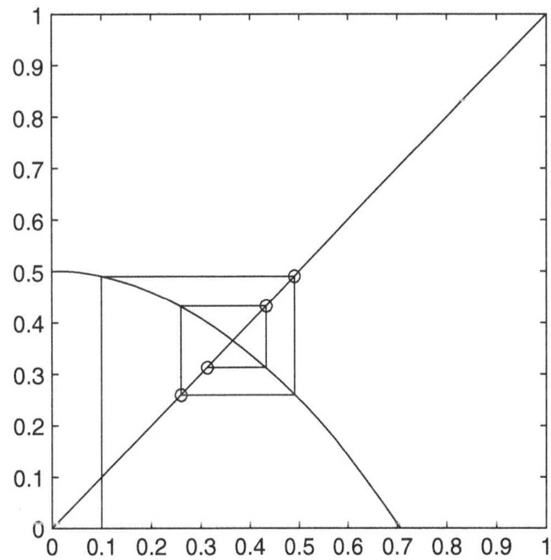

Figure 6.7 Cobweb diagram for Equation (6.12) with $N = 4$ iterations. A decreasing spiral indicates that the equilibrium $y = 0.3660$ is asymptotically stable.

Cobweb Diagram for Equation (6.12)

```
x0=0.1;          % Initial condition
N= input('number of iterations = ');
x(1) = x0;
for ic=1:N
 x(ic+1) = -x(ic)^2+0.5;
end
% plot the map function and the line y=x
clf;
t = 0:0.01:1;
plot(t,-t.^2+0.5,'k');hold on;
 axis('square'); axis([0 1 0 1]);
 set(gca,'XTick',(0:0.1:1),'YTick',(0:0.1:1))
 grid off;
 fplot(@(y)1.*y,[0 1],'k');
% plot the cobweb
 line([x(1) x(1)],[0 x(2)],'Color','k')
  plot(x(1), x(1))
 for ic=1:N-1
  line([x(ic) x(ic+1)],[x(ic+1) x(ic+1)],'Color','k')
  line([x(ic+1) x(ic+1)],[x(ic+1) x(ic+2)],'Color','k')
   plot(x(ic+1), x(ic+1),'ko');
 end
 plot(x(N+1),x(N+1),'ko');
 line([x(N) x(N+1)],[x(N+1) x(N+1)],'Color','k')
```

6.5.1 Cobweb Diagrams in Economics

Cobweb diagrams are used to analyze pricing of goods and services in business models. Cobweb economic theory was developed to explain why prices may fluctuate periodically in various market as a function of supply and demand.

Consider a hypothetical case in which a shortage of an agricultural product, say potatoes, causes prices to rise. As a result, farmers are likely to increase production the next year, which will result in a drop in price. The drop in price may cause the farmers to decrease production in the third year and so on, resulting in fluctuations in supply. Demand will also fluctuate with changes in prices.

Figure 6.8 shows typical supply and demand graphs, where for simplicity we have assumed linear supply and demand curves. The *equilibrium price* is

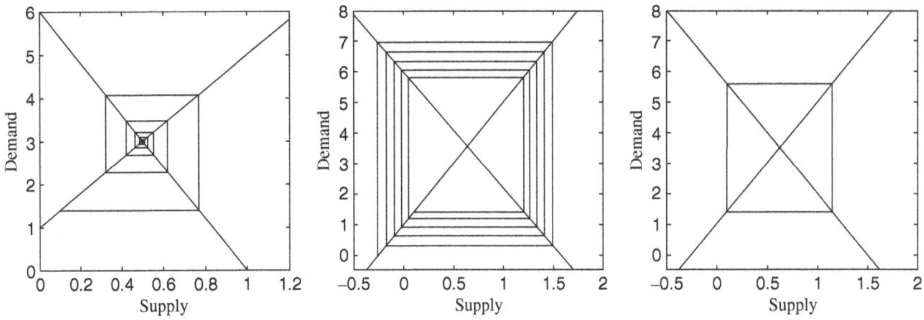

Figure 6.8 Cobweb diagrams for different slopes of the supply and demand curves showing stable equilibrium (left), unstable equilibrium (middle) and a periodic solution (right). The slopes of the supply curves are equal to 4 in all three cases. The slope of the demand curve is −6 in the left plot, −3.8 in the middle plot, and 4 in the right plot.

at the intersection where supply equals demand. The equilibrium price will be asymptotically stable if the slope of the supply curve is less than the absolute value of the demand curve, and unstable if the slope of the supply curve is greater than the absolute value of the demand curve. If the slopes are equal in absolute value, the price will fluctuate periodically.

Remark 6.2 Nonlinear supply and demand curves are also possible. The slope of the functions at the point of intersection will determine local stability properties of the equilibrium. Depending on how the slope changes away from the equilibrium, it is possible for the cobweb diagram to converge to a limit cycle. The details are omitted.

6.5.2 The Discrete Logistic Equation

The *discrete logistic equation* illustrates a fundamental difference between continuous-time differential equations and discrete-time difference equations. Suppose in our population studies that we sample a population only at discrete times, say yearly. Let x_0 represent the population of a species at the beginning of the year and suppose we sample the population at the beginning of each subsequent year. Then, we can relate the population at year $k + 1$ to the population at year k as

$$x_{k+1} = rx_k(1 - x_k). \tag{6.13}$$

Equation (6.13) is known as the *discrete logistic equation*. It is easy to see that there are in general two equilibrium points for the logistic equation, namely, $x = 0$ and $x = (r - 1)/r$ (Problem 6.12). The behavior of the logistic equation is different depending on the value of the parameter r.

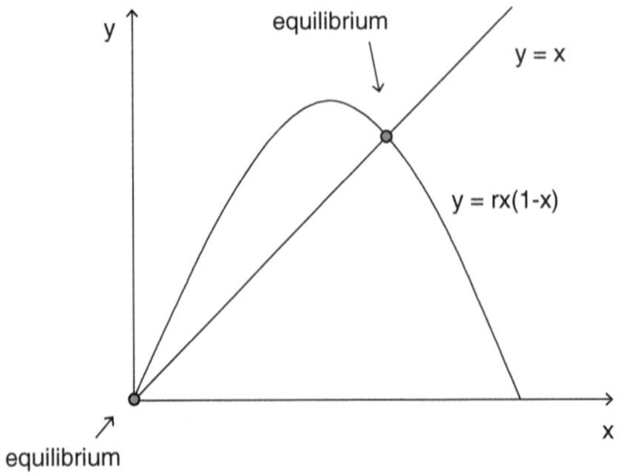

Figure 6.9 Equilibrium points of the discrete logistic equation at 0 and $(r-1)/r$.

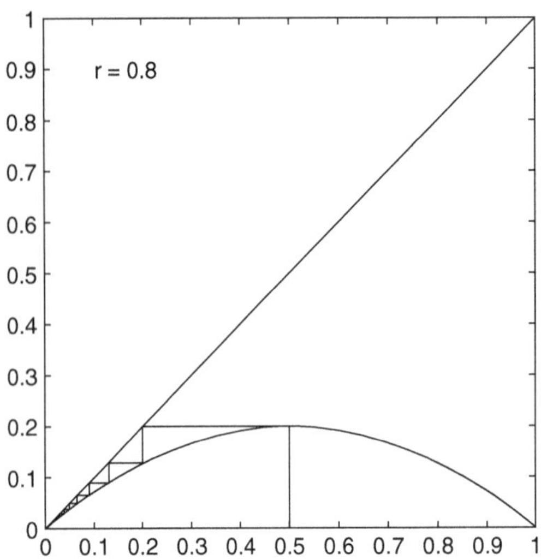

Figure 6.10 Cobweb diagram for $r \leq 1$. In this case the only equilibrium is at the origin and all solutions converge to zero.

We can use a cobweb diagram to plot the behavior of the equation. When the parameter r is less than or equal to 1, then for any initial starting value, the population will always tend towards 0 in future generations (Figure 6.10). That is, no population can sustain itself.

When r is between 1 and 2, then the nonzero equilibrium point is between 0 and 1/2. In this case all initial populations with the exceptions of 0 and 1 will converge to a population of $(r-1)/r$ (Figure 6.11).

A qualitative change occurs when r exceeds 2 (Figure 6.12). The slope of the function $y = rx(1-x)$ at the fixed point changes from positive to negative at $r = 2$. In this case, the iterates will still converge to the $(r-1)/r$ fixed point, but they no longer approach from either the positive side or the negative side, as they did when

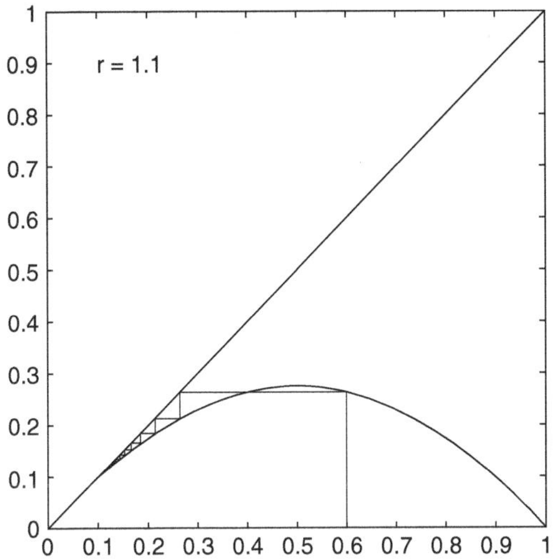

Figure 6.11 Cobweb diagram for $r = 1.1$. In this case, the solutions converge to the nonzero equilibrium.

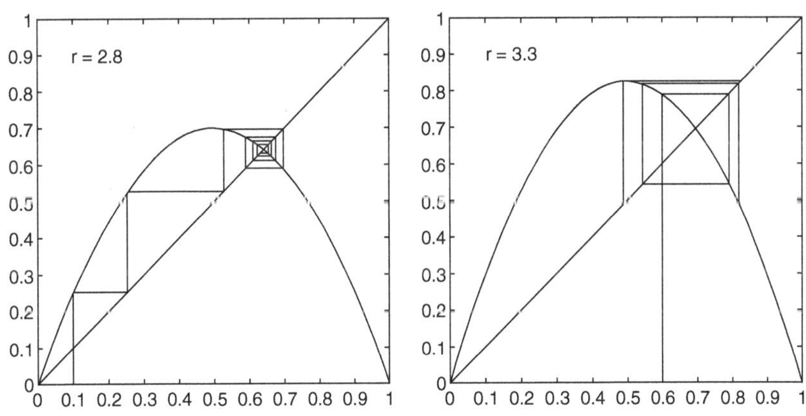

Figure 6.12 Stable equilibrium for $r = 2.8$ (left) and unstable equilibrium for $r = 3.3$ (right).

r was between 1 and 2. Instead, because of the downward slope at the fixed point they flip-flop on alternate sides of the fixed point.

Another change occurs when r becomes greater than 3. For r greater than 3, the point $(r - 1)/r$ is still a fixed point, but the slope at this fixed point becomes steeper than -1. This means that populations close to the fixed point will oscillate back and forth around the fixed point, but they will tend to get farther away from the

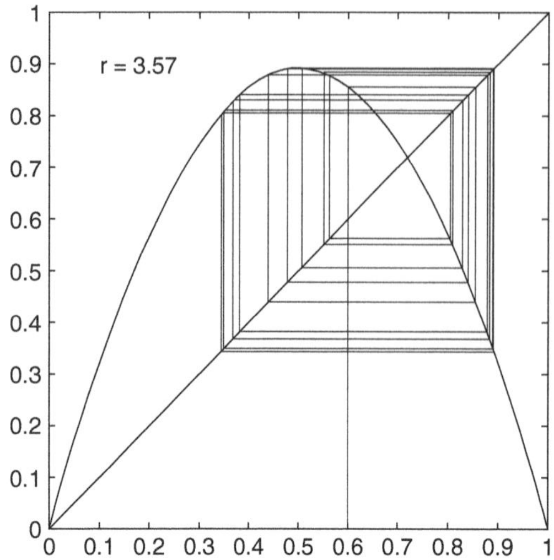

Figure 6.13 Chaotic solution for $r = 3.57$.

fixed point on every iteration. In other words, the fixed point becomes unstable for $r > 3$. This is another example of a bifurcation – a qualitative change in behavior as a parameter varies.

Increasing the parameter r further results in behavior that neither converges nor is periodic. When $r = 3.57$ the solution exhibits chaos (Figure 6.13).

The continuous-time logistic equation cannot exhibit chaotic behavior. In fact, no first- or second-order autonomous nonlinear differential equation can exhibit chaos. In discrete time even a first-order system can have chaotic solutions.

Problems

6.1 Consider the system in Equation (6.5). Show that the eigenvalues of the coefficient matrix A are $\lambda_i = 1/2 \pm \sqrt{5}/2$. The number $\lambda = 1/2 + \sqrt{5}/2$ is called the *golden ratio*. Show that the ratio $y(k+1)/y(k)$ in Equation (6.4) approaches the golden ratio as $k \to \infty$.

6.2 Consider the linear nonhomogeneous equation

$$8y(k+2) - 6y(k+1) + 2y(k) = 1, \quad \text{with } y(1) = 2,\ y(2) = 0.$$

Write a Matlab script to simulate this system and plot the results.

6.3 Consider the discrete-time system

$$y_{k+3} + 0.2y_{k+2} + 0.5y_{k+1} - 0.3y_k = 0.$$

Define a state vector x and write this system as

$$x_{k+1} = Ax_k.$$

Write a Matlab script to simulate and plot the solution of the system starting at a given initial condition x_0.

6.4 Assuming $y(0) = y(1) = y(2) = 0$, find by direct numerical recursion the values of $y(k)$, $k = 3,4,5$, satisfying the nonlinear difference equation.

$$y(k+3) - y(k+2) + [y(k+1) + y(k)]^2 = 1.$$

6.5 (Repeated Roots to the Characteristic Polynomial) Consider the second-order difference equation

$$y(k+2) - 2ay(k+1) + a^2y(k) = 0.$$

(a) Show that both $y(k) = a^k$ and $y(k) = ka^k$ are solutions.
(b) Find the solution to this equation with initial conditions $y(0) = 1$ and $y(1) = 0$.

6.6 Consider the second-order difference equation

$$y(k+2) + y(k) = 0.$$

(a) Write the system in state as $x(k+1) = Ax(k)$.
(b) What are the eigenvalues of the coefficient matrix A?
(c) Write a Matlab script to simulate the system for various initial conditions and plot the solution. Note the analogy to the harmonic oscillator differential equation.

6.7 Consider the discrete-time system $x(k+1) = Ax(k)$ where the matrix A is

$$A = \begin{bmatrix} 0.5 & 0 & 0 \\ 0 & 0 & 1 \\ 0 & -0.5 & -0.6 \end{bmatrix}.$$

Is the equilibrium (0,0) stable or unstable? Explain. Now consider the continuous-time system $\dot{x} = Ax$, where A is the same matrix as above. Is the equilibrium (0,0) stable or unstable? Explain.

6.8 Suppose in the discrete-time system $x(k+1) = Ax(k)$ that 1 is an eigenvalue of the matrix A. What can you say about the equilibrium states and stability in this case?

6.9 Consider the discrete-time system

$$x(k+1) = Ax(k), \quad \text{where } A = \begin{bmatrix} a & 1 \\ 0 & \frac{1}{a} \end{bmatrix}.$$

Are there values of the constant a that make the equilibrium $\bar{x} = 0$ asymptotically stable? Explain.

6.10 Consider the discrete-time system

$$x(k+1) = ax(k)^n,$$

where n is an integer. Find the equilibrium points and investigate stability of the system in terms of a and n. Investigate both analytically and experimentally using Matlab.

6.11 Consider the discrete-time system

$$x(k+1) = -2x(k)^3 + 0.4.$$

Find the equilibrium points and plot five points of a cobweb diagram for the system.

6.12 Show that $x = 0$ and $x = (r-1)/r$ are both equilibria for the discrete-time logistic equation.

7

Numerical Methods

7.1 Introduction

In this chapter we discuss numerical methods in simulation [32]. We discuss numerical approximation of both derivatives and integrals of functions and we derive iterative algorithms for numerical solution of ordinary differential equations (ODEs). Matlab has a number of built-in functions for numerical solution of ODEs. We discuss the most popular of these functions and give some examples of their use. We also give a brief introduction to *Simulink*, which is Matlab's graphical block-diagram program for constructing and simulating systems.

Example 7.1 As a first example of an iterative algorithm, suppose that we want to compute the square root of a real number N. Note that the square root of N is a solution in x of the equation

$$x^2 - N = 0. \tag{7.1}$$

Dividing both sides of equation (7.1) by x, we have

$$0 = x - \frac{N}{x} = 2x - x - \frac{N}{x}.$$

Thus, the square root of N satisfies the implicit equation

$$x = \frac{1}{2}\left(x + \frac{N}{x}\right). \tag{7.2}$$

We can use Equation (7.2) to define an iterative algorithm

$$x_{k+1} = \frac{1}{2}\left(x_k + \frac{N}{x_k}\right), \quad k = 0, 1, 2, \ldots, \tag{7.3}$$

to approximate the square root, starting at an initial guess x_0. The Matlab script below calculates the square root of N using this equation, accurate to four decimal places.

Introduction to Modeling and Simulation: A Systems Approach, First Edition. Mark W. Spong.

Companion Website: www.wiley.com/go/spong/engineeringmodeling

Iterative Algorithm to calculate the square root of *N*

```
N=input('Input a real number N: ');
x0=1;    % initial guess
x=x0;    % initialize x
while 1
    if abs(x*x-N) < 0.001
        break
    else
        x = 0.5*(x + N/x)
    end
end
fprintf('The square root of %4.4f is %4.4f',N,x)
```

Starting at the initial guess $x_0 = 1$, the Matlab output for $N = 2$ is

```
x = 1.5000
x = 1.4167
x = 1.4142
The square root of 2.0000 is 1.4142.
```

The algorithm defined by Equation (7.3) is called *Newton's Method*; we will revisit it in Chapter 8 when we discuss optimization methods.

7.2 Numerical Differentiation

There are several ways to approximate the derivative of a function using so-called *finite-difference approximations*. Such approximations are the basis for numerical solution of differential equations. We present three such approximation methods, the *forward*, *backward*, and *middle* difference approximations.

For $f(x)$ differentiable at $x = a$ the *forward difference quotient*

$$\frac{f(a+h) - f(a)}{h} \tag{7.4}$$

converges to the derivative $f'(a)$ in the limit as $h \to 0$. Thus, for small values of $h > 0$, the above difference quotient approximates the derivative $f'(a)$. The closeness of this approximation can be quantified using the Taylor series expansion of $f(x)$.

The reader should first review the definitions of the order notation $O(h^n)$ and the Taylor series approximation for a smooth function in Appendix B. The Taylor series approximates the value of a function $f(x)$ at a point $b = a + h$ in terms of the value of the function and its derivatives at a nearby point a as follows:

$$f(b) = f(a) + hf'(a) + \frac{h^2}{2!}f''(a) + \cdots + \frac{h^{n-1}}{(n-1)!}f^{(n-1)}(a) + O(h^n).$$

Keeping only the first two terms gives

$$f(b) = f(a) + hf'(a) + O(h^2).$$

From the above with $b = a + h$ we see that the error

$$\left|\frac{f(a+h) - f(a)}{h} - f'(a)\right| = O(h).$$

The approximation given by the forward difference quotient in Equation (7.4) is called the *forward difference approximation* of $f'(x)$ at $x = a$.

Example 7.2 Suppose $f(x) = x^2$. Then a simple calculation shows that

$$\frac{f(x+h) - f(x)}{h} = \frac{(x+h)^2 - x^2}{h} = \frac{x^2 + 2xh + h^2 - x^2}{h} = 2x + h,$$

showing that the difference quotient approximates the true derivative $f'(x) = 2x$ with an order $O(h)$ error.

Next, the *backward difference approximation* of $f'(x)$ at a point a is given by

$$\left|\frac{f(a) - f(a-h)}{h} - f'(a)\right| = O(h).$$

Combining the forward and backward approximations leads to the *middle difference approximation*

$$\frac{f(a+h) - f(a-h)}{2h} = f'(a) + O(h^2).$$

Note that the middle difference approximation gives a more accurate $O(h^2)$ approximation of the derivative. Figure 7.1 illustrates these approximations. The slopes

Figure 7.1 Illustrating the forward, backward, and middle difference approximations to $f'(a)$.

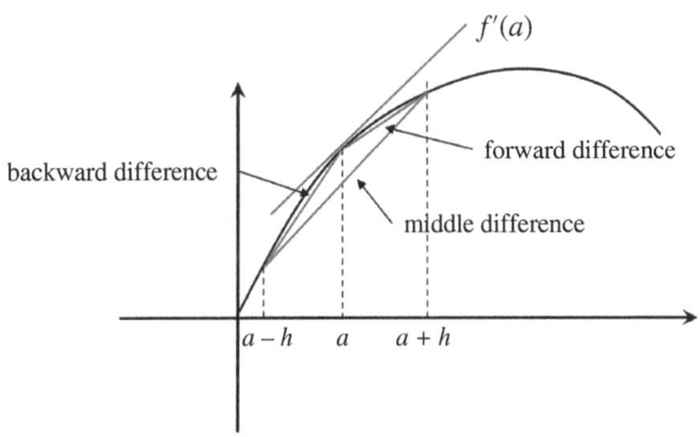

of the line segments indicated approximate the true derivative at $x = a$, which is the slope of the line tangent to $f(x)$ at a.

We may also approximate the second derivative $f''(x)$ of a smooth function as the difference quotient

$$\left|\frac{f(a+h) - 2f(a) + f(a-h)}{h^2} - f''(a)\right| = O(h^2). \tag{7.5}$$

To show this, we write the Taylor series expansions of $f(x)$ about $x + a$ and $x - a$ as

$$f(a+h) = f(a) + hf'(a) + \frac{h^2}{2}f''(a) + \frac{h^3}{6}f'''(a) + O(h^4)$$
$$f(a-h) = f(a) - hf'(a) + \frac{h^2}{2}f''(a) - \frac{h^3}{6}f'''(a) + O(h^4).$$

Therefore, combining these two equations and dividing by h^2 gives

$$\frac{f(a+h) - 2f(a) + f(a-h)}{h^2} = f''(a) + O(h^2).$$

Thus the difference quotient in Equation (7.5) approximates the second derivative with an $O(h^2)$ error.

Example 7.3 Let $f(x) = x^3 - 2x^2 + 2x$. Then the derivative of $f(x)$ at $x = 3$ is $f'(3) = 17$. Using the formulae for the forward, backward, middle difference approximations with $h = 0.1$ gives

```
forward = 17.7100
backward = 16.3100
middle = 17.0100.
```

Also, the second derivative at $x = 3$ is $f''(3) = 14$ and the difference quotient (7.5) gives

```
second derivative = 14.0000.
```

Example 7.4 The Matlab function `diff` computes the difference of adjacent elements of a vector and can be used to approximate the derivative of a function.

Compute the derivative of sin(t) using `diff`

```
h = 0.01;             % step size
t = 0:h:2*pi;         % domain
f = sin(t);           % range
y = diff(f)/h;        % derivative
plot(t,f,'k',t(:,1:length(y)),y,'--k')
```

Running the above script yields the plot in Figure 7.2.

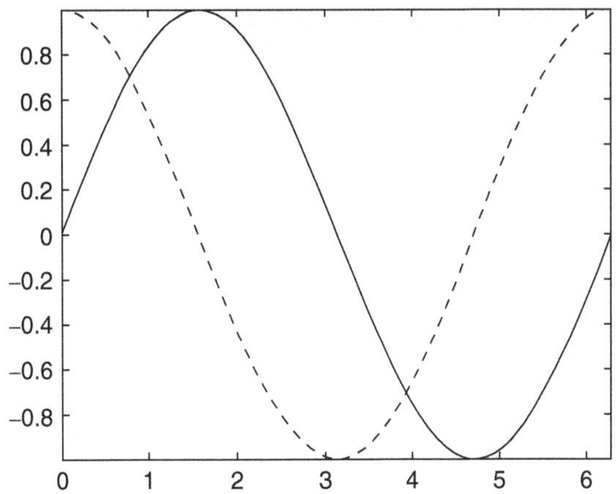

Figure 7.2 Numerical computation of the derivative of sin(t) using the Matlab command `diff`. The dotted line is the approximate derivative, representing cos(t).

In addition, the command `diff(f,2)/`h^2 computes the second derivative of the function f and is equivalent to `diff(diff(f))/`h^2 (Problem 7.5).

7.3 Numerical Integration

In this section we give algorithms for computing definite integrals of functions of a single variable.

Left-Endpoint Integration

To approximate a definite integral $\int_a^b f(x)dx$, we subdivide the interval $[a, b]$ into n subintervals $[x_0 = a, x_1, \dots, x_n = b]$ and use the finite summation

$$\int_a^b f(x)dx \approx \sum_{i=0}^{n-1} f(x_i)\Delta x_i,$$

where $\Delta x_i = x_{i+1} - x_i$. For simplicity, we assume a uniform partition, i.e., that the intervals Δx_i are all equal to $h = (b - a)/n$. The left-endpoint integration method is illustrated in Figure 7.3.

The summation $\sum_{i=0}^{n-1} f(x_i)\Delta x_i$ is called a *Riemann sum*. If the function $f(x)$ is continuous (almost everywhere) on $[a, b]$, then the limit

$$\lim_{n\to\infty} \sum_{i=0}^{n-1} f(x_i)\Delta x_i$$

exists and is equal to $\int_a^b f(x)dx$. This limit is called the *Riemann integral* of $f(x)$.

Right-Endpoint and Midpoint Integration

Alternate ways to approximate the integral of $f(x)$ are the *right-endpoint* and *midpoint* integration methods. The former is expressed as

$$\int_a^b f(x)dx \approx \sum_{i=0}^{n-1} f(x_{i+1})\Delta x_i,$$

and the latter as

$$\int_a^b f(x)dx \approx \sum_{i=0}^{n-1} f\left(\frac{1}{2}(x_i + x_{i+1})\right)\Delta x_i.$$

These are also illustrated in Figure 7.3.

The sum of the areas of the rectangles defined by the partition of the interval $[a, b]$ approximates the total area under the graph of the function. The Matlab script below implements the left-endpoint integration for the function $f(x) = x^2$ from $a = 0$ to $b = 1$. With the given partition size, the calculation is accurate to four decimal places.

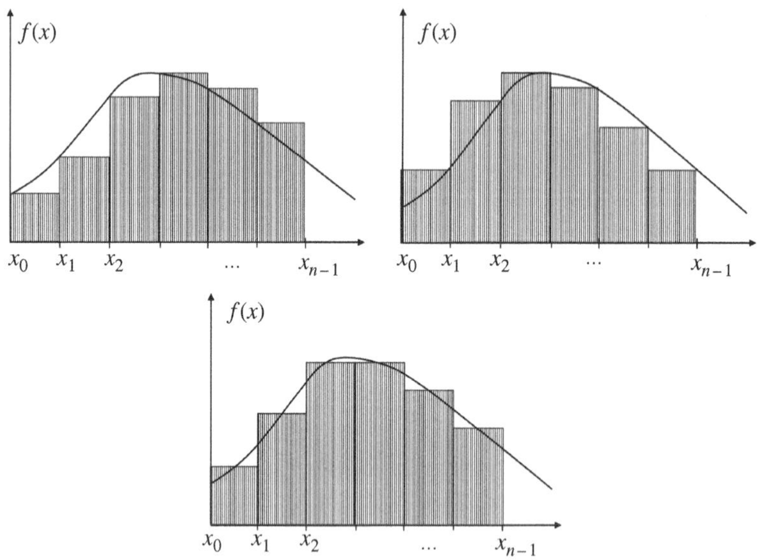

Figure 7.3 Illustrating the left-endpoint, right-endpoint and midpoint integration methods. The sum of the areas of the rectangles approximates the total area under the curve.

Left-Endpoint Integration of $f(x) = x^2$ from $x = 0$ to $x = 1$.

```
h=0.0001;     % partition size
N = 1/h;      % number of rectangles
x=0; area=0;
for i=0:N-1
f=x*x;
x=x+h;
area = area + f*h;
end
```

Note that it is not necessary to use a uniform partition of the interval. The key property is that the widths of the subintervals Δx_i approach zero as the number of subintervals increases. If the function $f(x)$ takes on both positive and negative values we define the integral as the area above the x-axis minus the area below the x-axis. It is left as an exercise (Problem 7.6) to modify the above Matlab code to implement the right-endpoint and midpoint integration methods.

Trapezoidal Integration

Another useful approximation of the integral, shown in Figure 7.4, is the *trapezoidal approximation* given by

$$\int_a^b f(x)dx \approx \sum_{i=0}^{n-1} \frac{1}{2}(f(x_i) + f(x_{i+1}))\Delta x_i.$$

Error Analysis

The closeness of the approximation of any of these integration methods is a function of the width of the interval $h = x_{i+1} - x_i$. We can calculate the error as

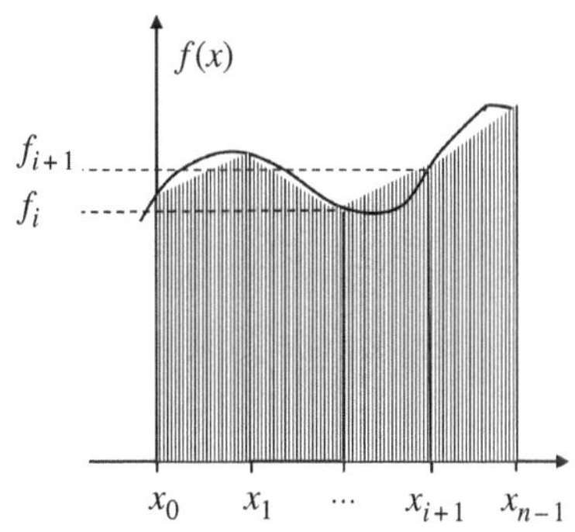

Figure 7.4 Illustrating trapezoidal integration. The area under a representative trapezoid is equal to $\frac{1}{2}(f(x_i) + f(x_{i+1}))$.

a function of h using the Taylor series expansion. We illustrate this error analysis with the trapezoidal integration method.

Let us denote $f(x_i)$ by f_i for $i = 0, 1, \dots, n-1$. For a given x_i, we can write

$$f(x) = f_i + f_i' \cdot (x - x_i) + \frac{1}{2}f_i'' \cdot (x - x_i)^2 + O((x - x_i)^3).$$

With $h = x_{i+1} - x_i$ we integrate both sides of the above to obtain

$$\int_{x_i}^{x_{i+1}} f(x)dx \approx \int_{x_i}^{x_{i+1}} f_i dx + \int_{x_i}^{x_{i+1}} f_i' \cdot (x - x_i)dx + \int_{x_i}^{x_{i+1}} \frac{1}{2}f_i'' \cdot (x - x_i)^2 dx$$
$$= f_i h + f_i' \frac{h^2}{2} + f_i'' \frac{h^3}{6}.$$

Next, starting with

$$f_{i+1} = f_i + f_i' h + f_i'' \frac{h^2}{2},$$

we rearrange terms to approximate the derivative term $f_i' = f'(x_i)$ as

$$f_i' \approx \frac{f_{i+1} - f_i}{h} - \frac{h}{2}f_i''.$$

Substituting this into the expression for the integral, we get

$$\int_{x_i}^{x_{i+1}} f_i dx = f_i h + f_i' \frac{h^2}{2} + f_i'' \frac{h^3}{6}$$
$$= f_i h + \left(\frac{f_{i+1} - f_i}{h} - \frac{h}{2}f_i'' \right) \frac{h^2}{2} + f_i'' \frac{h^3}{6}$$
$$= \frac{h}{2}(f_i + f_{i+1}) - \frac{h^3}{12}f_i'',$$

which is the trapezoidal approximation $\frac{h}{2}(f_i + f_{i+1})$ plus the error term $-\frac{h^3}{12}f_i''$. Summing over the interval $[a, b]$ gives

$$\int_a^b f(x)dx = \sum_{i=0}^{n-1} \int_{x_i}^{x_{i+1}} f(x)dx$$
$$= \sum_{i=0}^{n-1} \left[\frac{h}{2}(f_i + f_{i+1}) - \frac{h^3}{12}f_i'' \right].$$

Therefore, the total error E is given by

$$E = \left| \sum_{i=0}^{n-1} \frac{h^3}{12} f_i'' \right|.$$

Assuming that the second derivative is bounded as $|f''(x)| \leq M$ on $[a, b]$, we can write

$$E \leq \frac{h^3}{12} \sum_{i=0}^{n-1} M = \frac{h^3}{12} MN = \frac{h^2}{12} M(b - a) = O(h^2) \quad \text{since } Nh = b - a.$$

Thus, the trapezoidal approximation is an $O(h^2)$ approximation of the definite integral.

Remark 7.1 Note that the trapezoidal approximation

$$\int_a^b f(x)dx \approx \sum_{i=0}^{n-1} \frac{1}{2}(f(x_i) + f(x_{i+1}))\Delta x_i = \sum_{i=0}^{n-1} \frac{1}{2}(f_i + f_{i+1})\Delta x_i$$

can be written as

$$\int_a^b f(x)dx \approx (f_1 + \cdots + f_{n-1})h + \frac{1}{2}(f_0 + f_n)h, \tag{7.6}$$

which simplifies the calculation of the integral.

Example 7.5 Suppose we want to compute the natural logarithm ln(2) to a certain number of decimal places. Note that $\ln(2) = \int_1^2 \frac{1}{x}dx$. Therefore, we can use numerical integration methods to compute ln(2). The Matlab script below uses the trapezoidal approximation from Equation (7.6) to compute this integral.

Trapezoidal Rule to Compute ln(2)

```
h = 0.005;              % step size
x_init = 1;
x_final = 2;
x = x_init;             % Initial value of x
n = 20;                 % number of subintervals
sum_f = 0;
for i = 1:n-1
x = x+h;
f - 1/x;
sum_f = sum_f+f;
x1(:,i) = [x f];
end
ln2 = h*(sum_f+(1/2)*(1/x_init+1/x_final))
```

Running this script gives ln(2) = 0.6931.

Simpson's Rule

Simpson's rule is a method of numerical integration that is based on approximating the integrand $f(x)$ by a quadratic function. Let $P(x) = Ax^2 + Bx + C$ be a quadratic approximation of $f(x)$ passing through the three points

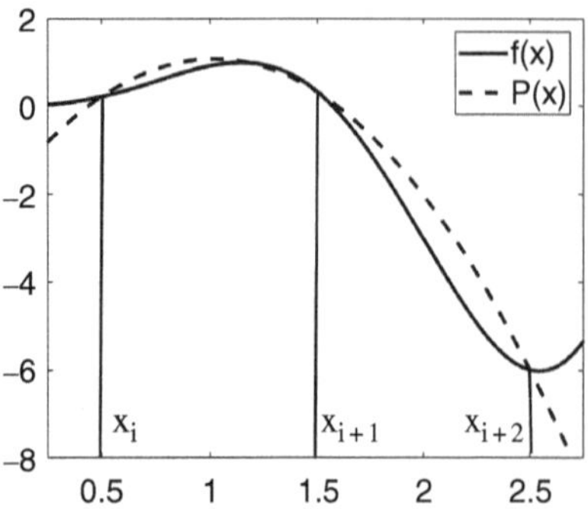

Figure 7.5 Illustrating Simpson's rule. The function $f(x)$ is approximated by a quadratic function $P(x)$.

$(x_i, f_i), (x_{i+1}, f_{i+1}), (x_{i+2}, f_{i+2})$ as shown in Figure 7.5. So the coefficients A, B, C should satisfy

$$\begin{aligned} f_i &= Ax_i^2 + Bx_i + C, \\ f_{i+1} &= Ax_{i+1}^2 + Bx_{i+1} + C, \\ f_{i+2} &= Ax_{i+2}^2 + Bx_{i+2} + C. \end{aligned}$$

This gives us three equations in the three unknowns A, B, C, which can be solved to determine the approximating quadratic polynomial.

Note that if $f(x) = Ax^2 + Bx + C$ then a straightforward calculation (Problem 7.7) shows that

$$\int_{-h}^{h} f(x)dx = \frac{h}{3}[f(-h) + 4f(0) + f(h)]. \tag{7.7}$$

It now follows that, if we choose n points $x_0, \ldots, x_{n-1}$ in the interval $[a, b]$, with $n =$ even and $x_{i+1} - x_i = h$, then

$$\int_a^b f(x)dx = \frac{h}{3}[f_0 + 4f_1 + 2f_2 + 4f_3 + 2f_4 + \cdots + 4f_{n-1} + f_n].$$

The error with Simpson's rule is approximately $E = \frac{h^4}{180}M(b-a)$, where M is a bound on the fourth derivative of f (see [59]).

Simpson's Rule to Compute $\int_0^1 (x^3 + 2x)dx$

```
f = @(x) x.^3+2*x;
a = 0; b= 1;
n = 10;
h = (b-a)/n;
s = f(a) + f(b);
for i = 1:2:n-1
    s = s +4*f(a+i*h);
```

```
end
for i = 2:2:n-2
    s = s+2*f(a+i*h);
end
I = h/3*s;
fprintf('The area equals %4.4f',I)
```

7.4 Numerical Solution of ODEs

We now turn to numerical solution of ordinary differential equations (ODEs). Consider the initial-value problem

$$\dot{x} = f(x), \; x(0) = x_0, \quad x \in \mathbb{R}^n,$$

and recall the forward difference approximation

$$\dot{x} \approx \frac{x(t+h) - x(t)}{h} \approx f(x).$$

Then, with $x(t_k) = x_k$ and $x(t_k + h) = x_{k+1}$, we define an iterative expression for the solution of the ODE, called *Euler's method*,

$$x_{k+1} = x_k + hf(x_k). \tag{7.8}$$

Since this expression is an approximation of the solution $x(t)$, there is an error at each iteration k. For Euler's method to succeed, the step size h should be small to avoid having these errors build up excessively.

Example 7.6 Consider the first-order linear equation

$$\dot{x} = ax, \quad x(0) = x_0.$$

Figure 7.6 shows the solution using Euler's method for two different values of the step size.

For this problem, and in general for unstable trajectories, the solution using Euler's method will eventually diverge from the true solution for any step size. As we will see, more sophisticated methods address this problem with variable step sizes and error checking routines.

Example 7.7 This next example illustrates the importance of using physical intuition to interpret the results of a numerical simulation. Let us use Euler's method with a step size of $h = 0.1$ to simulate the response of the simple pendulum model.

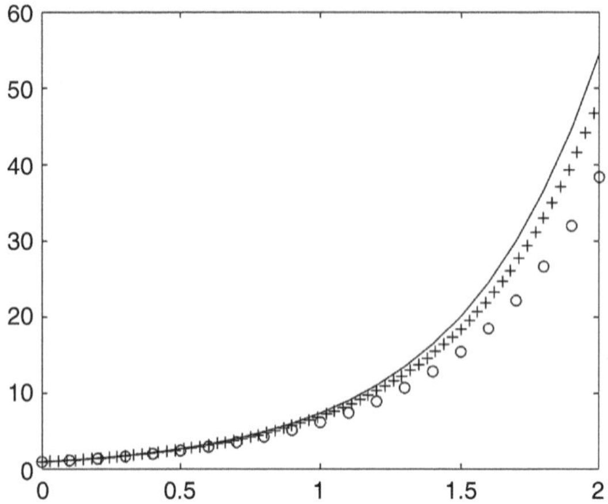

Figure 7.6 Euler's method for $\dot{x} = ax$ with $a = 2$, $x(0) = 1$, and step sizes $h = 0.1$ (∘) and $h = 0.03$ (+). The solid line is the exact solution $x(t) = e^{2t}$. The smaller step size gives a more accurate estimate of the solution at the expense of more computation.

Euler's Method for the Simple Pendulum

```
f = @(x) [x(2);-sin(x(1))];   % pendulum vector field
x0 = [0.2;0];                 % initial conditions
t0 = 0;                       % start time
tf = 20;                      % end time
h = 0.1;                      % step size
N = (tf-t0)/h;                % number of steps
t = t0;
x = x0;
s1 = x0;
for i = 1:N
    x = x+h*f(x);             % Euler step
    s1 = [s1,x];
end
```

Running the above script yields the plot in Figure 7.7.

Explicit and Implicit Methods

Equation (7.8) is called an *explicit* integration method since x_{k+1} is explicitly defined by the right-hand side, $x_k + hf(x_k)$. Let us now consider the backward difference approximation

$$\dot{x}(t+h) \approx \frac{x(t+h) - x(t)}{h} \approx f(x(t+h)).$$

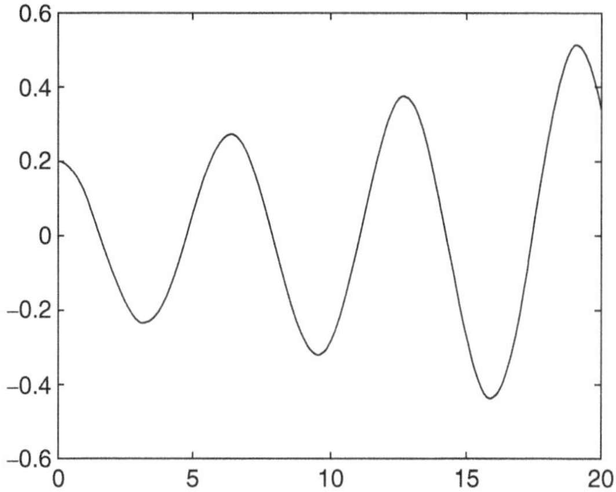

Figure 7.7 Response of the simple pendulum model using Euler's method with a step size of $h = 0.1$. Since we know that the trajectories of the simple pendulum are periodic, we conclude that the increasing amplitude of oscillation in the response is due to numerical errors and does not represent the true dynamics of the system. Therefore, the step size should be reduced to improve the accuracy of the simulation (Problem 7.14).

This leads to the iterative procedure

$$x_{k+1} = x_k + hf(x_{k+1}). \tag{7.9}$$

Equation (7.9) is called an *implicit* integration method since x_{k+1} appears on both sides of the equation and hence is only defined implicitly by Equation (7.9). Implicit methods often have advantages over explicit methods as we show in the next example.

Example 7.8 Consider the first-order linear equation

$$\dot{x} = -ax, \quad x(0) = x_0, \quad a > 0. \tag{7.10}$$

Using the forward difference equation, the explicit Euler method gives

$$x_{k+1} = x_k - ahx_k = (1 - ah)x_k, \quad x(0) = x_0. \tag{7.11}$$

The solution of this difference equation is

$$x_k = (1 - ah)^k x_0.$$

This discrete-time system is stable provided $|1 - ah| < 1$, which requires that the step size h satisfy $0 < h < 2/a$. On the other hand, using the backward difference approximation, the implicit Euler method for this equation gives

$$x_{k+1} = x_k - ahx_{k+1}, \tag{7.12}$$

which can be solved for x_{k+1} as

$$x_{k+1} = (1 + ah)^{-1} x_k. \tag{7.13}$$

The discrete-time system (7.13) has the solution

$$x_{k+1} = (1 + ah)^{-k} x_0,$$

which is stable for all values of $h > 0$ since $|(1 + ah)^{-1}| < 1$ for all $h > 0$. Thus, in this case, the discrete-time solution is insensitive to the choice of step size h.

In the case of the simple linear system (7.10) it was easy to solve the implicit equation (7.12) for x_{k+1}. In the general case, solving the implicit expression

$$x_{k+1} = x_k + hf(x_{k+1})$$

for x_{k+1} is more difficult. Thus the tradeoff between the explicit and implicit methods is that the former may require smaller step size but is generally easier to compute.

7.4.1 Euler Predictor–Corrector Method

We next discuss a modification to the Euler method, that falls into the category of so-called *predictor–corrector* methods. This is a two-step procedure that uses an explicit method in the first step (the predictor) and an implicit method in the second step (the corrector). The advantage of the predictor–corrector method is that we do not have to solve an implicit equation for x_{k+1}. Starting with the equation

$$\dot{x} = f(x),$$

we integrate both sides from t to $t + h$

$$\int_t^{t+h} \dot{x}(s)ds = \int_t^{t+h} f(x(s))ds.$$

The left-hand side is just $x(t + h) - x(t)$. We can approximate the right-hand- side integral using the trapezoid approximation

$$\int_t^{t+h} f(x(s))ds \approx \frac{h}{2}[f(x(t)) + f(x(t + h))],$$

which results in

$$x(t + h) = x(t) + \frac{h}{2}[f(x(t)) + f(x(t + h))].$$

Thus, with $x_k = x(t)$ and $x_{k+1} = x(t + h)$, we have the iterative formula

$$x_{k+1} = x_k + \frac{h}{2}[f(x_k) + f(x_{k+1})].$$

This is an implicit formula for x_{k+1}, and, since we do not know the value of x_{k+1}, we will compute an estimated value, $\tilde{x}_{k+1}$, for x_{k+1} via the Euler method

$$\tilde{x}_{k+1} \approx x_k + hf(x_k).$$

Thus, we finally have the complete formula

$$\tilde{x}_{k+1} = x_k + hf(x_k) \tag{7.14}$$

$$x_{k+1} = x_k + \frac{1}{2}h(f(x_k) + f(\tilde{x}_{k+1})). \tag{7.15}$$

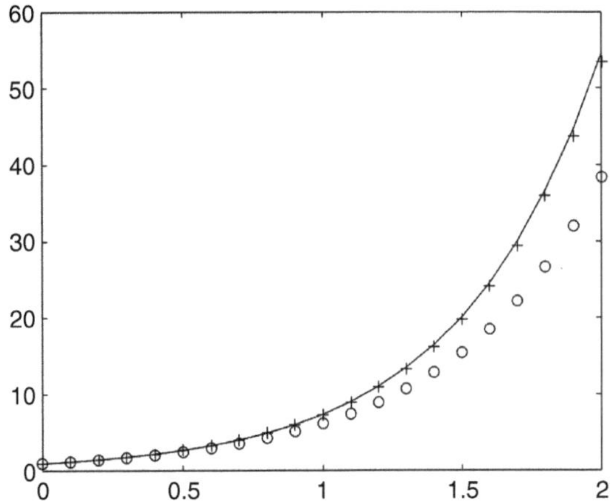

Figure 7.8 Euler's method (∘) and the Euler predictor–corrector method (+) for $\dot{x} = ax$ with $a = 2$, $x(0) = 1$, and step size $h = 0.1$. The predictor–corrector method shows better performance without having to decrease the step size.

Equation (7.14) is the *predictor* step and Equation (7.15) is the *corrector* step. The Matlab script below shows a comparison of Euler's method and the predictor–corrector method for the linear equation $\dot{x} = ax$. The results of the simulation are shown in Figure 7.8.

Predictor–Corrector Method for $\dot{x} = ax$

```
% Euler predictor-corrector example
a = 2;
f = @(x)a*x;
t0 = 0; tf = 2;
x0 = 1;    % initial condition
h = 0.1;   % step size
x = x0;    % initialize x
xout = x; % initialize output
for t = t0: h: tf-h
   xtilde = x+h*f(x);                      % predictor step
     x = x + 1/2*h*(f(x) + f(xtilde)); % corrector step
     xout = [xout;x];
end
% Compare with the explicit Euler method
y = x0;
yout = y;
for t = t0: h: tf-h
     y = y+h*a*y;
```

(Continued)

(Continued)

```
    yout = [yout;y];
end
t = t0:h:tf;
plot(t,exp(a*t),'k',t,xout,'k+',t,yout,'ko')
```

7.4.2 Runge–Kutta Methods

The previous predictor–corrector method is also called *Heun's method* and is an example of a so-called second-order *Runge–Kutta method*. In this section we consider more general second-order and fourth-order Runge–Kutta methods. The idea behind Runge–Kutta methods is to match the Taylor series expansion of $f(x)$ up to a certain order without having to compute the derivatives of $f(x)$ explicitly.

A second-order Runge–Kutta algorithm starts by writing $x(t+h)$ as

$$x(t+h) = x(t) + ak_1 + bk_2 \tag{7.16}$$

with

$$k_1 = hf(x(t)) \quad \text{and} \quad k_2 = hf(x(t) + \beta k_1),$$

where a, b, β are free parameters chosen to match Equation (7.16) with a Taylor series expansion of $f(x)$ up to second order. To this end we expand $x(t)$ in a Taylor series as

$$\begin{aligned} x(t+h) &= x(t) + h\dot{x}(t) + \frac{1}{2}h^2\ddot{x}(t) + O(h^3) \\ &= x(t) + hf(x(t)) + \frac{1}{2}h^2\frac{\partial f}{\partial x}f(x(t)) + O(h^3). \end{aligned}$$

Next, we expand the term k_2 in a Taylor series as

$$\begin{aligned} k_2 &= hf(x(t) + \beta k_1) \\ &= hf(x(t)) + \frac{\partial f}{\partial x}\beta k_1 + O(h^3). \end{aligned}$$

Combining the above equations results in

$$x(t+h) = x(t) + (a+b)hf(x(t)) + h^2 b\beta\frac{\partial f}{\partial x}f(x(t)).$$

Comparing terms in the above gives

$$a + b = 1, \quad \beta b = \frac{1}{2}.$$

Since there are two equations in three unknowns, there are infinitely many solutions. Choosing $a = b = 1/2$ and $\beta = 1$ gives the classical second-order Runge–Kutta algorithm

$$\begin{aligned} x(t+h) &= x(t) + \frac{1}{2}(k_1 + k_2) \\ k_1 &= hf(x(t)) \\ k_2 &= hf(x(t) + k_1). \end{aligned}$$

Higher-Order Runge–Kutta Methods

The same procedure can be used to derive higher-order Runge–Kutta methods, which match more terms in the Taylor series expansion of $f(x)$. Perhaps the most popular and widely used is the fourth-order Runge–Kutta (RK4) method. This method can be summarized as:

$$\begin{aligned} x(t+h) &= x(t) + \frac{1}{6}(k_1 + 2k_2 + 2k_3 + k_4) \\ k_1 &= hf(x(t)) \\ k_2 &= hf(x(t) + k_1/2) \\ k_3 &= hf(x(t) + k_2/2) \\ k_4 &= hf(x(t) + k_3). \end{aligned}$$

The above RK4 formula determines $x(t+h)$ by adding to $x(t)$ a weighted average of the four terms $k_1, \dots, k_4$.

- k_1 is based on the slope of the function at the beginning of the interval $[t, t+h]$. Thus, if k_2, k_3 and k_4 are zero, then this is equivalent to Euler's method.
- k_2 and k_3 are based on the slope of the function at the midpoint of the interval.
- k_4 is based on the slope of the function at the end of the interval.

Remark 7.2 For a nonautonomous equation

$$\dot{x} = f(x(t), t),$$

the RK4 method should be modified as

$$\begin{aligned} x(t+h) &= x(t) + \frac{1}{6}(k_1 + 2k_2 + 2k_3 + k_4) \\ k_1 &= hf(x(t), t) \\ k_2 &= hf(x(t) + k_1/2, t + h/2) \\ k_3 &= hf(x(t) + k_2/2, t + h/2) \\ k_4 &= hf(x(t) + k_3, t + h). \end{aligned}$$

Example 7.9 Figure 7.9 shows a comparison of a fourth-order Runge–Kutta method with Euler's method for a nonautonomous system $\dot{x} = 1/t^3 - x/2$.

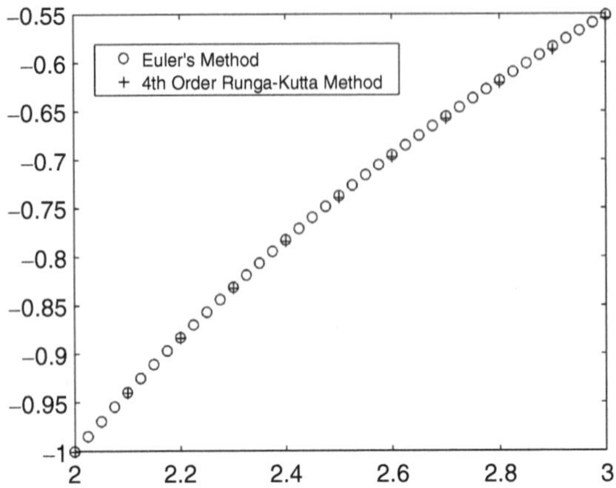

Figure 7.9 Comparison of Euler's method (∘) with a fourth-order Runge–Kutta method (+) for the nonautonomous system $\dot{x} = 1/t^3 - x/2$. The fourth-order Runge–Kutta method produces a good result with larger step size.

Comparison of Euler's Method and a Fourth-Order Runge–Kutta Method

```
f = @(t,x)1/t.^3-x/2;
xinit = -1; tinit = 2; tfinal = 3;
h = 0.025;                  % Euler step size
N = (tfinal-tinit)/h;       % number of steps
s1 = xinit; x = xinit; t = tinit;
for i = 1:N
    x = x+h*f(t,x);   % Euler step
    s1 = [s1;x];
    t = t+h;
end
% 4th-order Runge-Kutta
h = 0.1;                  % RK step size
N = (tfinal-tinit)/h;
s3 = xinit; x3 = xinit; t3 = tinit;
for i = 1:N
    k1 = h*f(t3,x3);
    k2 = h*f(t3+h/2,x3+k1/2);
    k3 = h*f(t3+h/2,x3+k2/2);
    k4 = h*f(t3+h,x3+k3);
    x3 = x3+1/6*(k1+2*k2+2*k3+k4);
    s3 = [s3;x3];
    t3 = t3+h;
end
```

We note that the Runge–Kutta methods are explicit methods and so the choice of step size h for stability is important. Matlab's `ode45` algorithm uses a modified RK4 with a variable step size. Both a fourth-order and a fifth-order Runge–Kutta step are computed and compared. If their difference is larger than the specified error tolerance, the step size is reduced and the Runge–Kutta steps are recomputed [19]. Further information about numerical integration and Matlab's ODE suite is contained in [60–63]. Refer also to the Matlab documentation for further details.

7.5 Stiff Systems

Stiffness is an important property to consider in numerical solution of ODEs. The definition of what constitutes a stiff system can be somewhat vague. Generally, systems with very large or very small parameters, such as singularly perturbed systems, or so-called differential-algebraic systems, discussed below, fall into the category of stiff systems. In the linear case, we say that a linear system is stiff if the ratio of the (absolute values of the) largest to the smallest eigenvalue, called the *stiffness ratio*, is large.

The Matlab ODE solvers `ode23s` and `ode15s` are designed to handle stiff systems. These are implicit methods in contrast to `ode45`, which is an explicit method.

Example 7.10 Consider the linear system

$$\begin{aligned} \dot{x}_1 &= x_2 \\ \dot{x}_2 &= -ax_1 - bx_2. \end{aligned}$$

For $a = 10$ and $b = 200.05$, the eigenvalues of the system matrix are $\lambda_1 = -0.005$ and $\lambda_2 = -200$ and so the stiffness ratio is $40,000$. The Matlab script below compares the elapsed simulation time for this system using `ode45` and `ode23s`.

A Comparison of `ode45` and `ode23s` for a Stiff Linear System

```
a = 10;
b = 200.05;
f = @(t,x) [x(2);-a*x(1)-b*x(2)];
y0 = [1;0];
tspan = [0 500];
% elapsed time using ode45
```

(Continued)

(Continued)

```
et45 = zeros(10,1);
for i = 1:10
tic
[t,x] = ode45(f,tspan,y0);
et45(i) = toc;
end
filename = 'data.xlsx';
writematrix(et45,filename,'Sheet','1','Range','A1:A10');
% elapsed time using ode23s
et23s = zeros(10,1);
for j = 1:10
tic
[t,x] = ode23s(f,tspan,y0);
et23s(j) = toc;
end
filename = 'data.xlsx';
writematrix(et23s,filename,'Sheet',1,'Range','B1:B10');
```

The results of ten simulation runs are stored in the Excel file data.xlsx so that an average simulation time can be computed. A typical average execution time for a simulation of 500 seconds is 0.330 seconds using `ode45` and 0.009 seconds using `ode23s`. Thus we see that the stiff equation solver `ode23s` is more than an order of magnitude faster than the nonstiff solver `ode45` even for this very simple system.

Example 7.11 As another example, consider the one-dimensional nonlinear system

$$\dot{y} = y^2 - y^3.$$

This is a simple model of *flame propagation*. The variable y represents the radius of a ball of flame, which starts at a small size δ and grows slowly until it expands explosively to a constant equilibrium $y = 1$, which represents a balance between the amount of oxygen being consumed by the ball and the amount available at the surface of the ball. Near the equilibrium $y = 1$ the equation becomes "stiff" because solutions starting near the equilibrium very rapidly approach the equilibrium. This means that a discretization of the solution can overshoot and oscillate about the equilibrium so that very small step sizes are required to keep the solution near the equilibrium. Figure 7.10 shows this phenomenon using `ode45`. More than 120,000 steps were required to compute the solution, most of them around $y = 1$.

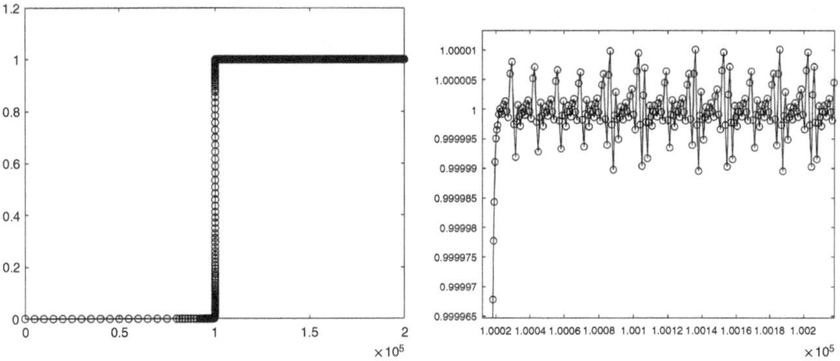

Figure 7.10 Solution of the stiff system using `ode45`. The expanded view on the right shows that `ode45` takes multiple steps near $y = 1$ in order to satisfy the error tolerances.

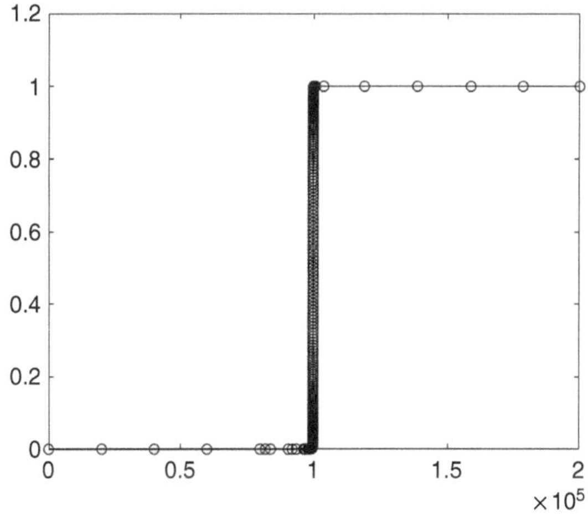

Figure 7.11 Solution of the stiff system using `ode23s`. As an implicit method, `ode23s` can "look ahead" at each step to reduce the number of steps needed around $y = 1$.

Figure 7.11 shows the response using the stiff equation solver `ode23s`. The implicit solution method, which effectively can "look ahead" and anticipate the overshoot, results in many fewer steps, in this case 241, to compute the solution.

Differential-Algebraic Equations

A particular example of the use of stiff equations solvers is for so-called *differential-algebraic equations* (DAEs). A differential-algebraic equation is one that combines differential equations and algebraic equations, for example, the system

$$\dot{x}_1 = f(x_1, x_2) \tag{7.17}$$

$$0 = g(x_1, x_2). \tag{7.18}$$

DAE systems arise in systems subject to constraints, in singular perturbations, and in other systems. If the Jacobian $\partial g/\partial x_2$ is nonsingular, the algebraic equation (7.18) can, in principle, be solved as

$$x_2 = h(x_1).$$

as a consequence of the *implicit function theorem*. In this case the system (7.17)–(7.18) is said to have *index* 1, and (7.17) reduces to

$$\dot{x}_1 = f(x_1, h(x_1)).$$

The Matlab ode solver `ode15s` can solve DAE systems with index 1, without the user having to specify the function $h(x_1)$. First, we write the system (7.17)–(7.18) as

$$\begin{bmatrix} 1 & 0 \\ 0 & 0 \end{bmatrix} \begin{bmatrix} \dot{x}_1 \\ \dot{x}_2 \end{bmatrix} = \begin{bmatrix} f(x_1, x_2) \\ g(x_1, x_2) \end{bmatrix}.$$

The matrix

$$M = \begin{bmatrix} 1 & 0 \\ 0 & 0 \end{bmatrix}$$

is called the *mass matrix*. The mass matrix is specified as an option in the `odeset` command, for example, as

Specifying the mass matrix *M*

```
options =  odeset('Mass',M)
[t,x] = ode15s(f,tspan,y0,options).
```

Example 7.12 Consider the system

$$\dot{x}_1 = x_2$$
$$0 = -x_1^3 - x_2.$$

The Matlab script below simulates this system using `ode15s`. Figure 7.12 shows the response of the system.

Differential-Algebraic System

```
f = @(t,x)[x(2);-x(1)^3-x(2)];  % DAE model
M = [1 0; 0 0];                 % Mass matrix
y0 = [-1;1];                    % Initial conditions
tspan = [0 5];                  % Simulation time
options = odeset('Mass',M);
[t,x] = ode15s(f,tspan,y0,options);
```

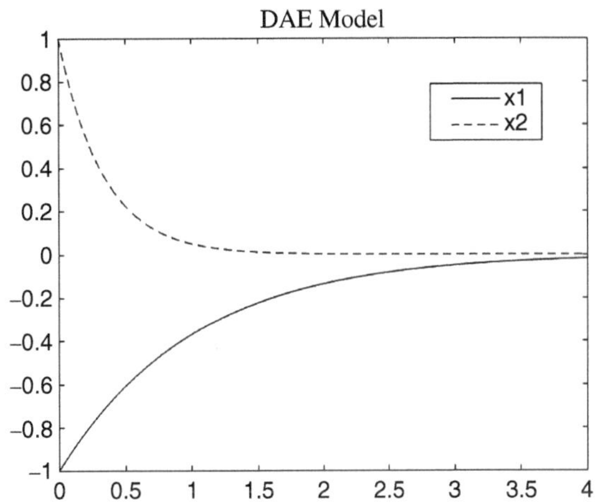

Figure 7.12 Solution of the DAE system using `ode15s`. `ode15s` should be used when the mass matrix is singular. For a nonsingular mass matrix, another ode solver, such as `ode23s`, may also be used.

Example 7.13 Consider next the SI epidemic model with a constant population size K:

$$\begin{aligned} \dot{S} &= -SI \\ \dot{I} &= SI \\ S + I &= K. \end{aligned}$$

Because of the constant population size, only one of the two differential equation is needed, and we can model this as the DAE system with index 1 given by

$$\begin{aligned} \dot{S} &= -SI \\ 0 &= S + I - K. \end{aligned}$$

The Matlab script below simulates this as a DAE system with the response shown in Figure 7.13.

SI Epidemic Model as a Differential-Algebraic System

```
K=100;                          % The population size
beta=0.01;                      % contact probability
f = @(t,x)[-beta*x(1)*x(2);x(1)+x(2)-K]; % SI model
M = [1 0;0 0];                 % The mass matrix
x0 = [99;1];                    % initial conditions
tspan=[0 12];                   % time span
options = odeset('Mass',M);
[t,x] = ode15s(f,tspan,x0,options);
```

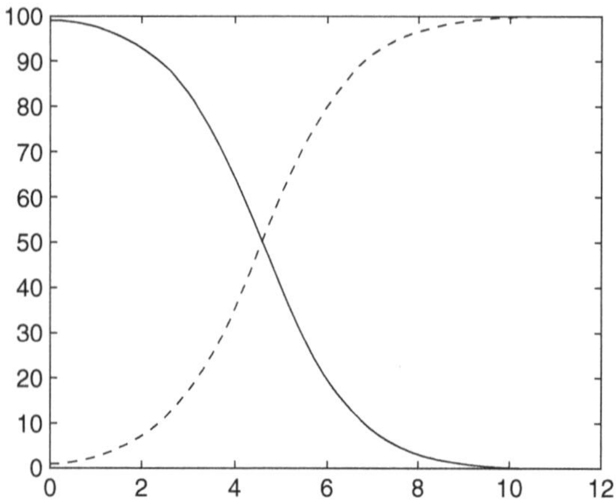

Figure 7.13 Response of the DAE model of the SI-epidemic system.

7.6 Event Detection

The ability to detect discrete events in simulation has several important applications. For example, to simulate a walking robot one needs to know when a foot comes into contact with the ground. In manipulation one needs to know when a robot hand touches an object, and so on. Event detection can also be used to determine the period of a periodic trajectory. The event detection feature in Matlab is controlled using the function `odeset`, as we show next.

Example 7.14 Consider an object at rest at height $x_0 > 0$ above the ground, starting with zero velocity and falling under the influence of gravity. If we include air resistance, then a simple model is given by

$$\ddot{x} = -g + \dot{x}^2, \quad x(0) = x_0, \quad \dot{x}(0) = 0.$$

Suppose we wish to know at what time t the object hits the ground. Thus, the event we are looking for is $x(t) = 0$ and we want Matlab to determine the time t at which the event happens. The following Matlab code solves this problem. First we define the differential equation to be integrated.

Falling Object with Air Resistance

```
function xdot = f(t,x)
xdot = [x(2); -9.8 + x(2)^2];
```

Now we need to define a function $g(x)$ for the event.

Falling Object with Air Resistance

```
function [xevent,isterminal,direction] = g(t,x)
xevent = x(1);
isterminal = 1;
direction = -1;
```

The first output, `xevent`, is the value we want to be zero, in this case $x(1) = x$. The second argument, `isterminal` should be +1 if we want Matlab to terminate the integration when the event is detected, and 0 otherwise. The third argument, `direction` = −1, tells Matlab to compute the zero value when $x(1)$ is decreasing.

With these two functions defined, we can now compute the solution and determine when the object strikes the ground.

Falling Object with Air Resistance

```
options = odeset('events', @g); % g defines the event
x0=[10;0];                      % the initial height
[t,y,tfinal] = ode45(@f, [0,inf], x0,options);
```

Note that we set the final integration time to `inf` since we do not know a priori the time at which the event will occur. Running this code gives the response shown in Figure 7.14. We note that the time at which the object strikes the ground is identified as `tfinal` = 3.4158 seconds.

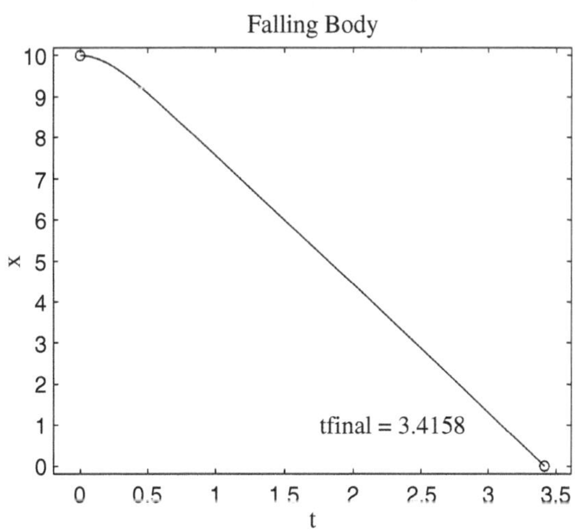

Figure 7.14 Response of the falling object. The simulation terminates when $x_1 = 0$ and returns the time of termination as 3.4158 seconds.

Example 7.15 ***(Bouncing Ball)***
As a more detailed example of event detection, let us consider a bouncing ball of mass $m = 1$. Ignoring the air resistance, the equation of motion is

$$\ddot{x} = -9.8. \tag{7.19}$$

If the coefficient of restitution is $\rho < 1$, then each time the ball strikes the ground there will be an instantaneous change in velocity according to

$$v^{+} = -\rho v^{-}, \tag{7.20}$$

where v^{-} is the velocity just prior to the impact with the ground and v^{+} is the velocity just after impact. This is a special case of a *hybrid* or *switched* system, in which Equation (7.19) is the continuous-time dynamics and Equation (7.20) is called a discrete *reset map*. We define the dynamics and the event as before with the following script.

Bouncing Ball

```
function xdot = f(t,x)
xdot = [x(2);-9.8];
end
function [xevent,isterminal,direction] = g(t,x)
xevent = x(1);
isterminal = 1;
direction = -1;
end
[t,x] = ode45(@f,[tstart,tfinal],x0,options);
```

Since the simulation will terminate when the ball strikes the ground, we restart the simulation using the new initial conditions given by the reset map in order to continue the simulation to the next event.

Bouncing Ball

```
tstart=0;
tfinal=20;
x0=[10;0]; % set the initial height above the ground
options = odeset('events',@g);
tout=tstart;
xout= x0.';
```

Figure 7.15 The bouncing ball starting at initial height $x_0 = 10$ with coefficient of restitution $\rho = 0.8$.

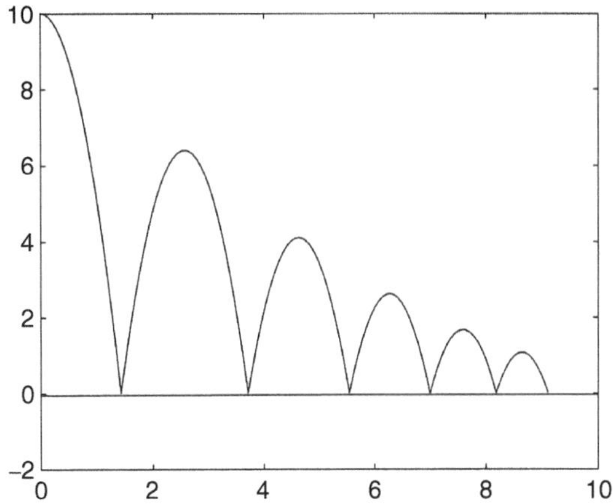

Then we call the `ode45` simulation in a `for` loop as follows to compute, in this case, six bounces with coefficient of restitution $\rho = 0.8$ as shown in Figure 7.15.

Bouncing Ball

```
for i=1:6
[t,x,te,ye,ie] = ode45(@f,[tstart,tfinal],x0,options);
nt = length(t);
tout = [tout; t(2:nt)];
xout = [xout; x(2:nt,:)];
%
x0=[0;-.8*x(nt,2)]; % new initial condition with rho = 0.8
tstart = t(nt);
end
x1=xout(:,1);
plot(tout,x1)
```

7.7 Simulink

Simulink is a block-diagram graphical environment for Matlab simulation. Simulink has an extensive library of blocks to create detailed simulation models. A few of the more commonly used blocks are shown in Figure 7.16.

Such blocks are interconnected to model differential and difference equations. As a first example, suppose that we wish to simulate the step response for the linear second-order equation

$$\ddot{y} + \dot{y} + 3y = 3u.$$

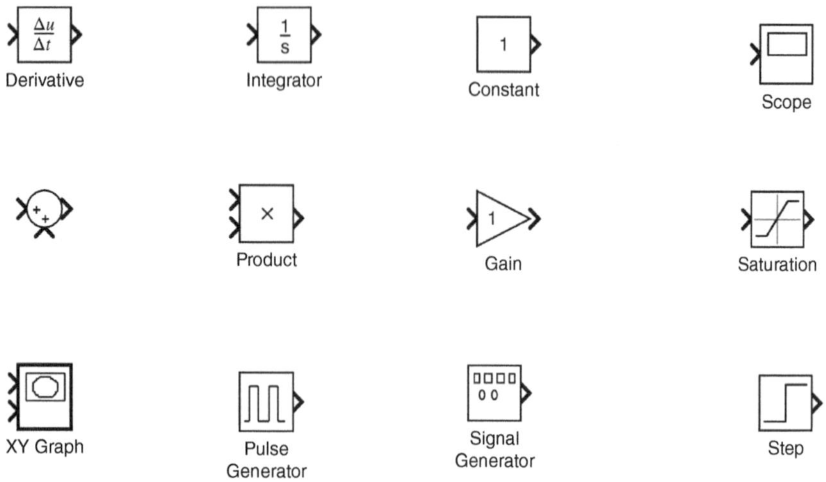

Figure 7.16 Simulink blocks from the Simulink library.

The key idea is to solve for the highest derivative, in this case $\ddot{y} = 3u - 3y - \dot{y}$, and successively integrate the derivatives until we reach the variable y.

Remark 7.3 The idea of successive integration of the highest derivative goes back to William Thomson (Lord Kelvin), who proposed the idea in his seminal work [67]. Using this method avoids having to compute numerical derivatives of the output $y(t)$. Numerical differentiation is inherently subject to noise and round-off errors, whereas integration is a smoothing operation and thus less sensitive to noise.

Figure 7.17 shows the resulting Simulink diagram using two integrators. For an explanation of the notation $\frac{1}{s}$ to denote an integrator, refer to Section 11.2.

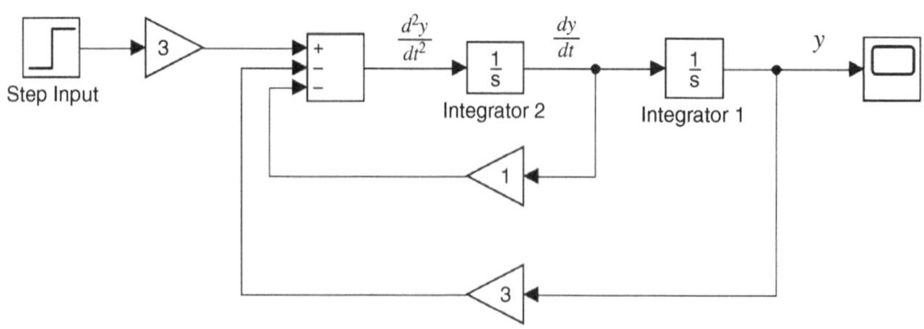

Figure 7.17 Realization of the second-order system $\ddot{y} + \dot{y} + 3y = 3u$ using two integrator blocks.

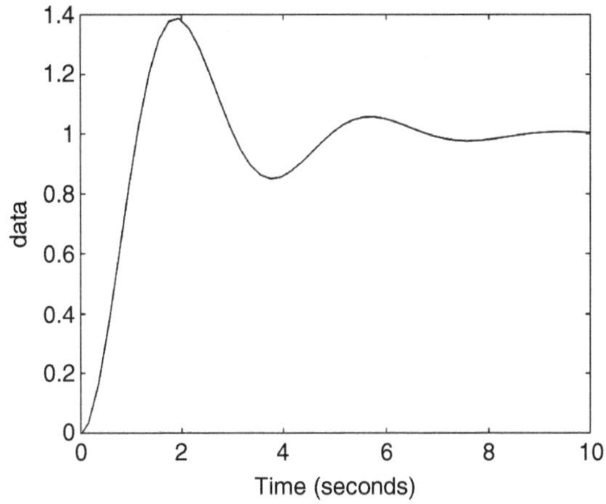

Figure 7.18 Step response of the system (7.21) starting with initial conditions $y(0) = 0$ and $\dot{y}(0) = 0$.

Figure 7.18 shows the step response of the system starting from zero initial conditions.

The Simulink Integrator

The Simulink integrator block, shown in Figure 7.19, treats the integrator as a dynamical system

$$\dot{x} = u(t), \quad x(0) = x_0$$
$$y = x(t),$$

where $u(t)$ is the block input and $y(t)$ is the block output. Thus, various methods for numerical integration of ODEs may be used to compute the integral of the input. These can be specified by the user depending on the particular application.

How to Make a Differentiator From an Integrator

Suppose that we want to estimate the derivative of a signal $y(t)$ in Simulink. Since numerical differentiation is subject to errors due to noise and roundoff, it is always desirable to use integrators instead of differentiators whenever possible. Consider the differential equation

$$\tau\dot{y} + y = \dot{u}. \tag{7.21}$$

Figure 7.19 Simulink integrator block.

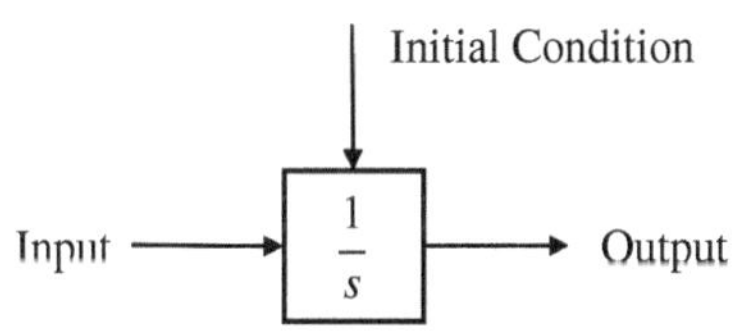

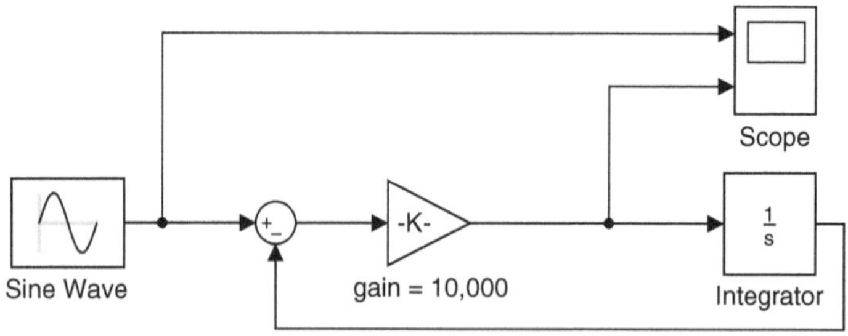

Figure 7.20 A practical differentiator built from an integrator. With input sin(t), the output approximates cos(t).

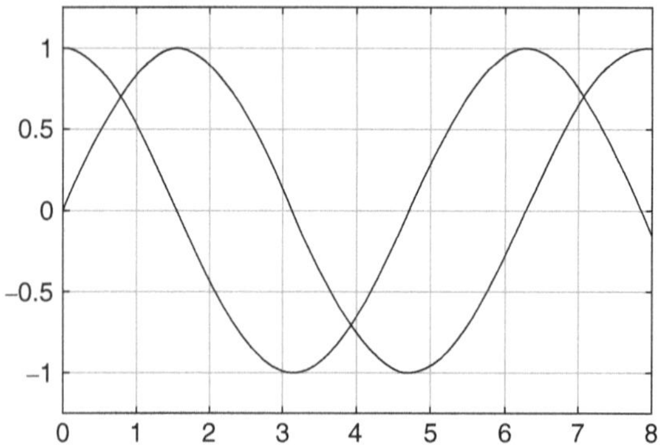

Figure 7.21 The practical differentiator with input sin(t) produces a signal that approximates cos(t). The accuracy of the approximation is determined by the time constant τ.

If the parameter τ in Equation (7.21), called the *time constant*, is small then y is approximately equal to $\dot{u}$, the time derivative of u. Introducing an intermediate variable z, we can write this equation as

$$\tau\dot{z} + z = u$$
$$y = \dot{z},$$

which can be realized as the Simulink model using only an integrator, as shown in Figure 7.20. The gain in the forward path is $1/\tau$. We call this a *practical differentiator* since it is constructed using only an integrator and the output approximates the derivative of the input.

As a test, take $u = \sin(t)$ and set $\tau = 10^{-4}$. The plot in Figure 7.21 shows the response of the practical differentiator to a sinusoidal input.

Nonlinear Systems in Simulink

Simulink also has blocks to include nonlinearities in the system equations. For example, the nonlinear system

$$\dot{x}_1 = x_2 \tag{7.22}$$

$$\dot{x}_2 = -2x_1 - 3x_2 - x_2^3 \tag{7.23}$$

can be simulated using a simple `product` block as shown in Figure 7.22.

However, for very complicated nonlinear systems, it is often easier to include the Matlab equations directly into Simulink using the Matlab function block. For example, the system in Figure 7.23 simulates the Van der Pol equation.

The Matlab function block contains Matlab code to define the equation to be simulated.

Matlab code used in the function block to define the Van der Pol equation.

```
function dxdt = vdpol(x)
mu=1.8;
dxdt=zeros(2,1);
dxdt = [x(2);-x(1)+mu*x(2)*(1-x(1)*x(1))];
end
```

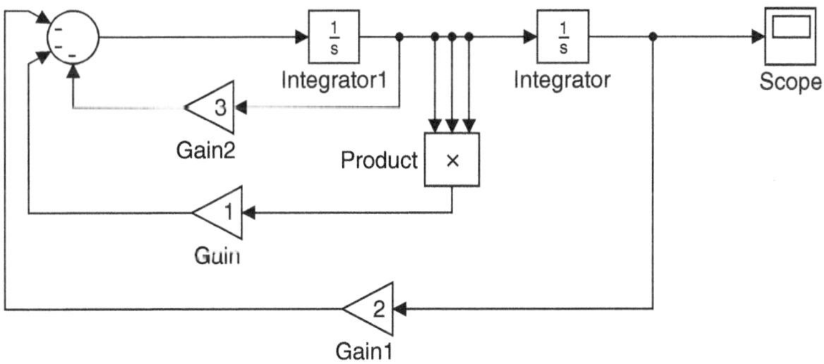

Figure 7.22 Simulink block diagram representation of the nonlinear system (7.22)–(7.23).

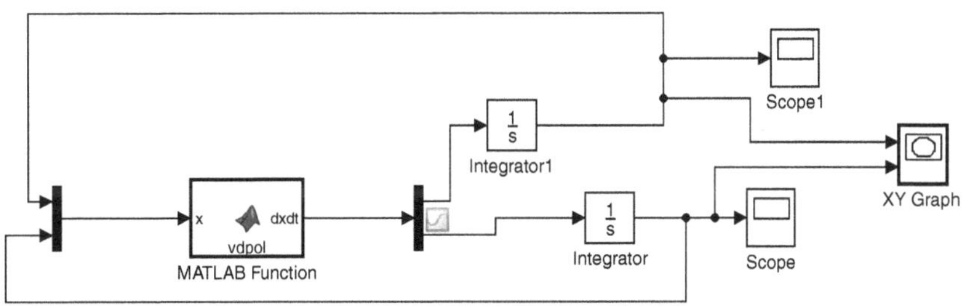

Figure 7.23 Simulink block diagram representation of the Van der Pol equation.

7.8 Summary

Numerical Differentiation Formulas

Forward Difference

$$f'(x) = \frac{f(x+h) - f(x)}{h}$$

Backward Difference

$$f'(x) = \frac{f(x) - f(x-h)}{h}$$

Midpoint Difference

$$f'(x) = \frac{f(x+h) - f(x-h)}{2h}$$

Second Derivative

$$f''(x) = \frac{f(x+h) - 2f(x) + f(x-h)}{h^2}$$

Numerical Integration Formulas

Left-Endpoint Method

$$\int_a^b f(x)dx \approx \sum_{i=0}^{n-1} f(x_i)\Delta x_i$$

Right-Endpoint Method

$$\int_a^b f(x)dx \approx \sum_{i=0}^{n-1} f(x_{i+1})\Delta x_i$$

Midpoint Method

$$\int_a^b f(x)dx \approx \sum_{i=0}^{n-1} f\left(\frac{1}{2}(x_i + x_{i+1})\right)\Delta x_i$$

Trapezoidal Method

$$\int_a^b f(x)dx \approx (f(x_1) + \cdots + f(x_{n-1}))h + \frac{1}{2}(f(x_0) + f(x_n))h$$

Simpson's Method

$$\int_a^b f(x)dx = \frac{h}{3}(f(x_0) + 4f(x_1) + 2f(x_2) + 4f(x_3) + 2f(x_4) + \cdots + 4f(x_{n-1}) + f(x_n))$$

Numerical Solutions of ODEs

Euler's Method

$$x_{k+1} = x_k + hf(x_k)$$

Predictor–Corrector (Heun's) Method

$$\begin{aligned}\tilde{x}_{k+1} &= x_k + hf(x_k)\\ x_{k+1} &= x_k + \frac{1}{2}h(f(x_k) + f(\tilde{x}_{k+1}))\end{aligned}$$

Second-Order Runge–Kutta Method

$$\begin{aligned}x_{k+1} &= x_k + \frac{1}{2}(k_1 + k_2)\\ k_1 &= hf(x_k)\\ k_2 &= hf(x_k + k_1)\end{aligned}$$

Fourth-Order Runge–Kutta Method

$$\begin{aligned}x_{k+1} &= x_k + \frac{1}{6}(k_1 + 2k_2 + 2k_3 + k_4)\\ k_1 &= hf(x_k)\\ k_2 &= hf(x_k + k_1/2)\\ k_3 &= hf(x_k + k_2/2)\\ k_4 &= hf(x_k + k_3).\end{aligned}$$

Problems

7.1 Experiment with the Matlab code in Example 7.1 using various values of N and initial guess x_0 to see how many iterations are needed to converge to the solution. Try to modify the code to compute the mth root of N, where m is a positive real number. Test your modified code with various values of m.

7.2 Write a Matlab script to estimate the derivative of the function $f(x) = x^3$ using the forward difference approximation.

7.3 Repeat the above problem using the backward difference approximation and the middle difference approximation.

7.4 The velocity of a rocket, measured at half-second intervals in meters per second, is

```
0 14.7 29.4 44.1 58.8 73.5 88.2 102.9 117.6 132.3 147.0
```

Use the middle difference method to estimate the acceleration of the rocket.

7.5 Write a Matlab function to input a function $f(t)$ and compute the first and second derivatives using `diff`.

7.6 Modify the Matlab code for left-endpoint integration to implement right-endpoint integration and midpoint integration.

7.7 Verify Equation (7.7) by direct calculation.

7.8 Write a Matlab script file to compute $\int_{x_0}^{x_f} f(x)dx$ using Simpson's rule. Test your file on the function

$$f(x) = x^3 + 2x \quad \text{with } x_0 = 0 \text{ and } x_f = 1.$$

7.9 Estimate the integral $\int_0^\infty e^{-x}dx$ using any of the numerical integration schemes discussed in this chapter.

7.10 Using the facts that $\frac{d}{dt}\sin(t) = \cos(t)$ and $\sin^2(t) + \cos^2(t) = 1$, it follows that the solution of the initial-value problem

$$\dot{y} = \sqrt{1 - y^2}, \quad y(0) = 0, \quad 0 \le t \le 1,$$

is $y(t) = \sin(t)$. Write a Matlab script using the Euler predictor–corrector method to solve this problem. Plot the value of $y(t)$ and compare it with $\sin(t)$.

7.11 Consider the initial-value problem

$$\dot{x} = \frac{1}{t^3} - \frac{x}{t}, \quad 2 \le t \le 3, \quad x(2) = -1.$$

Simulate this differential equation using:

- Euler's method with $h = 0.025$;
- Euler predictor–corrector method with $h = 0.05$;
- fourth-order Runge–Kutta method with $h = 0.1$.

Compare the output from the three methods.

7.12 Consider the nonautonomous initial-value problem

$$\dot{x} = -tx^2, \quad x(0) = 2.$$

Solve this problem on the interval $t = [0, 1]$ using Euler's method and the Euler predictor–corrector method and compare the results.

7.13 Consider the initial-value problem for the linear, nonautonomous system

$$\dot{x} = -2tx, \quad x(0) = 1.$$

Show that the solution of this problem is $x(t) = e^{-t^2}$. Solve the initial-value problem using various numerical integration methods and compare the results.

7.14 Consider the simple pendulum model $\ddot{y} + \sin(y) = 0$. Simulate this model for 20 seconds, starting at initial conditions $y(0) = 0.2, \dot{y}(0) = 0$, using Euler's method, the Euler predictor–corrector method, and a fourth-order Runge–Kutta method. Choose the step sizes carefully to obtain a physically meaningful result while keeping the number of steps as low as possible.

7.15 Consider a rope of length $L = 1$ with a weight attached to the end. If the end of the rope moves along a straight line, the path of the weight is governed by

$$\dot{x} = -\frac{x}{\sqrt{L^2 - x^2}}.$$

Compute and plot the path of the weight starting from $x(0) = L$ using the RK4 integration method.

7.16 We note that

$$\int_0^1 \frac{4}{1+t^2} dt = \pi.$$

Therefore, the initial-value problem

$$\dot{y} = \frac{4}{1+t^2}, \quad y(0) = 0, \quad 0 \leq t \leq 1.$$

can be used to compute the value of π. Compare the Matlab ODE solvers `ode23` and `ode45` for this problem. Which gives the better approximation of π? Experiment with the error tolerance using the Matlab `odeset` command.

7.17 Use the Matlab functions `int` and `integral` to compute the integral

$$\int_0^1 \frac{4}{1+t^2} dt$$

both symbolically and numerically.

7.18 Consider the initial-value problem

$$\ddot{y} - 4y = 0, \quad y(0) = 1, \quad \dot{y}(0) = 1.$$

Write the system in state space and compute the solution of the vector-valued equation for 2 seconds using the Euler predictor–corrector method with step size $h = 0.1$.

7.19 If $f(x, t)$ does not depend on x, then we have

$$\dot{x} = f(t),$$

so that $x(t) = \int_0^t f(u)du$. Show that, in this case, the RK4 algorithm reduces to Simpson's rule.

7.20 Using a Matlab function block, generate a Simulink model to simulate the SIR epidemic model.

7.21 Repeat the above problem for the predator–prey system. Use the Simulink XY-graph block to plot the solution in phase space.

7.22 Consider the Van der Pol equation

$$\begin{aligned}\dot{x}_1 &= x_2\\ \dot{x}_2 &= -x_1 + \mu x_2(1 - x_1^2),\end{aligned}$$

with $\mu = 2$. Use the Matlab event detection feature to estimate the period of the limit cycle.

7.23 Simulate the leaky bucket model from Chapter 2 using Matlab's event detection feature to detect when all the water has drained from the bucket.

8

Optimization

8.1 Introduction

In this chapter we present concepts and methods from optimization theory. The notation and derivations of the main concepts in this chapter are based largely on the books [16, 49], to which the reader can refer for additional details.

Optimization methods are used in many areas of science and engineering, including:

- operations research;
- supply-chain management;
- risk management;
- reliability;
- economics and game theory;
- financial engineering;
- optimal control;
- neural networks and machine learning.

In most problems in design, decision-making, or control we want to find the best solution. For example:

- in business, we may want to maximize profit, minimize loss, or minimize risk;
- in mechanical design, we may want to maximize strength, minimize weight, or minimize cost;
- in transportation, we may want to minimize travel time, or minimize fuel consumption.

Often a decision-making process involves constraints, such as:

- a speed should not exceed a value V;
- a temperature should be held at a constant value T;
- cost cannot exceed a fixed budget amount.

Introduction to Modeling and Simulation: A Systems Approach, First Edition. Mark W. Spong.

Companion Website: www.wiley.com/go/spong/engineeringmodeling

Example 8.1 Let us begin with a simple example. Suppose you own a large apartment complex with 500 units. You know that if you charge \$1000 per month rent you will have 100% occupancy. Historical data suggest that each \$100/month increase in rent will result in a 5% reduction in occupancy as people turn to less expensive options. How much rent should you charge to maximize your income?

For this problem, we can create a table showing the income generated by various values of the monthly rent in \$100 increments. We see that the total income increases with each \$100 increase rent up to \$1500 per month, after which the total income starts to decline. To prove analytically that \$1500 does indeed maximize your income, let x be the number of apartments rented and y be the rent per month. Then the monthly income is $f = xy$.

Now, it is not too difficult to see that x and y are related according to

$$\begin{aligned} x &= 500(100 - (y - 1000)0.05)/100 \\ &= 5(100 - (y - 1000)0.05) \end{aligned}$$

reflecting the fact that x decreases by 5% for each \$100 increase in y. Therefore, the monthly income f becomes

$$f = xy = 5y(100 - (y - 1000)0.05) = -0.25y^2 + 750y.$$

So the problem of maximizing the income can be stated as

$$\text{maximize} - 0.25y^2 + 750y.$$

We can solve this using elementary calculus. Computing the derivative of f gives $\frac{df}{dy} = -0.5y + 750 = 0$ and so $y = 1500$. Since $\frac{d^2f}{dy^2} = -0.5 < 0$ the value $y = 1500$ is the maximum. Thus, you should charge \$1500 per month, in which case your occupancy rate will be 75% and your income will be \$562,500, which agrees with Table 8.1.

The above exercise is an example of a nonlinear programming problem. A general nonlinear programming problem is of the form

$$\text{minimize} \quad f(x), \qquad x \in \mathbb{R}^n \tag{8.1}$$

$$\text{subject to} \quad h_i(x) = 0, \quad i = 1, \dots, k \tag{8.2}$$

$$g_j(x) \leq 0, \quad j = 1, \dots, p. \tag{8.3}$$

We begin with some definitions.

Definition 8.1

- The vector $x = (x_1, x_2, \dots, x_n) \in \mathbb{R}^n$ is an n-dimensional vector of independent variables, called *decision variables*, also called the *policy*.

Table 8.1 Income versus occupancy rate. The income increases to a maximum at a monthly rent of 1500 with 75% occupancy.

Rent	Occupancy rate	Number of units rented	Total income
\$1000	100	500	\$500,000
\$1100	95	475	\$522,500
\$1200	90	450	\$540,000
\$1300	85	425	\$552,500
\$1400	80	400	\$560,000
\$1500	75	375	\$562,500
\$1600	70	350	\$560,000
\$1700	65	325	\$552,500
\$1800	60	300	\$540,000
\$1900	55	275	\$522,500
\$2000	50	250	\$500,000

- The function $f : \mathbb{R}^n \to \mathbb{R}$ is called the *objective function* or *cost function.*
- The functions $h_i : \mathbb{R}^n \to \mathbb{R}$ and $g_j : \mathbb{R}^n \to \mathbb{R}$ are *equality constraints* and *inequality constraints*, respectively.
- For $x \in \mathbb{R}^n$, if the inequality constraint function $g_j(x)$ is equal to zero, the constraint is called an *active constraint*, while if $g_j(x)$ is strictly negative, the constraint is called *inactive*. The equality constraints $h_i(x) = 0$ are, therefore, always considered active. We will always assume that the functions defining the cost and constraint functions are differentiable as many times as needed.
- A vector $x \in \mathbb{R}^n$ is called a *feasible point* if it satisfies the constraints. The set of all feasible points, i.e., all points $x \in \mathbb{R}^n$ that satisfy the equality and inequality constraints, is called the *feasible set* Ω.
- A feasible point x^* that results in the smallest possible value of $f(x)$ for x in the feasible set Ω is called a *minimizer* of f, and we write $x^* = \text{argmin} f(x)$.

Remark 8.1 If the problem is to maximize an objective function $f(x)$ we simply state the problem as one of minimizing the function $-f(x)$. Thus our problem formulation (8.1)–(8.3) is without loss of generality. Moreover, Matlab's optimization functions assume that the problem is formulated as one of minimizing an objective function.

By exploiting special structure in the objective function and/or constraints, many techniques have been derived to achieve efficient solutions. By "efficient" we may mean fewer computations, less computer memory needed, faster convergence, or better scaling as the number of variables increases. We conclude this section by describing some special classes of the optimization problem in Equations (8.1)–(8.3).

Linear Programming

$$\begin{aligned} \text{minimize} \quad & c^T x \\ \text{subject to} \quad & Ax \leq b \\ & A_{\text{eq}} x = b_{\text{eq}} \\ & lb \leq x \leq ub, \end{aligned}$$

where both the objective function and constraints are linear in x. Here $c, b, b_{\text{eq}}, lb, ub$ are vectors and A, A_{eq} are matrices. The Matlab command

$$\mathtt{x = linprog(c, A, b, A_{eq}, b_{eq}, lb, ub, options)}$$

is designed to solve linear programming problems in this form. Linear programming is a vast subject in its own right. Several very efficient numerical algorithms are available to solve linear programming problems, the most famous one being the *simplex algorithm*, developed by George Dantzig in 1947. The simplex algorithm has been called one of ten most influential algorithms of the twentieth century for its impact on science and engineering. For information about linear programming, refer to [16].

Quadratic Programming

$$\begin{aligned} \text{minimize} \quad & \frac{1}{2} x^T Q x + c^T x \\ \text{subject to} \quad & Ax \leq b \\ & A_{\text{eq}} x = b_{\text{eq}} \\ & lb \leq x \leq ub, \end{aligned}$$

where Q is a symmetric, positive-definite matrix, $c, b, b_{\text{eq}}, lb, ub$ are vectors, and A, A_{eq} are matrices. The Matlab command

$$\mathtt{x = quadprog(c, A, b, A_{eq}, b_{eq}, lb, ub, x_0, options)}$$

is designed to solve quadratic programming problems in this form starting at the vector x_0. Quadratic programming problems are important for a couple of reasons. First, any suitably smooth nonlinear function can be approximated locally by a quadratic function, which therefore provides useful methods to search for minimizers for more general objective functions. Second, quadratic functions are so-called *convex functions*, which we will discuss later in this chapter. Convex optimization is an important subfield of optimization.

One-Dimensional Problems

$$\begin{aligned} &\text{minimize} \quad f(x),\ x \in \mathbb{R} \\ &\text{subject to} \quad a \leq x \leq b, \end{aligned}$$

where $f(x) : \mathbb{R} \to \mathbb{R}$ is a one-dimensional function and a, b are constants. The Matlab command

```
x = fminbnd(f,a,b,options)
```

can be used to solve one-dimensional problems in this form. If there are no constraints on the objective function $f(x)$, the Matlab command `fminunc()` may be used instead. One-dimensional problems often arise as part of multi-dimensional problems that use a so-called one-dimensional *line search* at each iteration.

8.2 Unconstrained Optimization

Optimization theory builds on the familiar notion from calculus that if the derivative of a function $f(x)$ of a single variable x is zero and the second derivative of $f(x)$ is positive (negative) at a point $x = x^*$, then x^* is a local minimum (maximum) of $f(x)$. In the multivariable case the roles of the first and second derivatives are played by the gradient and Hessian, respectively. The reader should review the definitions of the gradient and Hessian functions from Appendix B before continuing with the rest of this chapter.

We first discuss the unconstrained problem, which is to minimize an objective function $f(x)$ with no constraints, meaning that every point $x \in \mathbb{R}^n$ is a feasible point.

Definition 8.2 ***(Local and Global Minimizers)***

- Suppose $f : \mathbb{R}^n \to \mathbb{R}$ is defined on $\Omega \subset \mathbb{R}^n$. Then a point $x^* \in \Omega$ is a *local minimizer* if $f(x^*) \leq f(x)$ for $\|x - x^*\| < \epsilon$ for $x \in \Omega \backslash \{x^*\}$, for some $\epsilon > 0$.
- A point x^* is a *global minimizer* if $f(x^*) \leq f(x)$ for all $x \in \Omega \backslash \{x^*\}$.
- A minimizer is said to be *strict* if the inequalities in the above definitions are strict. This is illustrated in Figure 8.1.

If there are no constraints on the objective function $f(x)$ we may give simple conditions for local optimality in terms of the gradient and Hessian functions.

Definition 8.3 ***(Extreme Points)*** For the problem

$$\text{minimize } f(x),\ x \in \mathbb{R}^n,$$

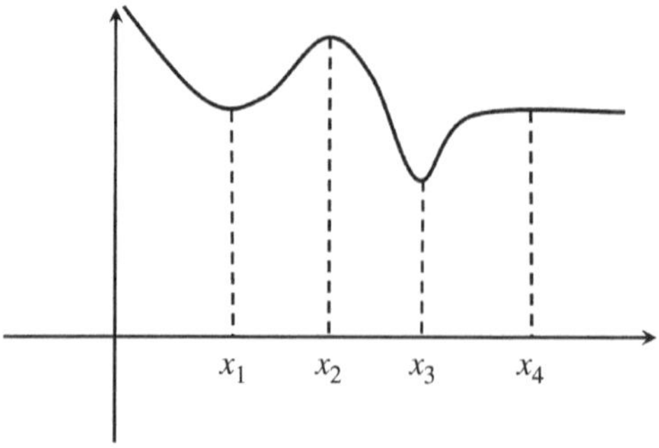

Figure 8.1 x_1 is a strict local minimizer; x_2 is a strict local maximizer; x_3 is a global minimizer; x_4 is a local maximizer (also a local minimizer) but not strict.

an *extreme point* or *extremum*, also called a *stationary point* of $f(x)$, is a vector x^* satisfying

$$\nabla f(x^*) = 0.$$

Proposition 8.1 A necessary condition for a vector $x^* \in \mathbb{R}^n$ to be a local minimizer of $f(x)$ is that x^* be a extremum of $f(x)$.

Proof: Let $\alpha \geq 0$ and let x be any vector in $\mathbb{R}^n$. Then by Taylor's theorem we can write

$$f(x^* + \alpha x) = f(x^*) + \alpha x^T \nabla f(x^*) + o(\alpha).$$

Thus, if x^* is a minimizer, we must have $\alpha x^T \nabla f(x^*) \geq 0$ for sufficiently small $\alpha > 0$. Since there are no constraints, the vector x is arbitrary, and therefore $\alpha(-x)^T \nabla f(x^*)$ must also be greater than or equal to zero. But the only way this can happen is if $\nabla f(x^*) = 0$.

The next proposition generalizes the condition that an extreme point $x^* \in \mathbb{R}$ in a one-dimensional problem is a minimum if the second derivative of $f(x)$ at x^* is positive.

Proposition 8.2 If x^* is an extremum of $f(x)$ then a sufficient condition for x^* to be a strict local minimizer of $f(x)$ is that the Hessian matrix of $f(x)$ is positive definite at $x = x^*$.

Proof: Suppose that

$$\nabla f(x^*) = 0 \quad \text{and} \quad H(x^*) > 0.$$

Since the Hessian $H(x^*)$ is positive definite at the extreme point x^*, we recall the identity from Equation (A.4) in Appendix A

$$\lambda_{\min}(H(x^*))\|x\|^2 \leq x^T H(x^*) x \leq \lambda_{\max}(H(x^*))\|x\|^2,$$

where $\lambda_{\min}(H(x^*))$ and $\lambda_{\max}(H(x^*))$ are the minimum and maximum eigenvalues, respectively, of $H(x^*)$. Then, for any vector $x \neq 0$ and $\|x\| = 1$, we have

$$x^T H(x^*) x \geq \lambda_{\min}(H(x^*)) > 0.$$

Using the second-order Taylor series expansion (since the gradient is zero at x^*)

$$\begin{aligned} f(x^* + \alpha x) &= f(x^*) + \frac{\alpha^2}{2} x^T H(x^*) x + o(\alpha^2) \\ &> f(x^*) + \frac{\alpha^2}{2} \lambda_{\min}(H(x^*)) + o(\alpha^2). \end{aligned}$$

Therefore, there exists $\overline{\alpha} > 0$, such that $f(x^* + \alpha x) > f(x^*)$, for any x, for $\alpha \in (0, \overline{\alpha})$. Therefore, x^* is a strict local minimizer of $f(x)$.

Example 8.2 Consider the problem

$$\text{minimize } f(x) = 2x_1^2 - 4x_1x_2 + 4x_2^2 - 4x_2.$$

Setting the gradient of $f(x)$ equal to zero gives

$$\begin{aligned} 4x_1 - 4x_2 &= 0 \\ -4x_1 + 8x_2 - 4 &= 0, \end{aligned}$$

which has the unique solution

$$x_1 = 1, \quad x_2 = 1.$$

The Hessian matrix is

$$H = \begin{bmatrix} 4 & -4 \\ -4 & 8 \end{bmatrix},$$

which is seen to be positive definite. Therefore, the point $x^* = (1, 1)^T$ is a strict local minimizer of $f(x)$. In this case, x^* is the global minimizer since it is the only extreme point of $f(x)$.

In the case where there are constraints the situation is, of course, more difficult. However, the gradient and Hessian still play an important role in solving constrained problems as we will see.

8.2.1 Iterative Search

In cases where it is difficult or impossible to solve an optimization problem analytically one must turn to numerical search or iterative methods. These typically take the form

$$x_{k+1} = x_k + \alpha_k d_k, \quad k = 0, 1, \ldots.$$

The vector d_k is called the *search direction* and the constant α_k is the *step size*. We start with an initial guess x_0 and compute a sequence of vectors, $x_1, x_2, \ldots$, in such a way that

$$\lim_{k\to\infty} x_k = x^*,$$

where x^* is a local minimizer of $f(x)$. Some key challenges of iterative search methods are as follows.

1) Choosing the search direction d_k and step size α_k so that $f(x_{k+1}) < f(x_k)$, i.e., so that the value of the objective function is decreased at each iteration. If the step size is too large, then the value of the objective function might not be decreased at each step. On the other hand, choosing a very small step size increases the number of iterations necessary to converge.
2) Choosing the initial guess x_0. Ideally, x_0 should be chosen close enough to the minimizer to guarantee convergence. Since, in general, there may be multiple local minima, the choice of x_0 determines which minimizer is found.
3) Deciding when to stop the iterations. Some possible stopping criteria are the following. With ϵ suitably small,
   a) $\|\nabla f(x_k)\| < \epsilon$: stop when the gradient is close to zero.
   b) $\|x_{k+1} - x_k\| < \epsilon$: stop when the iterates do not change very much.
   c) $\|f(x_{k+1}) - f(x_k)\| < \epsilon$: stop when the objective function does not decrease very much.

8.2.2 Gradient Descent

The first optimization methods that we discuss are so-called *gradient descent* methods. Since the gradient of a scalar function points in the direction of maximum increase of the function, a common approach to finding the minimum value of an objective function $f(x)$ is to choose the search direction d_k as the negative gradient of $f(x_k)$ at each iteration. In other words,

$$x_{k+1} = x_k - \alpha_k \nabla f(x_k), \quad k = 0, 1, \ldots .$$

Example 8.3 Let us return to the problem in Example 8.2, which is

$$\text{minimize } f(x) = 2x_1^2 - 4x_1x_2 + 4x_2^2 - 4x_2,$$

with the gradient

$$\nabla f(x_1, x_1) = [4x_1 - 4x_2, -4x_1 + 8x_2 - 4]^T.$$

The Matlab script below implements a gradient search for this problem using a fixed step size $\alpha = 0.1$ at each iteration.

Matlab script for Example 8.3 using gradient descent

```
f=@(x)2*x(1)^2-4*x(1)*x(2)+4*x(2)^2-4*x(2); % cost function
delf=@(x)[4*x(1)-4*x(2);-4*x(1)+8*x(2)-4];  % gradient
x=[0.5;0.5];                                 % initial guess
num_iterations=0;              % check the number of iterations
eps = 0.001;                   % used in the stopping criterion
alpha=0.1;                     % step size
while norm(delf(x)) > eps % the stopping criterion
     x=x-alpha*delf(x);
     num_iterations = num_iterations+1;
end
```

Gradient search algorithms, such as the above, are sensitive to the initial guess and the step size. Problem 8.1 investigates convergence to the optimum in the above problem for various values of initial guess and step size.

The Method of Steepest Descent

The method of *steepest descent* is a gradient descent method where the step size α_k at each iteration k is chosen according to

$$\alpha_k = \text{argmin}_\alpha\, f(x_k - \alpha\nabla f(x_k)), \quad \alpha \geq 0. \tag{8.4}$$

In this case, α_k is the largest step size for which $f(x_{k+1}) \leq f(x_k)$. Solving for α_k in Equation (8.4) is called a *line search* since α_k is one-dimensional. There are several one-dimensional iterative search methods that can be used to find α_k at each step, for example *Newton's method*, which is discussed below. Details of several other one-dimensional search methods can be found in [16].

Example 8.4 ***(Steepest Descent)***
Rather than detail specific line search methods, we will use the Matlab one-dimensional search function `fminsearch` in the previous example to find the optimal step size at each iteration.

Perform a Line Search to find Optimal Step Size

```
global x
x=[0.5,0.5];                       % initial value of x
alphak=@(a)f(x-a*delf(x));
a0=0.1;                            % initial step size guess
alfk = fminsearch(alphak,a0); % optimal step size
```

Running the above script at each iteration using the objective function and gradient from Example 8.3 finds the optimal step size at each step, which generally

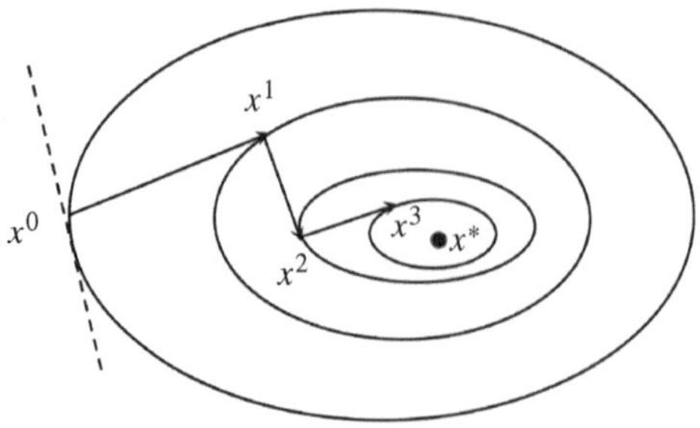

Figure 8.2 The search directions in gradient descent are orthogonal to level sets of the objective function. In the case of steepest descent, each new search direction is orthogonal to the previous search direction.

reduces the number of iterations needed to converge to a solution. The details are left to the reader in this example.

In the case of a quadratic cost function, the optimal step size α_k can be calculated in closed form so that no line search is needed. To show this, we need the following property of the steepest descent method (Figure 8.2).

Theorem 8.1 *If the sequence* $\{x_k\}_{k=0}^{\infty}$ *is computed from a steepest descent problem, then* $x_{k+1} - x_k$ *is orthogonal to the vector* $x_{k+2} - x_{k+1}$ *for each* k.

Proof: To prove this, set $g(x_k) = \nabla f(x_k)$ for notational simplicity. Then

$$x_{k+1} = x_k - \alpha^* g(x_k),$$

where α^* is determined by the minimizing $f(x_k - \alpha g(x_k))$ with respect to α. The value of α^* is found by setting the derivative of $f(x_k - \alpha g(x_k))$ with respect to α equal to zero,

$$\begin{aligned}\frac{df}{d\alpha}(x_k - \alpha g(x_k)) &= \nabla f(x_k - \alpha g(x_k))^T(-g(x_k)) \\ &= g(x_{k+1})^T g(x_k) = 0,\end{aligned}$$

which shows that the gradients at x_k and x_{k+1} are orthogonal.

Steepest Descent: The Quadratic Case

Quadratic programming problems are among the most useful and important problems to consider, primarily because near a local minimum we can approximate more general objective functions by quadratic functions using Taylor's theorem. Suppose that the cost function is quadratic of the form

$$f(x) = \frac{1}{2}x^T Qx + b^T x,$$

where Q is a symmetric, positive-definite $n \times n$ matrix and b is an n-dimensional vector. The gradient of $f(x)$ is

$$\nabla f(x) = Qx + b.$$

Note that the Hessian of the objective function is $Q > 0$ and therefore there is a global minimum at $x^* = -Q^{-1}b$. For large problems, an iterative search may be easier than inverting the matrix Q to find the exact solution. Therefore, considering a steepest descent search, we take as a search direction

$$d(x) = -Qx - b.$$

With the line search problem $\min_\alpha f(x_k - \alpha g(x_k))$ in the quadratic case, the orthogonality condition from Theorem 8.1,

$$\nabla f(x_k - \alpha^* g(x_k))^T(-g(x_k)) = 0,$$

becomes

$$(Q(x_k - \alpha^* g(x_k)) + b)^T g(x_k) = 0.$$

Multiplying this out gives

$$\begin{aligned}&(Qx_k + b)^T g(x_k) - \alpha^* g(x_k)^T Q g(x_k)\\ &\quad = g(x_k)^T g(x_k) - \alpha^* g(x_k)^T Q g(x_k) = 0.\end{aligned}$$

Solving this expression for α^*, we get

$$\alpha^* = \frac{g(x_k)^T g(x_k)}{g(x_k)^T Q g(x_k)}.$$

This gives us a closed-form expression for the optimal step size α^* without the need to solve a one-dimensional line search at each iteration. The method of steepest descent in the quadratic case is summarized below.

The Steepest Descent Algorithm in the Quadratic Case

$$\text{minimize } \frac{1}{2}x^T Q x + b^T x, \quad Q = Q^T > 0,\ b \in \mathbb{R}^n$$

The minimizer is found from the sequence

$$x_{k+1} = x_k - \alpha_k g(x_k)$$

where the gradient is $g(x_k) = Qx_k + b$ and the step size is

$$\alpha_k = \frac{g(x_k)^T g(x_k)}{g(x_k)^T Q g(x_k)}.$$

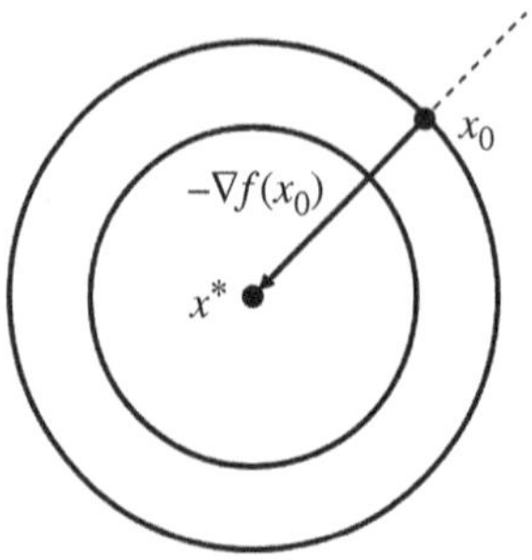

Figure 8.3 If the condition number of Q is 1, the level sets are concentric circles and the method of steepest descent will converge in one iteration.

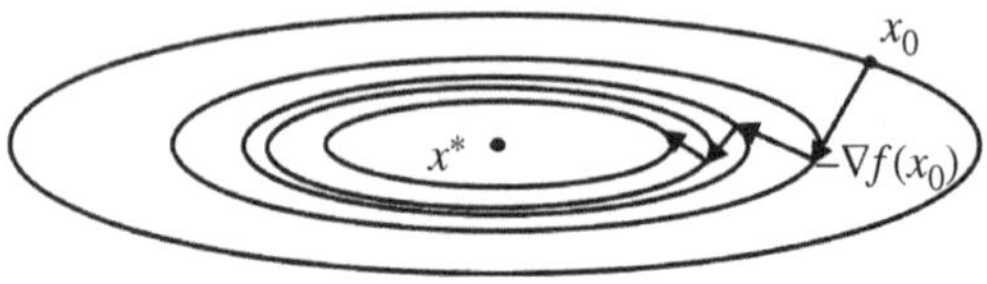

Figure 8.4 If the condition number of Q is large, the convergence of the method of steepest descent can be slow.

Condition Number of Q

Since the matrix Q is symmetric and positive definite, we again recall the inequality

$$\lambda_{\min}(Q)||x||^2 \leq x^T Qx \leq \lambda_{\max}(Q)||x||^2.$$

The ratio of the largest to smallest eigenvalue of Q,

$$r = \frac{\lambda_{\max}(Q)}{\lambda_{\min}(Q)},$$

is called the *condition number* of Q. If the condition number of Q is close to 1, convergence by the method of steepest descent will be fast. For example, if all the eigenvalues of Q are equal, so that $r = 1$, then the method of steepest descent will converge in one iteration (Figure 8.3). On the other hand, if the condition number is large, then convergence can be very slow (Figure 8.4).

8.2.3 Newton's Method

The gradient descent methods in the previous section are so-called first-order methods since they use only the first derivative information, i.e., the gradient. In this section we discuss Newton's method, which uses both first- and second-order information, i.e., both the gradient and the Hessian. We first consider minimization of a function of a single variable,

$$\text{minimize} \quad f(x), \quad f : \mathbb{R} \to \mathbb{R}.$$

The idea of Newton's method is to approximate the function $f(x)$ with a Taylor series expansion up to second order and then use the minimum of the quadratic approximation in an iterative search.

Given a function $f(x)$, a second-order Taylor series expansion of $f(x)$ at a point x_k gives the approximation

$$f(x) \approx q(x) = f(x_k) + f'(x_k)(x - x_k) + \frac{1}{2}f''(x_k)(x - x_k)^2.$$

We can minimize $q(x)$ by setting its derivative with respect to x equal to zero:

$$q'(x) = f'(x_k) + f''(x_k)(x - x_k) = 0.$$

Solving for x, the minimizer of the quadratic function $q(x)$, gives

$$x = x_k - \frac{f'(x_k)}{f''(x_k)}.$$

Thus, we can use this as an iterative method to minimize $f(x)$ by setting

$$x_{k+1} = x_k - \frac{f'(x_k)}{f''(x_k)}.$$

Example 8.5 Consider the problem

$$\text{minimize } x^4 - 14x^3 + 60x^2 - 70x.$$

The solution of this problem in the interval $[0, 2]$ is $x^* = 0.7809$. With first and second derivatives

$$f'(x) = 4x^3 - 42x^2 + 120x - 70, \quad f''(x) = 12x^2 - 84x + 120,$$

and starting with initial guess $x^0 = 0$, the Matlab script below implements Newton's method for this example.

Matlab script for Example 8.5 using Newton's Method

```
f=@(x) x.^4-14*x.^3+60*x^2-70*x;     % the objective function
delf=@(x)4*x.^3-42*x.^2+120*x-70;   % first derivative
Hf=@(x)12*x*2-84*x+120;              % second derivative
x=0.0;                              % the initial guess
num_iterations=0;                   % the number of iterations
eps = 0.001;                        %
while abs(delf(x))> eps             % the stopping criterion
    x=x-delf(x)/Hf(x);              % the Newton method iteration
    num_iterations = num_iterations+1;
end
```

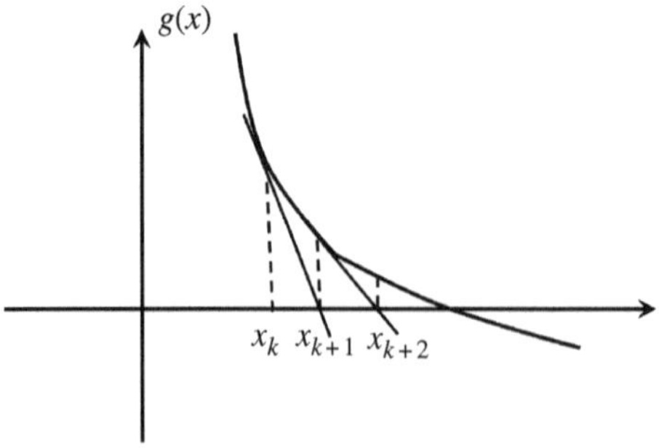

Figure 8.5 Newton's method for finding a root of $g(x)$.

Root Finding

Newton's method can also be used to find roots of a function $g(x)$ (Figure 8.5).

To find $g(x) = 0$, we fit a line at the current point and find where the line crosses zero:

$$\ell(x) = g(x_k) + g'(x_k)(x - x_k) = 0.$$

Solving for x leads to the iterative algorithm

$$x_{k+1} = x_k - \frac{g(x_k)}{g'(x_k)}.$$

Example 8.6 Consider the problem of finding a root of the polynomial $g(x) = 4x^3 - 42x^2 + 120x - 70$. Newton's method gives the following algorithm

$$x_{k+1} = x_k - \frac{4x_k^3 - 42x_k^2 + 120x_k - 70}{12x_k^2 - 84x_k + 120}.$$

Example of Newton's method for root finding

```
g=@(x)4*x^3-42*x^2+120*x-70; % objective function
dg=@(x)12*x*2-84*x+120;          % derivative
x=0.0;                       % the initial guess
num_iterations=0;            % the number of iterations
eps = 0.001;                 %
while abs(g(x)) > eps        % the stopping criterion
    x=x-g(x)/dg(x);          % Newton iteration
    num_iterations = num_iterations+1;
end
```

In the multi-dimensional case,

$$\text{minimize} \quad f(x), \quad x \in \mathbb{R}^n,$$

Newton's method takes the form

$$x_{k+1} = x_k - F(x_k)^{-1}\nabla f(x_k),$$

where $F(x_k)$ is the $n \times n$ Hessian matrix of $f(x)$ at x_k. The Hessian should be positive definite at each x_k for the iteration to reduce the value of the objective function. Since the inverse of the Hessian must be computed at each iteration, the method can be computationally expensive. Methods known as *quasi-Newton* methods have therefore been developed to avoid computation of the inverse Hessian. For example, one may use

$$x_{k+1} = x_k - \alpha F(x_0)^{-1}\nabla f(x_k),$$

thus keeping the Hessian constant at x_0 and introducing a step size α to avoid taking large steps. This can work if the Hessian does not change rapidly. Another approach is to use

$$x_{k+1} = x_k - \alpha(\epsilon I + F(x_k))^{-1}\nabla f(x_k),$$

where ϵ is chosen large enough to make $\epsilon I + F(x_k)$ positive definite. We note that, for large ϵ, this algorithm approximates a steepest descent algorithm. Several other quasi-Newton methods are discussed in [16].

8.3 Case Study: Numerical Inverse Kinematics

In this section, we present a case study to compare Newton's method and gradient descent for the problem of numerical inverse kinematics for a two-link robot. Figure 8.6 shows a so-called selective compliant articulated robot arm (SCARA), which is used in assembly operations.

Figure 8.7 shows a schematic of the first two links of this robot. The coordinates (x_1, x_2) of the end of the second link, called *end-effector* coordinates, are given in terms of the joint angles (θ_1, θ_2) by

$$x_1 = a_1 \cos\theta_1 + a_2 \cos(\theta_1 + \theta_2) \tag{8.5}$$

$$x_2 = a_1 \sin\theta_1 + a_2 \sin(\theta_1 + \theta_2), \tag{8.6}$$

where a_1 and a_2 are the lengths of the respective links. The function $f(\theta_1, \theta_2) : \mathbb{R}^2 \to \mathbb{R}^2$ defined by equations (8.5) and (8.6) is called the *forward kinematics*.

In order to plan the motion of a robot to complete a task, we need the *inverse kinematics*, which is to determine the joint variables θ_1, θ_2 in terms of given values of x_1 and x_2. In other words, given end-effector coordinates, x_1 and x_2, we want to solve equations (8.5) and (8.6) for θ_1 and θ_2. In this simple example, it is straightforward to solve equations (8.5) and (8.6) directly using elementary trigonometry [64].

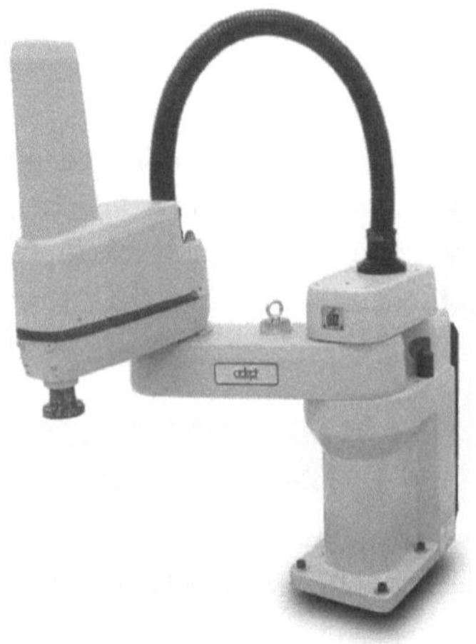

Figure 8.6 A selective compliant articulated robot arm. SCARA robots are useful for applications such as pick-and-place and table-top assembly.

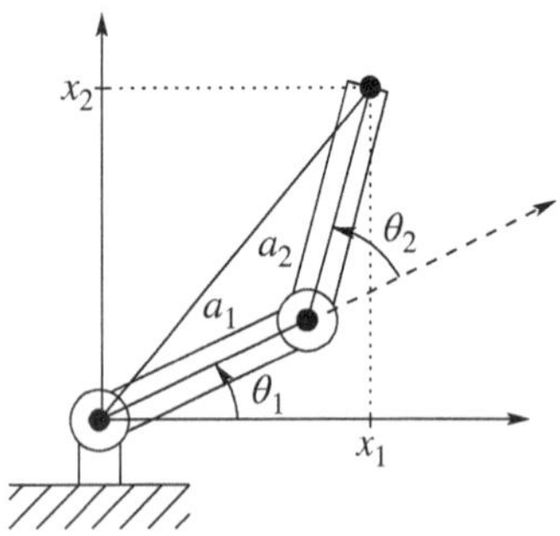

Figure 8.7 Schematic representation of a two-link robot. θ_1 and θ_2 are the joint variables. x_1 and x_2 are the end-effector coordinates and are related to the joint variables through the forward kinematic equations.

However, for a general robot with multiple links, the forward kinematic equations can be quite complicated and the inverse kinematics problem is difficult. In some cases, a closed-form solution to the inverse kinematics may not exist. For these reasons, numerical search techniques are frequently employed to find a solution.

Let $X^d = (x_1^d, x_2^d) \in \mathbb{R}^2$ be a vector specifying the desired end-effector coordinates and let $\Theta = (\theta_1, \theta_2)$ be the vector of joint coordinates. If we set

$$G(\Theta) = f(\Theta) - X^d, \tag{8.7}$$

where $f(\Theta)$ is the forward kinematics function, then a solution to the inverse kinematics problem is a root of the function $G(\Theta)$, in other words, a configuration Θ^d satisfying

$$G(\Theta^d) = f(\Theta^d) - X^d = 0.$$

Newton's Method

With X^d given as a desired robot configuration, we use a first-order Taylor series approximation

$$f(\Theta^d) \approx f(\Theta) + J(\Theta)(\Theta^d - \Theta), \tag{8.8}$$

where $J(\Theta)$ is the Jacobian of the forward kinematics map, $f(X)$, given by

$$J(\Theta) = J(\theta_1, \theta_2) = \begin{bmatrix} -a_1 \sin\theta_1 - a_2 \sin(\theta_1 + \theta_2) & -a_2 \sin(\theta_1 + \theta_2) \\ a_1 \cos\theta_1 + a_2 \cos(\theta_1 + \theta_2) & a_2 \cos(\theta_1 + \theta_2) \end{bmatrix}.$$

Assuming that the Jacobian matrix is invertible, we have

$$\Theta^d = \Theta + J^{-1}(\Theta)(X^d - f(\Theta)) \tag{8.9}$$

with $X^d = f(\Theta^d)$. To find a solution for Θ^d, we begin with an initial guess, Θ_0, and form a sequence of successive estimates, $\Theta_0, \Theta_1, \Theta_2, \ldots$, where

$$\Theta_{k+1} = \Theta_k + \alpha_k J^{-1}(\Theta_k)(X^d - f(\Theta_k)), \quad k = 0, 1, \ldots, \tag{8.10}$$

for a given step size $\alpha_k > 0$.

Remark 8.2 Equation (8.10) is based on Newton's method for root finding and is a first-order approximation of the inverse kinematics. In this case only local convergence can be expected. Also, since there are generally multiple solutions to the inverse kinematics, the particular configuration that results from running the algorithm is dependent on the initial guess.

Figure 8.8 shows the results obtained for the two-link SCARA robot. The algorithm converges to within 10^{-4} of the exact solution after 10 iterations with the given parameters.

Gradient Search Method

Next we use a gradient search method to compute the inverse kinematics for the above SCARA robot. To begin, we define an optimization problem

$$\min_{\Theta} F(\Theta) = \min_{\Theta} \frac{1}{2}(f(\Theta) - X^d)^T (f(\Theta) - X^d) \tag{8.11}$$

where, as above, X^d is the desired configuration and $f(\Theta)$ is the forward kinematics map. The gradient of the above cost function $F(\Theta)$ is given by

$$\nabla F(\Theta) = J^T(\Theta)(f(\Theta) - X^d). \tag{8.12}$$

A *gradient descent* algorithm to minimize $F(\Theta)$ is then

$$\Theta_{k+1} = \Theta_k - \alpha_k \nabla F(\Theta_k) = \Theta_k - \alpha_k J^T(\Theta_k)(f(\Theta_k) - X^d), \tag{8.13}$$

where, again, $\alpha_k > 0$ is the step size.

Remark 8.3 An advantage of the gradient method is that the Jacobian transpose is easier to compute than the Jacobian inverse and does not suffer from configuration singularities, which are configurations where the Jacobian matrix becomes singular. In general, however, the convergence, in terms of number of iterations, is slower with this method.

Figure 8.9 shows the response for the two-link SCARA robot for the same desired configuration and initial guess as in Figure 8.8. In this case, the algorithm converges after 30 iterations.

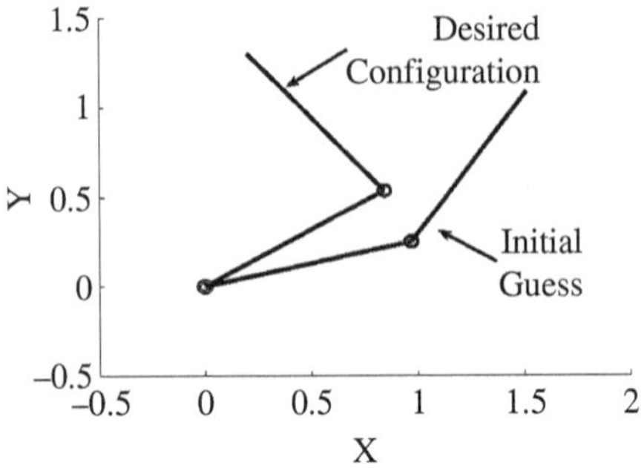

Iteration	θ_1	θ_2
1	−0.33284	2.6711
2	0.80552	2.1025
3	0.46906	1.9316
4	0.53554	1.7697
5	0.55729	1.7227
6	0.56308	1.7104
7	0.56455	1.7073
8	0.56492	1.7065
9	0.56501	1.7063
10	0.56503	1.7062

Figure 8.8 Inverse kinematics solution using Newton's method. Desired end-effector coordinates are $X^d = (0.2, 1.3)$. The joint variables corresponding to x^d are $\theta_1 = 0.5650$, $\theta_2 = 1.7062$. The initial guess is $\theta_1 = 0.25, \theta_2 = 0.75$. The step size α was chosen as 0.75.

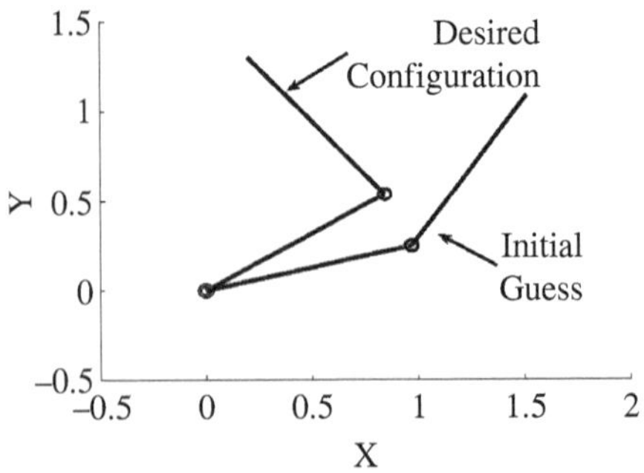

Iteration	θ_1	θ_2
1	1.8362	1.3412
2	0.4667	1.1025
3	1.1215	1.6233
4	0.45264	1.415
5	0.83519	1.7273
⋮	⋮	⋮
26	0.56522	1.7063
27	0.56492	1.7061
28	0.56514	1.7063
29	0.56498	1.7062
30	0.5650	1.7062

Figure 8.9 Inverse kinematics solution using gradient descent. Desired end-effector coordinates are $X^d = (0.2, 1.3)$. The joint variables corresponding to X^d are $\theta_1 = 0.5650$, $\theta_2 = 1.7062$. The initial guess is $\theta_1 = 0.25, \theta_2 = 0.75$. The step size α was chosen as 0.75.

8.4 Constrained Optimization

So far we have considered only unconstrained optimization problems. In this section we study problems with both equality and inequality constraints. For these problems, it is not necessary that the gradient of the objective function is equal zero at the optimum. In the case of equality constraints, we introduce the concept of *Lagrange multipliers*, which provides elegant necessary and sufficient conditions for optimality. For inequality constraints, the conditions for optimality are given by the *Karush–Kuhn–Tucker* (KKT) conditions.

8.4.1 Equality Constraints

In this subsection we consider an optimization problem with one or more equality constraints,

$$\begin{aligned} &\text{minimize} && f(x), \quad x \in \mathbb{R}^n \\ &\text{subject to} && h_i(x) = 0, \quad i = 1, \dots, m. \end{aligned}$$

The key concept for this problem is the notion of *Lagrange multipliers*. To introduce this concept, let us consider a two-dimensional problem with one constraint

$$\begin{aligned} &\text{minimize} && f(x_1, x_2) \\ &\text{subject to} && h(x_1, x_2) = c, \end{aligned}$$

and examine the graphs of the level sets of f in Figure 8.10.

The minimum occurs when the level curve $f(x_1, x_2) = d_1$ and the constraint curve $h(x_1, x_2) = c$ are tangent, because shrinking the level curve any further would violate the constraint. At this point, the gradient of f and the gradient of h

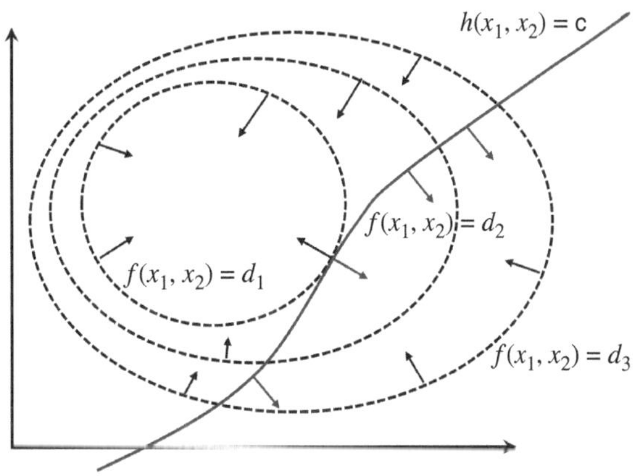

Figure 8.10 Level sets of the objective function $f(x)$ with $d_1 < d_2 < d_3$ and the graph of the constraint $h(x) = c$. The particular level curve that is tangent to the constraint curve determines the minimum value.

are proportional,

$$\nabla f(x_1, x_2) = -\lambda \nabla h(x_1, x_2),$$

which we can write as

$$\frac{\partial f}{\partial x_1} + \lambda \frac{\partial h}{\partial x_1} = 0$$
$$\frac{\partial f}{\partial x_2} + \lambda \frac{\partial h}{\partial x_2} = 0.$$

Definition 8.4 ***(Lagrangian)*** The *Lagrangian* or *Lagrangian function* is defined as

$$\mathcal{L}(x_1, x_2, \lambda) = f(x_1^*, x_2^*) + \lambda(h(x_1^*, x_2^*) - c), \tag{8.14}$$

where the scalar λ is called a *Lagrange multiplier*.

The gradient $\nabla\mathcal{L}$ of the Lagrangian $\mathcal{L}(x_1, x_2, \lambda)$ is given in terms of the partial derivatives

$$\frac{\partial \mathcal{L}}{\partial x_1} = \frac{\partial f}{\partial x_1} + \lambda \frac{\partial h}{\partial x_1}$$
$$\frac{\partial \mathcal{L}}{\partial x_2} = \frac{\partial f}{\partial x_2} + \lambda \frac{\partial h}{\partial x_2}$$
$$\frac{\partial \mathcal{L}}{\partial \lambda} = h(x_1, x_2) - c.$$

This leads to the following necessary condition for an extreme point.

Theorem 8.2 *A necessary condition for a minimizer (also for a maximizer)* (x_1^*, x_2^*) *is that there exists* λ^* *such that* $(x_1^*, x_2^*, \lambda^*)$ *is a stationary point of the Lagrangian, i.e.,*

$$\frac{\partial \mathcal{L}}{\partial x_1}(x_1^*, x_2^*, \lambda^*) = 0$$
$$\frac{\partial \mathcal{L}}{\partial x_2}(x_1^*, x_2^*, \lambda^*) = 0$$
$$\frac{\partial \mathcal{L}}{\partial \lambda}(x_1^*, x_2^*, \lambda^*) = 0.$$

Remark 8.4 The first two equations express the fact that the gradients of the objective function and constraint function are proportional, and the third equation just says that the point (x_1^*, x_2^*) satisfies the constraint. The Lagrange multiplier approach essentially turns the constrained problem into an unconstrained problem.

Example 8.7 Consider the problem

$$\begin{aligned} &\text{minimize } x - x^2 - 2y^2 \\ &\text{subject to } x^2 + y^2 = 1. \end{aligned}$$

The Lagrangian is therefore

$$\mathcal{L} = x - x^2 - 2y^2 + \lambda(x^2 + y^2 - 1).$$

The necessary conditions for an optimum are

$$\begin{aligned} \frac{\partial \mathcal{L}}{\partial x} &= 1 - 2x + 2\lambda x = 0 \\ \frac{\partial \mathcal{L}}{\partial y} &= -4y + 2\lambda y = 0 \\ \frac{\partial \mathcal{L}}{\partial \lambda} &= x^2 + y^2 - 1 = 0. \end{aligned}$$

There are multiple solutions of these equations:

$$\begin{aligned} &x = 1, \quad y = 0, \quad \lambda = 1/2 \\ &x = -1, \quad y = 0, \quad \lambda = 3/2 \\ &x = -1/2, \quad y = \sqrt{3}/2, \quad \lambda = 2 \\ &x = -1/2, \quad y = -\sqrt{3}/2, \quad \lambda = 2. \end{aligned}$$

We can evaluate the objective function at all of these points to find the minimizer. The choice $x^* = -1/2, y^* = \pm\sqrt{3}/2, \lambda^* = 2$ gives the minimum value of the objective function of -2.25.

Interpretation of the Lagrange multiplier

We note from Equation (8.14) that

$$\lambda = -\frac{\partial \mathcal{L}}{\partial c}.$$

Thus $-\lambda$ is the rate of change of the quantity being optimized as a function of the constraint parameter c. We say that λ^* is the *marginal cost* of the constraint. In economics this is referred to as the *shadow price*.

The shadow price may be interpreted as the change in the objective function resulting from a unit change in the constraint value c. For example, in Example 8.7, $\lambda^* = -2$ and so a unit increase in the constraint value as $x^2 + y^2 = 2$ results in the minimum of the objective function decreasing by 2 from -2.25 to -4.25 (Problem 8.8).

The shadow price λ is called a *dual* variable. Duality theory, which is covered in more advanced treatments of optimization theory, provides additional decision-making ability. For example, in a business application, one may have a constraint that workers should only work 40 hours per week. A shadow price of

\$20 means that one should be willing to pay at most \$20 for an additional hour of work.

Multiple Equality Constraints

Consider the more general case of multiple equality constraints

$$\begin{aligned} \text{minimize} \quad & f(x) \\ \text{subject to} \quad & h_1(x) = 0 \\ & h_2(x) = 0 \\ & \vdots \\ & h_m(x) = 0, \end{aligned}$$

where $f, h_1, \ldots, h_m : \mathbb{R}^n \to \mathbb{R}$ with $m < n$. The Lagrangian function in this case is

$$\begin{aligned} \mathcal{L} &= f(x) + \lambda_1 h_1(x) + \lambda_2 h_2(x) + \cdots + \lambda_m h_m(x) \\ &= f(x) + \lambda^T h(x), \end{aligned}$$

where $\lambda^T = (\lambda_1, \lambda_2, \ldots, \lambda_m)$ is a vector of Lagrange multipliers and $h = (h_1, h_2, \ldots, h_m)^T$ is the constraint vector.

Definition 8.5 ***(Regular Point)*** A vector x^* satisfying the constraints $h_i(x^*) = 0$ is called a *regular point* if the gradient vectors $\nabla h_i(x^*)$ are linearly independent at x^*.

Example 8.8 Suppose $h_1 = 3x_1 + x_2 - 1$, $h_2 = x_1 - 2x_2$. Then $\nabla h_1 = [3, 2]^T$ and $\nabla h_2 = [1, -2]^T$ are linearly dependent at each point $[x_1, x_2]^T$. Thus every point in $\mathbb{R}^2$ is a regular point of these constraints.

The next two theorems give necessary and sufficient conditions for a solution to the minimization problem. The proofs are omitted.

Theorem 8.3 *Consider the constrained optimization problem*

$$\begin{aligned} \text{minimize} \quad & f(x), \quad x \in \mathbb{R}^n \\ \text{subject to} \quad & h_i(x) = 0, \quad i = 1, \ldots, m, \end{aligned}$$

and suppose that x^* is a regular point of the constraints. If x^* is a minimizer of $f(x)$ subject to the constraints $h_i(x) = 0$, $i = 1, \ldots, m$, then there is a vector $\lambda^* = [\lambda_1^*, \ldots, \lambda_m^*]$ such that

$$\frac{\partial \mathcal{L}}{\partial x}(x^*, \lambda^*) = \nabla f(x^*) + \sum_{i=1}^{m} \lambda_i^* \nabla h_i(x^*) = 0 \tag{8.15}$$

$$\frac{\partial \mathcal{L}}{\partial \lambda}(x^*, \lambda^*) = h(x^*) = 0. \tag{8.16}$$

In order to state sufficient conditions for a minimizer, we examine second-order terms using the Hessian matrices. To this end, we define the matrix function

$$L(x, \lambda) = F(x) + \lambda_1 H_1(x) + \cdots + \lambda_m H_m(x), \tag{8.17}$$

where $F(x)$ and $H_i(x)$ are the Hessian matrices of $f(x)$ and $h_i(x)$, respectively. We also define the set M, called the *tangent space* of the constraints at a point x, as

$$M = \left\{ y \in \mathbb{R}^n \mid \nabla h_i(x)^T y = 0,\ i = 1, \dots, m \right\}.$$

The tangent space M is the space spanned by all vectors y that are orthogonal to the gradients of the equality constraints $h_i(x)$. Then we can state the following theorem.

Theorem 8.4 *If x^*, λ^* satisfy equations (8.15) and (8.16), a sufficient condition for x^* to be a strict local minimum is that the matrix $L(x^*, \lambda^*)$ in Equation (8.17) is positive definite on the tangent space M.*

Example 8.9 Consider the problem

$$\begin{aligned} &\text{minimize} \quad -x_1 x_2 \\ &\text{subject to} \quad x_1 + x_2 = 2. \end{aligned}$$

The Lagrangian for this problem is given by

$$\mathcal{L} = -x_1 x_2 + \lambda(x_1 + x_2 - 2).$$

It is straightforward to compute the partial derivatives and to show that the point $(x^*, y^*, \lambda^*) = (1, 1, 1)$ is the unique point that satisfies the first-order necessary conditions. Since the gradient of the constraint function, $h(x_1, x_2) = x_1 + x_2 - 2$, is $\nabla h = [1, 1]^T$, it follows that the tangent space M is

$$M = \left\{ y \in \mathbb{R}^2 \mid [1, 1] \begin{bmatrix} y_1 \\ y_2 \end{bmatrix} = 0 \right\} = \left\{ y \in \mathbb{R}^2 \mid y_2 = -y_1 \right\}.$$

The Hessians of the objective function, $f(x_1, x_2) = -x_1 x_2$, and constraint function, $h(x_1, x_2)$, are

$$F = \begin{bmatrix} 0 & -1 \\ -1 & 0 \end{bmatrix}, \quad H = \begin{bmatrix} 0 & 0 \\ 0 & 0 \end{bmatrix}.$$

and so $L = F + \lambda H = F$. It follows that, on the tangent space M,

$$y^T L y = [y_1, -y_1] \begin{bmatrix} 0 & -1 \\ -1 & 0 \end{bmatrix} \begin{bmatrix} y_1 \\ -y_1 \end{bmatrix} = 2y_1^2 > 0.$$

Therefore, $(x^*, y^*, \lambda^*) = (1, 1, 1)$ is a strict local minimum. Since there are no other candidate minimizers, (x^*, y^*, λ^*) is actually the global minimizer.

Remark 8.5 It is important to note that the matrix L above is not a positive-definite matrix on $\mathbb{R}^2$. It is enough that L is positive definite on the space M.

8.4.2 Inequality Constraints

Suppose now that we have both equality and inequality constraints

$$\begin{aligned} \text{minimize} \quad & f(x), \quad x \in \mathbb{R}^n \\ \text{subject to} \quad & h_1(x) = 0, \ g_1(x) \leq 0 \\ & h_2(x) = 0, \ g_2(x) \leq 0 \\ & \quad \vdots \qquad\qquad \vdots \\ & h_m(x) = 0, \ g_p(x) \leq 0, \end{aligned}$$

which we can express in vector notation as

$$\begin{aligned} \text{minimize} \quad & f(x), \quad x \in \mathbb{R}^n \\ \text{subject to} \quad & h(x) = 0, \ g(x) \leq 0, \quad h \in \mathbb{R}^m, \ g \in \mathbb{R}^p. \end{aligned}$$

Recall our definition of active and inactive constraints.

Definition 8.6 ***(Active Constraints)*** An inequality constraint $g_i(x) \leq 0$ is said to be *active* at a feasible point x if $g_i(x) = 0$, and *inactive* if $g_i(x) < 0$.

Definition 8.7 ***(Regular Point)*** Let x^* be a point satisfying the constraints

$$h(x^*) = 0, \ g(x^*) \leq 0$$

and let J be the set of indices j for which $g_j(x^*) = 0$. Then x^* is a *regular point* of the constraints if the gradients $\nabla h_i(x^*)$, $\nabla g_j(x^*)$ are linearly independent for $1 \leq i \leq k$, $j \in J$.

Note that only the active constraints are included in the definition of a regular point. The next theorem is known as the *Karush–Kuhn–Tucker theorem* and the conditions are known as the *KKT conditions.*

Theorem 8.5 ***(KKT Conditions)*** *Let x^* be a local minimum point for the problem*

$$\begin{aligned} \text{minimize} \quad & f(x), \quad x \in \mathbb{R}^n \\ \text{subject to} \quad & h(x) = 0, \ g(x) \leq 0, \ h \in \mathbb{R}^m, \ g \in \mathbb{R}^p \end{aligned}$$

and suppose that x^* is a regular point for the constraints. Then there exist a vector $\lambda^* \in \mathbb{R}^m$ and a vector $\mu^* \in \mathbb{R}^p$ with $\mu^* \geq 0$ such that

$$\begin{aligned} \nabla f(x^*) + \lambda^{*T} \nabla h(x^*) + \mu^{*T} \nabla g(x^*) &= 0 \\ \mu^{*T} g(x^*) &= 0. \end{aligned}$$

The condition $\mu^{*T}g(x^*) = 0$ with $\mu^* \geq 0$ is called a *complementary slackness* condition. The implication of this equality is that if $g_j(x^*) < 0$, then it must be true that $\mu_j^* = 0$. Similarly, if $\mu_j^* > 0$, then the corresponding constraint must be active, i.e., $g_j(x^*) = 0$.

Second-Order Necessary and Sufficient Conditions

In the case of inequality constraints, the *tangent space* of the constraints at x^* is defined as

$$M = \{y \in \mathbb{R}^n \mid \nabla h_i(x^*)^T y = 0, \text{ for all } i, \text{ and } \nabla g_j(x^*)^T y = 0, \text{ for all } j \in J\},$$

where $J = \{j \mid g_j(x^*) = 0, \mu_j \geq 0\}$. In other words, M is the space spanned by all vectors y that are orthogonal to the gradients of the equality constraints $h_i(x^*)$ and the gradients of the active inequality constraints $g_j(x^*) = 0$.

Second-Order Necessary Conditions

Proposition 8.3 Suppose that x^* is a regular point of the constraints. If x^* is a relative minimum, then there exist a λ^* and $\mu^* \geq 0$ such that the Hessian

$$L(x^*) = F(x^*) + \sum_{i=1}^{m} \lambda_i^* H_i(x^*) + \sum_{i=1}^{p} \mu_i^* G_i(x^*)$$

is positive definite on the tangent space M.

This result essentially follows from the equality constraint case because a local minimum over the constraints is a local minimum for the problem over the active constraints taken as equality constraints.

Second-Order Sufficient Conditions

Theorem 8.6 *A sufficient condition that a regular point x^* be a strict local minimum is that there exist λ^*, μ^* such that*

$$\mu^* \geq 0, \quad \mu^{*T} g(x^*) = 0$$
$$\nabla f(x^*) + \lambda^{*T} \nabla h(x^*) + \mu^{*T} \nabla g(x^*) = 0,$$

and that the Hessian

$$L(x^*) = F(x^*) + \sum_{i=1}^{m} \lambda_i^* H_i(x^*) + \sum_{i=i}^{p} \mu_i^* G_i(x^*)$$

is positive definite on the tangent space M.

Example 8.10 Consider the problem

$$\begin{aligned} \text{minimize} \quad & f(x) = (x_1 - 1)^2 + x_2 - 2 \\ \text{subject to} \quad & h(x) = x_2 - x_1 - 1 = 0 \\ & g(x) = x_1 + x_2 - 2 \leq 0. \end{aligned}$$

From the KKT conditions, we get

$$\begin{aligned} \nabla f(x) + \lambda \nabla h(x) + \mu \nabla g(x) = [2x_1 - 2 - \lambda + \mu, 1 + \lambda + \mu]^T &= 0 \\ \mu(x_1 + x_2 - 2) &= 0 \\ \mu &\geq 0 \\ x_2 - x_1 - 1 &= 0 \\ x_1 + x_2 - 2 &\leq 0. \end{aligned}$$

We first try $\mu > 0$, which implies that $x_1 + x_2 - 2 = 0$. In this case, the KKT conditions become

$$\begin{aligned} 2x_1 - 2 - \lambda + \mu &= 0 \\ 1 + \lambda + \mu &= 0 \\ x_2 - x_1 - 1 &= 0 \\ x_1 + x_2 - 2 &= 0, \end{aligned}$$

which has solution

$$(x_1, x_2, \lambda, \mu) = \left(\frac{1}{2}, \frac{3}{2}, -1, 0\right), \tag{8.18}$$

which contradicts the assumption that $\mu > 0$. We therefore take $\mu = 0$ in which case the KKT conditions are

$$\begin{aligned} 2x_1 - 2 - \lambda &= 0 \\ 1 + \lambda &= 0 \\ x_2 - x_1 - 1 &= 0, \end{aligned}$$

since the inequality constraint is not active. The solution of these equations is the same solution (8.18), which is now a feasible candidate solution.

Next, we check the second-order sufficient conditions on the candidate minimizer

$$(x_1^*, x_2^*, \lambda^*, \mu^*) = \left(\frac{1}{2}, \frac{3}{2}, -1, 0\right),$$

which gives

$$L(x^*, \lambda^*) = \begin{bmatrix} 2 & 0 \\ 0 & 0 \end{bmatrix} + (-1) \begin{bmatrix} 0 & 0 \\ 0 & 0 \end{bmatrix} = \begin{bmatrix} 2 & 0 \\ 0 & 0 \end{bmatrix}.$$

The tangent space to the active constraint $h = 0$ is

$$\left\{y \in \mathbb{R}^2 \mid [-1,1]y = 0\right\} = \left\{[a,a]^T \mid a \in \mathbb{R}\right\}.$$

Therefore, $y^T L y = 2a^2 > 0$ for $a \neq 0$ and so $(x_1^*, x_2^2) = \left(\frac{1}{2}, \frac{3}{2}\right)$ is a strict local minimizer.

Example 8.11 ***(Linear Programming)***
Consider the problem

$$\begin{aligned} \text{maximize} \quad & f(x) = 2x_1 + x_2 \\ \text{subject to} \quad & x_1 + x_2 \leq 8 \\ & 0 \leq x_1 \leq 6 \\ & 0 \leq x_2 \leq 4. \end{aligned}$$

This is an example of a *linear programming problem*, since the objective function and constraints are linear in the decision variables. The feasible region for this problem is shown in Figure 8.11.

We first note that the gradient of the objective function, $\nabla f = [2,1]^T$, is never equal to zero. This means that no point in the interior of the feasible region can be an extreme point for $f(x)$ and, therefore, the solution of this problem must lie on the boundary of the feasible set, where one or more constraints are active. It can be shown that the maximum occurs at one of the corner points of the feasible region. Therefore, we need only evaluate the objective function at the points $(0,0)$, $(0,4)$, $(4,4)$, $(6,2)$ and $(0,6)$, which gives the maximum value of 14 at $x_1 = 6$, $x_2 = 2$.

Two-dimensional linear programming problems may be solved graphically in this fashion. For higher-dimensional problems, the simplex method provides

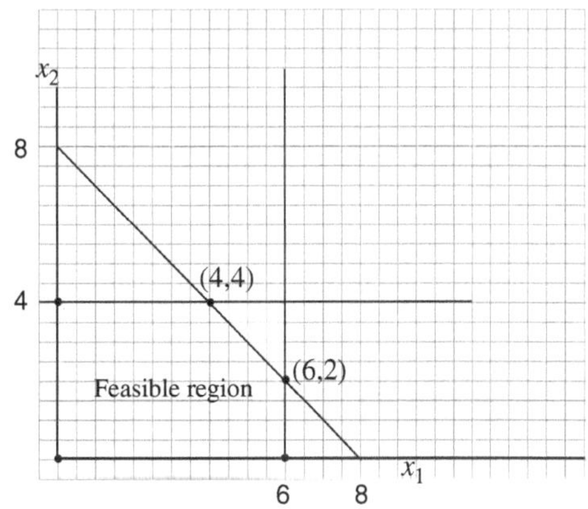

Figure 8.11 Feasible region for the linear programming problem.

a systematic way to check corner points of the feasible region to arrive at the optimal solution. See [16].

8.5 Convex Optimization

An important class of optimization problems are *convex problems*. A convex optimization problem is one where the objective function and constraints are convex functions (Figure 8.12).

Definition 8.8 ***(Convexity)*** A convex function is one that satisfies

$$f(\alpha x + \beta y) \leq \alpha f(x) + \beta f(y)$$

for all $x, y \in \mathbb{R}^n$ and all $\alpha,\ \beta \in \mathbb{R}$ with $\alpha + \beta = 1, \alpha \geq 0, \beta \geq 0$.

Note that linear functions are convex (Figure 8.13) and, therefore, convexity is a more general condition than linearity.

Definition 8.9 ***(Convex Set)*** A set $S \subset \mathbb{R}^n$ is convex if for all $u, v \in S$, the line segment between u and v is contained in S. In other words, if u and v are in S then $\alpha u + (1 - \alpha)v$ is in S for all $\alpha \in [0, 1]$.

It is easy to show that the intersection of any collection of convex sets is convex. As a consequence, if a constraint set Ω is defined by functional constraints,

$$\Omega = \{x \in \mathbb{R}^n : h_i(x) = 0,\ i = 1, \ldots, m,\ g_j(x) \leq 0,\ j = 1, \ldots, p\},$$

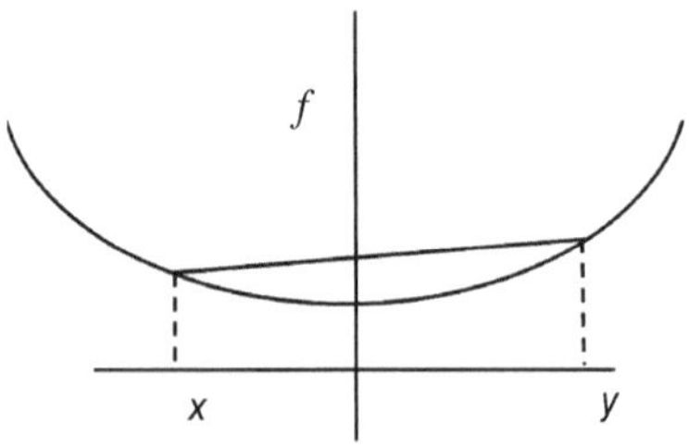

Figure 8.12 Illustrating the notion of a convex function. The convexity condition is that the graph of the function $f(x)$ between points x and y should be below the line segment connecting $f(x)$ and $f(y)$.

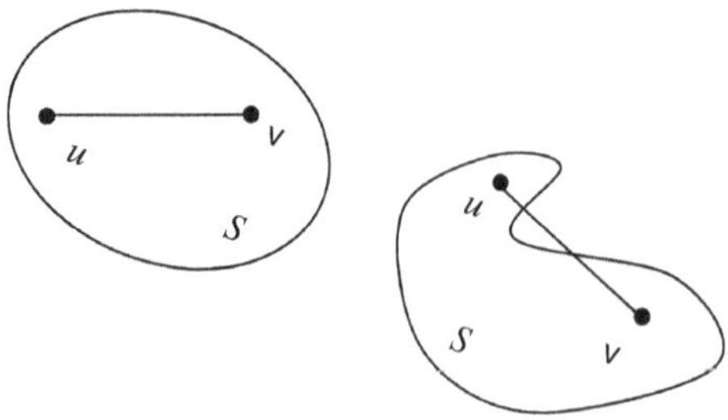

Figure 8.13 The set S on the left is convex. The set S on the right is nonconvex since the line segment connecting u and u is not wholly contained within S.

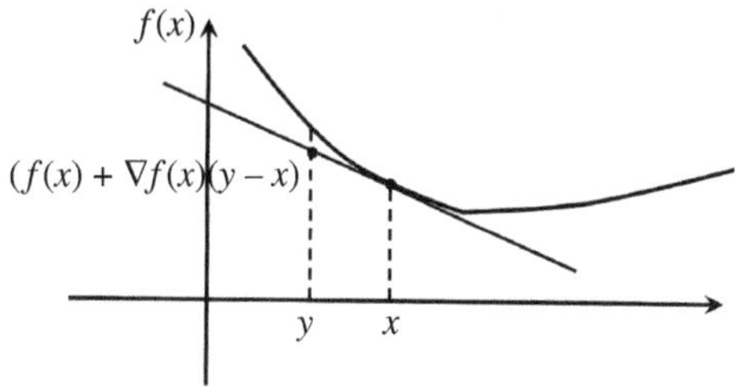

Figure 8.14 Illustrating Proposition 8.4. The condition (8.19) means that the tangent to the function at x is below the function.

then Ω is a convex set if the constraints are convex functions. Another characterization of a convex function is given in the following proposition (see also Figure 8.14).

Proposition 8.4 Let $f : \Omega \to \mathbb{R}$ be a continuously differentiable function defined on an open convex set $\Omega \subset \mathbb{R}^n$. Then f is convex on Ω if and only if, for all $x, y \in \Omega$,

$$f(y) \geq f(x) + \nabla f(x)^T (y - x). \tag{8.19}$$

And, finally, we have the following characterization of convexity.

Proposition 8.5 Let $f : \Omega \to \mathbb{R}$ be a twice continuously differentiable function defined on an open convex set $\Omega \subset \mathbb{R}^n$. Then f is convex on Ω if and only if, for each $x \in \Omega$, the Hessian matrix $F(x)$ is positive semidefinite, i.e.,

$$y^T F(x) y > 0, \quad \text{for all } y \in \mathbb{R}^n.$$

As a result of this property of convex functions, one can show that the previous necessary conditions for optimality are also sufficient in the convex case. An important consequence of convexity in optimization problems is the following result.

Theorem 8.7 *Let $f : \Omega \to \mathbb{R}$ be a convex function defined on a convex set $\Omega \subset \mathbb{R}^n$. Then a point x^* is a global minimizer of f on Ω if and only if x^* is a local minimizer of f on Ω.*

Proof: A global minimizer is obviously a local minimizer. To prove that a local minimizer is a global minimizer, let us suppose the x^* is not a global minimizer and show that it cannot be a local minimizer.

If x^* is not a global minimizer, then for some $y \in \Omega$ we have $f(y) < f(x^*)$. Since f is convex, then for all $\alpha \in (0, 1)$,

$$f(\alpha y + (1 - \alpha)x^*) \leq \alpha f(y) + (1 - \alpha) f(x^*).$$

Because $f(y) < f(x^*)$, we have

$$\alpha f(y) + (1-\alpha)f(x^*) = \alpha(f(y) - f(x^*)) + f(x^*) < f(x^*).$$

Therefore,

$$f(\alpha y + (1-\alpha)x^*) < f(x^*) \quad \text{for all } \alpha \in (0,1).$$

This means that there are points arbitrarily close to x^* that have lower objective

Next, consider the constrained optimization problem

$$\begin{array}{ll} \text{minimize} & f(x), \quad x \in \mathbb{R}^n \\ \text{subject to} & h_i(x) = 0, \; i = 1, \dots, m, \end{array}$$

and assume that the objective function and constraints are convex.

Theorem 8.8 *Let $f : \mathbb{R}^n \to \mathbb{R}$ be a continuously differentiable, convex function on the constraint set $\Omega = \{x \in \mathbb{R}^n : h_i(x) = 0, \; i = 1, \dots, m\}$. Suppose that there exists $x^* \in \Omega$ and $\lambda^* \in \mathbb{R}^m$ such that*

$$\nabla f(x^*) + \lambda_1^* \nabla h_1(x^*) + \cdots + \lambda_m^* \nabla h_m(x^*) = 0.$$

Then x^* is a global minimizer of f on Ω.

Finally, consider the general problem

$$\begin{array}{ll} \text{minimize} & f(x), \quad x \in \mathbb{R}^n \\ \text{subject to} & h_i(x) = 0, \; i = 1, \dots, m \\ & g_j(x) \le 0, j = 1, \dots, p. \end{array}$$

As before, assume that the objective function and constraints are convex. Suppose there exist $x^* \in \Omega$ and $\lambda^* \in \mathbb{R}^m$, $\mu^* \in \mathbb{R}^p$, such that

$$\mu^* \ge 0$$

$$\nabla f(x^*) + \sum_i \lambda_i^* \nabla h_i(x^*) + \sum_j \mu_j g_j(x^*) = 0$$

$$\mu^{*T} g(x^*) = 0.$$

Then x^* is a global minimizer of f over Ω. Thus, in all these cases, the necessary conditions for a local minimum become necessary and sufficient for a global minimum.

Examples

Here we give two additional examples of the use of Matlab optimization functions to solve constrained optimization problems.

Example 8.12 Let us consider the nonlinear programming problem on $\mathbb{R}^3$ given by

$$\begin{aligned} \text{minimize} \quad & 2x_1^2 - x_1x_2 + 4x_2^2 + x_3^2 \\ \text{subject to} \quad & x_1 - x_2 = 6 \\ & x_3 + x_1 - x_2^2 = 0. \end{aligned}$$

The Matlab script below uses `fmincon` to solve this problem. Running this script gives the optimal values `xopt` = (4.0339, −1.9661, −0.1684) and Lagrange multiplier vector $\lambda = (-18.4385, 0.3368)$. The optimal cost is `fval` = 55.9663.

fmincon example with nonlinear equality constraints

```
x0=[0.1; 0.1;0.1]; % initial guess
%
[xopt,fval,exit,output,lambda] =
fmincon(@objective,x0,[],[],[],[],[],[],@nonlcon);

function f= objective(x)     % the objective function
f = 2*x(1)^2-x(1)*x(2)+4*x(2)^2+x(3)^2;
end
%
function [c,ceq]=nonlcon(x) % the equality constraints
ceq= [x(1)-x(2)-6;x(3)+x(1)-x(2)^2];
c = [];
end
```

Example 8.13 Suppose, in the above example, that we have an inequality constraint

$$x_1 \leq 4$$

in addition to the equality constraints. In this case we simply include the inequality constraint in the function `nonlcon(x)` as follows.

```
function [c,ceq]=nonlcon(x) % the equality constraints
ceq= [x(1)-x(2)-6;x(3)+x(1)-x(2)^2];
c = x(1)-4;
end
```

Now, the minimum value returned is `xopt` = [4, −2, 0] with $\lambda = [-20, 0]$. and $\mu = 2$. The positive value of μ indicates that the inequality constraint $x_1 \leq 4$ is active at the optimum. The value of the object function increases to 56.

Example 8.14 Consider the quadratic optimization problem

$$\begin{aligned} \text{minimize} \quad & \frac{1}{2}x^T Q x + c^T x \\ \text{subject to} \quad & Ax \leq b \\ & A_{\text{eq}}x = b_{\text{eq}} \\ & LB \leq x \leq UB. \end{aligned}$$

This is a convex problem if $Q \geq 0$, i.e. positive semidefinite. Matlab has the function `quadprog()` for this problem. As a concrete example, consider the quadratic programming problem

$$\begin{aligned} \text{minimize} \quad & 2x_1^2 - 2x_1x_2 + x_2^2 + x_1 - x_2 \\ \text{subject to} \quad & x_1 + x_2 \leq 1. \end{aligned}$$

Then we define Q, c, A, b as below and call `quadprog`.

Quadratic program example

```
Q= [ 4 -2; -2 2];
c=[1;-1];
b= 1;
A=[1 1];
[x,fval,ex]=quadprog(Q,c,A,b);
```

This returns `x` $= (0, 1/2)$, `fval` $= -0.25$, `ex` $= 1$. The exit flag 1 means that the first-order optimality conditions are satisfied.

Problems

8.1 Copy the script in Example 8.3 into a Matlab live script and experiment with the initial guess and step size. Draw conclusions from your experiments.

8.2 Consider the following problem, illustrated in the figure below. Suppose that you want to reach an island offshore by running along the beach for a certain distance x and then swimming from that point to the island. Suppose that you can run at a speed of 6 miles per hour (mph) and swim at a speed of 3 mph. At which point x should you start swimming in order to minimize the total travel time to the island? How long will it take to reach the island?

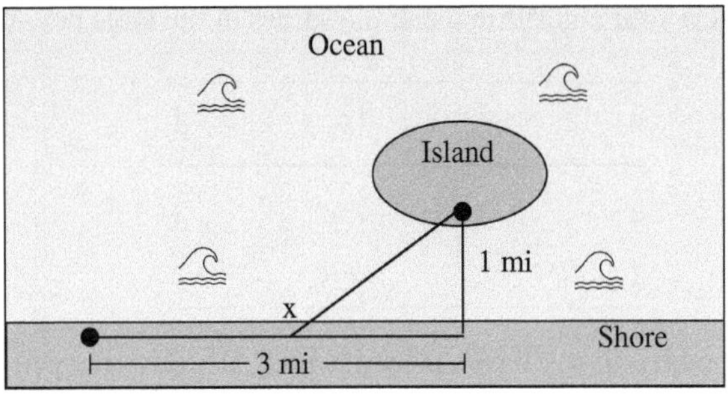

8.3 Consider the two-link planar robot shown in the figure below.

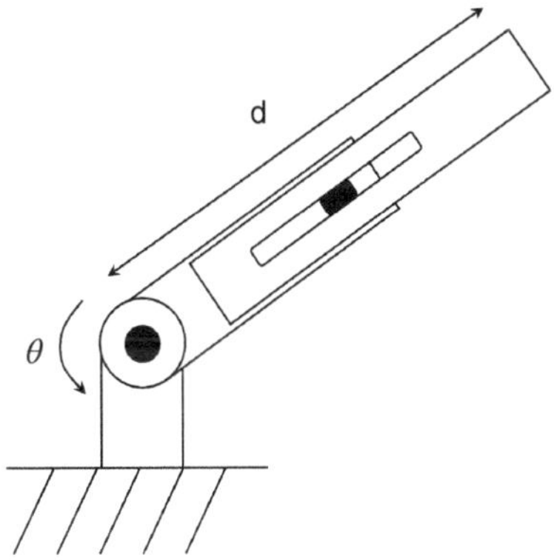

In contrast to the SCARA robot, this robot has the second joint *prismatic* or *telescoping*. The two joint variables are therefore θ and d, as shown in the figure. The forward kinematic equations for this robot are

$$x = d\cos(\theta)$$

$$y = d\sin(\theta).$$

1. Compute the Jacobian matrix $J(\theta, d)$ as a function of the variables θ and d.
2. Write a Matlab program to compute the inverse kinematics using Newton's method. Your program should input the desired endpoint coordinates (x, y) and an initial guess (θ_0, d_0) and output both the variables θ and d corresponding to the desired endpoint coordinates and the number of iterations needed to converge to the solution within the specified error tolerances.
3. Experiment with both the error tolerances and the step size until you get a satisfactory solution.

4. Test your algorithm using the values in the table below.

x	y	θ_0	d_0
2	3	0	1
1	5	$\pi/4$	2

8.4 Suppose that you have a piece of wood of length c and you want to construct a frame with sides of length x_1 and x_2 as shown below.

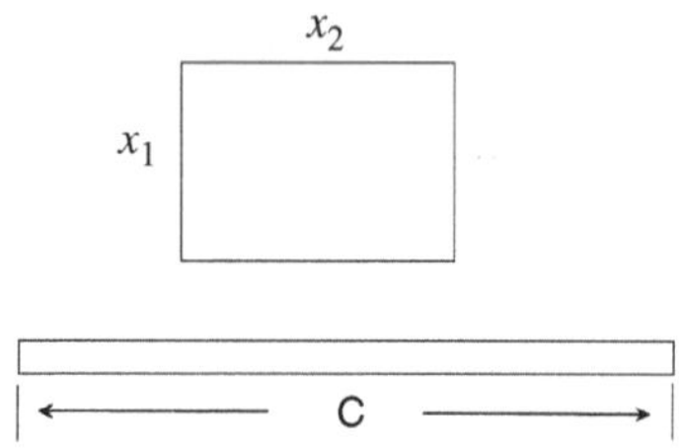

What should the dimensions x_1 and x_2 be to maximize the enclosed area inside the frame? Formulate and solve a constrained optimization problem using the method of Lagrange multipliers to answer this question.

8.5 The US Post Office requires that any box used for shipment must have the sum of its length and girth at most 108 inches. The girth is the maximum distance around the package perpendicular to the length, while, for a rectangular box, the length is the largest of the three dimensions. Find the dimensions of the largest acceptable box with square front and back.

8.6 Find the maximum and minimum values of the function $f(x, y) = 4x^2 + 10y^2$ on the circle $x^2 + y^2 = 4$.

8.7 Consider the problem

$$\begin{aligned} \text{minimize} \quad & x_1^2 + x_1x_2 + x_2^2 \\ \text{subject to} \quad & 3x_1 - x_2 = 2. \end{aligned}$$

Find a local minimizer for this problem. Is it a global minimizer? Justify your answer.

8.8 Verify the statement after Example 8.7 that increasing the constraint value as $x^2 + y^2 = 2$, results in the minimum of the objective function decreasing from -2.25 to -4.25.

8.9 Consider the quadratic function $f(x) = x^T Qx$. Show that $f(x)$ is convex on a set $\Omega \subset \mathbb{R}^n$ if and only if

$$(y - x)^T Q(y - x) \geq 0$$

for all $x, y \in \Omega$.

8.10 Solve the following problem using Lagrange multipliers:

$$\begin{aligned} \text{maximize} \quad & x_1x_2 + x_2x_3 + x_1x_3 \\ \text{subject to} \quad & x_1 + x_2 + x_3 = 3. \end{aligned}$$

8.11 Consider the problem

$$\begin{aligned} \text{minimize} \quad & x_1x_2 \\ \text{subject to} \quad & x_1 + x_2 \geq 2 \\ & x_1 \leq x_2 \end{aligned}$$

1. Write the KKT conditions for this problem.
2. Find all points and KKT multipliers satisfying the KKT conditions.

8.12 Compute the tangent space to the following constraint functions at $x = (1, 2)$:

$$\begin{aligned} h_1 &= x_1 + x_2 - 1 \\ h_2 &= x_1^2 - 2x_2 - 3. \end{aligned}$$

8.13 Write a Matlab script to solve the following problem by the method of steepest descent:

$$\text{minimize} \quad x_1^2 + 4x_1x_2 + 6x_2^2 - 2x_1 + 3x_2.$$

8.14 Use `fmincon` to solve the following problem:

$$\begin{aligned} \text{minimize} \quad & x_1^2 - x_1x_2/3 + x_2^2 \\ \text{subject to} \quad & x_1 + x_2 \leq 2 \\ & x_1 + x_2/4 \leq 1 \\ & -x_1 + x_2 \geq -1. \end{aligned}$$

8.15 Consider the following problem:

$$\begin{aligned} \text{minimize} \quad & 2x_1^2 + 2x_2^2 + 8x_1 + 6x_2 \\ \text{subject to} \quad & x_1 + x_2 \leq 2 \\ & x_1 + 3x_2 \leq 3. \end{aligned}$$

Solve this problem using the KKT necessary and sufficient conditions. Verify your solution using `quadprog`.

9

System Identification

9.1 Introduction

In this chapter we give an introduction to *system identification*. Roughly speaking, system identification involves the use of measured data to determine the best model for a given system. We will primarily focus on *regression* and *parameter estimation*. In this case we assume that we have a model of our system expressed in terms of certain parameters and the task is to estimate the values of the parameters from input/output measurements.

Some key questions to consider are:

- How to generate "sufficiently rich input data." In other words, how to choose inputs so that the system can be identified accurately.
- What are the assumptions on the model structure? It may be linear or nonlinear, parametric or nonparametric.
- How to validate the model? How good a fit is the model? Can it predict outputs for new inputs?

Before continuing to the next section, the reader should review the concepts of *range* and *null space* of a linear transformation in Appendix A.

9.2 Least Squares

The method of least squares has a rich and interesting history [53]. The best-known anecdote is the story of the mathematician Karl Friedrich Gauss's prediction of the location of the asteroid Ceres. On New Year's Day in 1801, the Italian astronomer Giuseppe Piazzi discovered Ceres and was able to track its path for 40 days until it passed behind the Sun. The problem then posed by astronomers was how to predict the location of Ceres when it emerged from behind the Sun. Gauss's prediction, using the method of least squares, enabled the

Introduction to Modeling and Simulation: A Systems Approach, First Edition. Mark W. Spong.

Companion Website: www.wiley.com/go/spong/engineeringmodeling

Hungarian astronomer Franz Xaver von Zach to relocate Ceres. The 24-year-old Gauss became an instant celebrity. Because of this, Gauss is often credited as the inventor of the method of least squares. However, the method was also discovered independently by the French mathematician, Legendre, and the Irish-American mathematician, Robert Adrian.

Consider a system of linear equations $Ax = b$ with

$$A \in \mathbb{R}^{n \times m}, \quad x \in \mathbb{R}^m, \quad b \in \mathbb{R}^n,$$

with $m < n$ so that A is a so-called *tall matrix*, meaning it has more rows (n) than columns (m). We assume that the matrix A has full rank m. Since A has rank m, the range space $\mathcal{R}(A)$ of A is an m-dimensional subspace of $\mathbb{R}^n$. If $b \notin \mathcal{R}(A)$ then there is no solution to the equation $Ax = b$. Such a system is called *overdetermined*.

Example 9.1 Suppose we are given a linear equation $Ax = b$ where

$$A = \begin{bmatrix} 1 & 0 \\ 0 & 1 \\ 1 & 1 \end{bmatrix}, \quad x = \begin{bmatrix} x_1 \\ x_2 \end{bmatrix}, \quad b = \begin{bmatrix} b_1 \\ b_2 \\ b_3 \end{bmatrix}.$$

Then $Ax = b$ corresponds to the equations

$$\begin{aligned} x_1 &= b_1 \\ x_2 &= b_2 \\ x_1 + x_2 &= b_3, \end{aligned}$$

so there can be no solution unless $b_3 = b_1 + b_2$.

If $b_3 \neq b_1 + b_2$ there is no exact solution but we can look for a vector $y \in \mathbb{R}^n$ that is closest to the vector b in the sense that $y = Ax$ minimizes the norm $f(x) = ||y - b||^2 = ||Ax - b||^2$. This is called the *least-squares* solution. We note that the least-squares problem to minimize $f(x)$ is an unconstrained quadratic optimization problem since we have

$$f(x) = ||Ax - b||^2 = (Ax - b)^T(Ax - b) = x^T A^T Ax - 2b^T Ax + b^T b.$$

Therefore, we can find the least-squares solution by computing the gradient of the function $f(x)$ and setting it equal to zero. This gives

$$\nabla f = 2A^T Ax - 2A^T b = 0 \Rightarrow x^* = (A^T A)^{-1} A^T b,$$

provided that $A^T A$ is nonsingular, which we show next.

Proposition 9.1 Let $A \in \mathbb{R}^{n \times m}$ with $m < n$. Then $\text{rank}(A) = m$ if and only if $\text{rank}(A^T A) = m$.

Proof: First we recall that the dimensions of the range space $\mathcal{R}(A)$ and null space $\mathcal{N}(A)$ of an $n \times m$ matrix A satisfy

$$\dim \mathcal{N}(A) + \dim \mathcal{R}(A) = m.$$

Thus, rank$(A) = m$ if and only if the null space $\mathcal{N}(A) = \{0\}$. Therefore, the result will follow if we show that

$$\mathcal{N}(A) = \mathcal{N}(A^T A).$$

To this end let $x \in \mathcal{N}(A)$. Then $Ax = 0$ and so $A^T Ax = 0$. Therefore, $\mathcal{N}(A) \subset \mathcal{N}(A^T A)$. Next, let $x \in \mathcal{N}(A^T A)$. Then $A^T Ax = 0$, which implies that $x^T A^T Ax = 0$. Thus, $\|Ax\| = 0$, which implies that $Ax = 0$. Hence, $\mathcal{N}(A^T A) \subset \mathcal{N}(A)$. Putting these together, we have shown that $\mathcal{N}(A^T A) = \mathcal{N}(A)$.

Example 9.2 In Example 9.1 suppose $b = [1,1,3]^T$. Then the least-squares solution of the equation $Ax = b$ is

$$x^* = \left(\begin{bmatrix} 1 & 0 & 1 \\ 0 & 1 & 1 \end{bmatrix} \begin{bmatrix} 1 & 0 \\ 0 & 1 \\ 1 & 1 \end{bmatrix} \right)^{-1} \begin{bmatrix} 1 & 0 & 1 \\ 0 & 1 & 1 \end{bmatrix} \begin{bmatrix} 1 \\ 1 \\ 3 \end{bmatrix} = \begin{bmatrix} 4/3 \\ 4/3 \end{bmatrix}$$

and $y^* = Ax^*$ is given by

$$y^* = \begin{bmatrix} 1 & 0 \\ 0 & 1 \\ 1 & 1 \end{bmatrix} \begin{bmatrix} 4/3 \\ 4/3 \end{bmatrix} = \begin{bmatrix} 4/3 \\ 4/3 \\ 8/3 \end{bmatrix}.$$

Interpretation of the least-squares solution as a projection

If x^* is the least-squares solution of $Ax = b$, then the vector $y^* = Ax^*$ is the perpendicular projection of the vector b onto the range space of A as illustrated in Figure 9.1.

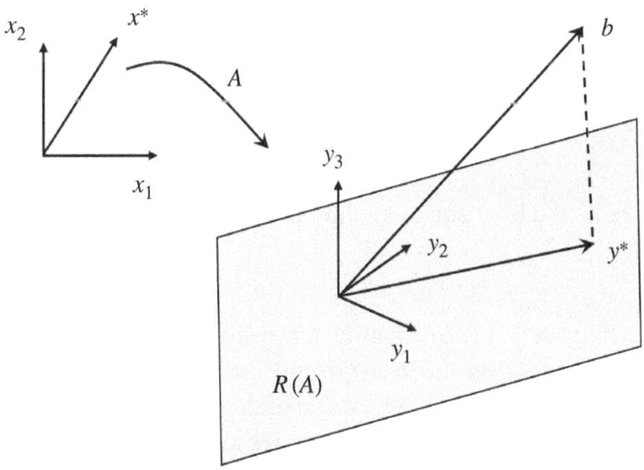

Figure 9.1 The range $R(A)$ of A is an n-dimensional subspace of $\mathbb{R}^n$. If b is not in the range of A, then the least-squares solution x^* of $Ax = b$ is such that Ax^* is the perpendicular projection of b onto $R(A)$.

9.3 Regression

We next consider the problem of regression and show how the method of least squares applies to this problem. The term *regression* was originally coined by Francis Galton [25] in reference to the biological phenomenon of *regression to the mean*. For example, descendants of taller than average individuals tend to regress toward the average height of the population. Regression now refers to the general statistical method of computing relationships among dependent and independent variables.

We first consider a linear input/output model of the form

$$y = a_1 u_1 + \cdots + a_n u_n = u^T a, \tag{9.1}$$

where $u = [u_1, \ldots, u_n]$ is a vector of inputs and $a = [a_1, \ldots, a_n]$ is a vector of constant but unknown parameters. The procedure to estimate these unknown parameters is to take a number $N \geq n$ of measurements, i.e., apply known inputs, and then measure the resulting outputs. Since we assume that the inputs and outputs are related according to Equation (9.1), we equate the measured input/output pairs as

$$\begin{aligned} y(1) &= a_1 u_1(1) + \cdots + a_n u_n(1) = u(1)^T a \\ &\vdots \\ y(N) &= a_1 u_1(N) + \cdots + a_n u_n(N) = u(N)^T a, \end{aligned}$$

which we can write as

$$Y = Ua, \quad \text{where } Y = \begin{bmatrix} y(1) \\ \vdots \\ y(N) \end{bmatrix}, \; U = \begin{bmatrix} u(1)^T \\ \vdots \\ u(N)^T \end{bmatrix}, \; a = \begin{bmatrix} a_1 \\ \vdots \\ a_n \end{bmatrix}.$$

Our best guess for the unknown parameters, in the least-squares sense, is to choose the parameter vector a as the solution of

$$\min_{a \in \mathbb{R}^n} \|Y - Ua\|^2.$$

If the matrix U is full rank, the least-squares solution is therefore

$$a^* = (U^T U)^{-1} U^T Y.$$

Example 9.3 Suppose that we want to identify the resistance R of a linear resistor by applying a voltage across the resistor and measuring the current through the resistor. By Ohm's law, $V = RI$. We take N measurement pairs $(I(1), V(1)), \ldots, (I(N), V(N))$ and compute the least-squares solution for R as

$$R = \text{argmin}_R \sum_{i=1}^{N} |RI(i) - V(i)|^2.$$

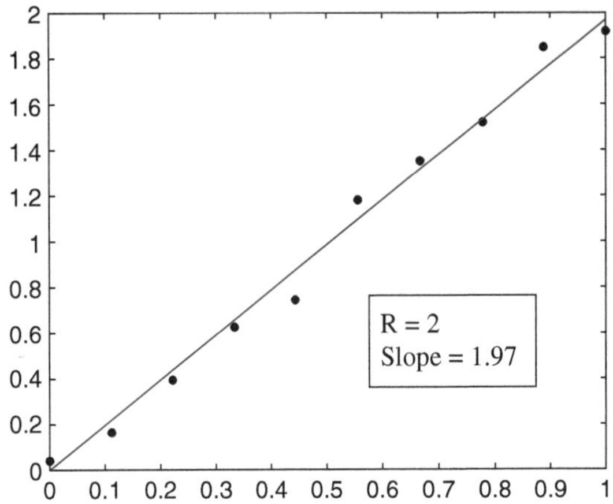

Figure 9.2 Estimate of the resistance R from 10 measurements. The horizontal axis is the current I and the vertical axis is the voltage V.

We define

$$A = \begin{bmatrix} I(1) \\ \vdots \\ I(N) \end{bmatrix}, \quad b = \begin{bmatrix} V(1) \\ \vdots \\ V(N) \end{bmatrix}.$$

Then the least-squares solution for R is given by

$$R = (A^T A)^{-1} A^T b = \frac{\sum I(i)V(i)}{\sum I(i)^2}.$$

Figure 9.2 shows the results of an experiment taking $N = 10$ measurements. The true value of the resistance is $R = 2$ and the value computed by the least-squares solution from the given data set is $R = 1.97$.

The previous example of least-squares optimization is an example of *linear regression*. The more general linear regression problem is to find both the slope m and y-intercept c to determine a straight line $y = mx + c$ that best fits a set of measured data $(x_1, y_1), \ldots, (x_n, y_n)$. With each data pair ideally related as $y_i = mx_i + c$, we seek the values of m and c to minimize the sum of squares

$$\sum_{i=1}^{n} (mx_i + c - y_i)^2. \tag{9.2}$$

Note that we can write expression (9.2) as

$$\sum_{i=1}^{n} (mx_i + c - y_i)^2 = ||Au - y||^2,$$

where

$$A = \begin{bmatrix} x_1 & 1 \\ x_2 & 1 \\ \vdots & \vdots \\ x_n & 1 \end{bmatrix}, \quad u = \begin{bmatrix} m \\ c \end{bmatrix}, \quad y = \begin{bmatrix} y_1 \\ y_2 \\ \vdots \\ y_n \end{bmatrix}.$$

Example 9.4 Suppose we have measured data as shown in the table below.

Observation	x	y
1	2	3
2	3	4
3	4	15

Then the Matlab script below generates the least-squares solution plotted in Figure 9.3.

Linear Regression Example

```
xdata=[2;3;4];
ydata=[3;4;15];
plot(xdata,ydata,'ko','MarkerFaceColor','k','MarkerSize',4)
hold on
A=[2 1;3 1;4 1];
u=(A'*A)\A'*ydata;
m=u(1);c=u(2);
x=linspace(0,5);
y=m*x +c;
plot(x,y)
```

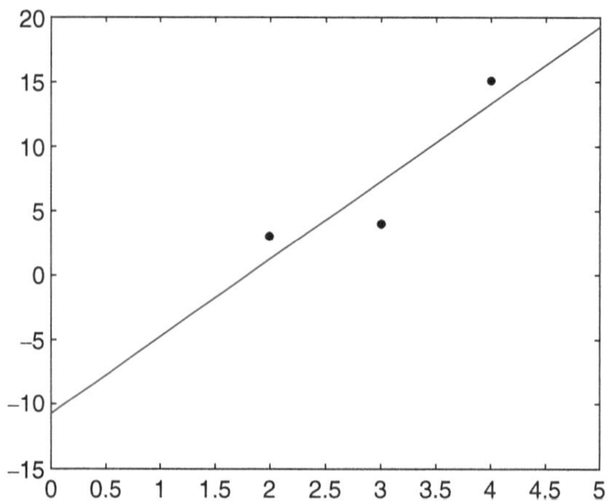

Figure 9.3 Linear regression example. The straight line is estimated from three data pairs.

Remark 9.1 When solving systems of equations $Ax = b$, it is recommended to use Matlab's `mldivide` operator `A\b` rather than the matrix inverse operator `inv(A)*b`, since the `mldivide` operator is faster and more accurate. In fact, if the matrix A is not square, $A\backslash b$ computes the least-squares solution and so is equivalent to $(A' * A)\backslash A' * b$. Matlab also provides the function `lsqminnorm` for solving systems of linear equations via the least-squares method.

Higher-Order Polynomial Fitting

We may also consider fitting higher-order polynomials, exponentials, or other functions to a given data set. As is well known, one can exactly fit a line to two points, a quadratic polynomial to three points, and so on. In general, a polynomial of degree $n - 1$ can exactly fit n data points. Since the measurements are always corrupted by noise, it is not desirable to try to fit the n data points exactly with an $(n - 1)$-degree polynomial. The choice of what function, polynomial or other, to chose as a parameterized model is an important consideration.

Therefore, given N data points, take $m < N$ and define a polynomial model of degree m as

$$y(k) = a_0 + a_1 u(k) + a_2 u(k)^2 + \cdots + a_m u(k)^m = \phi^T(k)a,$$

where

$$\phi^T(k) = \left[1, u(k), u(k)^2, \cdots, u(k)^m\right], \quad a = \left[a_0, a_1, \cdots, a_m\right]^T.$$

Therefore, we have

$$\begin{bmatrix} y(1) \\ y(2) \\ \vdots \\ y(N) \end{bmatrix} = \begin{bmatrix} 1 & u(1) & u(1)^2 & & u(1)^m \\ 1 & u(1) & u(1)^2 & \cdots & u(1)^m \\ \vdots & \vdots & \vdots & \ddots & \vdots \\ 1 & u(1) & u(1)^2 & \cdots & u(1)^m \end{bmatrix} \begin{bmatrix} a_0 \\ a_1 \\ \vdots \\ a_m \end{bmatrix},$$

which we write in vector notation as

$$Y = \Phi a.$$

We seek to minimize over N observations the objective

$$\min_{a \in \mathbb{R}^n} \frac{1}{2} \|Y - \Phi \cdot a\|^2,$$

which, if $\Phi^T\Phi$ is invertible, is given by the least-squares solution

$$a^* = (\Phi^T\Phi)^{-1}\Phi^T Y.$$

Example 9.5 Using the data set below, we can compare both a linear and a quadratic polynomial fit to the data. The results are shown in Figure 9.4.

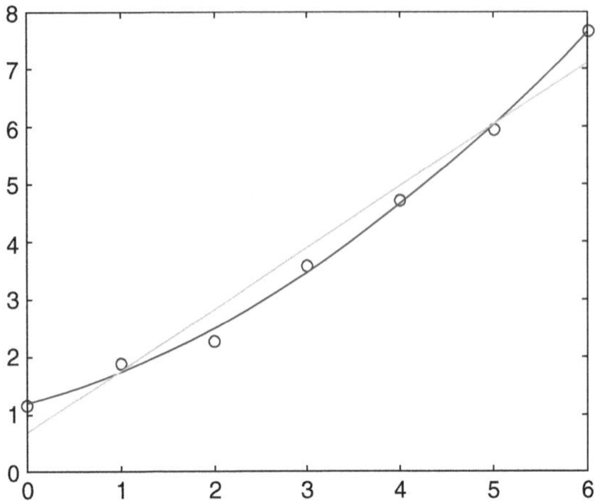

Figure 9.4 Linear and quadratic polynomial fitting. The choice of which model is best for the given data can be understood by examining the so-called *residuals*.

u	0	1	2	3	4	5	6
y	1.1538	1.8834	2.2741	3.5862	4.7319	5.9692	7.6566

Overfitting and Underfitting

An important question to ask is what order polynomial one should use to model a given data set. There are several statistical methods that can be used to address this question. One of the simplest is to examine the so-called *residuals*.

Definition 9.1 ***(Residuals and Mean-Squared Error)*** The *residual* R is the difference between the output y from the data and an output $\hat{y}$ computed from the fitted model

$$R = y - \hat{y}.$$

The *mean-squared error* (MSE) is defined as

$$MSE = \frac{||R||^2}{n} = \frac{||y - \hat{y}||^2}{n},$$

where n is the size of the data set.

We can define the best-fit model, therefore, as the one giving the smallest mean-squared error. In the above example, it is easy to compute the mean-squared error for the linear model as 0.1424 and the mean-squared error for the quadratic model as 0.0141 (Problem 9.4). We conclude that the quadratic polynomial is the better model for the given data.

Matlab Curve-Fitting Functions

Matlab has several functions for fitting curves and surfaces to data, regression and least-squares analysis. A few of them are listed below:

- `polyfit` – fit polynomial to data (uses least squares);
- `fit` – fit a curve or surface to data;
- `lsqlin` – constrained linear least squares;
- `lsqnonlin` – nonlinear least-squares problems;
- `lsqcurvefit` – nonlinear least-squares problems.

9.4 Recursive Least Squares

We next discuss recursive least squares. The idea is that, as new data come in, we can update the least-squares solution using the previous solution together with the new data rather than computing everything over again. Suppose that $x_0^* = (A_0^T A_0)^{-1} A_0^T b_0$ is the least-squares solution to the problem

$$\min \|A_0 x - b_0\|^2,$$

and new information arrives in the form of additional rows of data A_1, b_1, so that we now have the problem

$$\min \left\| \begin{bmatrix} A_0 \\ A_1 \end{bmatrix} x - \begin{bmatrix} b_0 \\ b_1 \end{bmatrix} \right\|^2,$$

where the rows of A_1 and b_1 are appended to A_0 and b_0, respectively. The least-squares solution to this problem is

$$x_1^* = G_1^{-1} \begin{bmatrix} A_0 \\ A_1 \end{bmatrix}^T \begin{bmatrix} b_0 \\ b_1 \end{bmatrix} = G_1^{-1}(A_0^T b_0 + A_1^T b_1),$$

where

$$\begin{aligned} G_1 &= \begin{bmatrix} A_0 \\ A_1 \end{bmatrix}^T \begin{bmatrix} A_0 \\ A_1 \end{bmatrix} \\ &= A_0^T A_0 + A_1^T A_1 = G_0 + A_1^T A_1, \end{aligned}$$

and where $G_0 = A_0^T A_0$ has already been computed in the previous step. We can then express the recursive solution as

$$\begin{aligned} x_1^* &= G_1^{-1}(A_0^T b_0 + A_1^T b_1) \\ &= G_1^{-1}(G_0 G_0^{-1} A_0^T b_0 + A_1^T b_1) \\ &= G_1^{-1}(G_0 x_0^* + A_1^T b_1), \end{aligned}$$

using the fact that $x_0^* = G_0^{-1}A_0^Tb_0$. Let us rewrite this to emphasize the dependence on the new data, noting that $G_0 = G_1 - A_1^TA_1$. Thus we have

$$\begin{aligned}x_1^* &= G_1^{-1}((G_1 - A_1^TA_1)x_0^* + A_1^Tb_1)\\ &= x_0^* + G_1^{-1}A_1^T(b_1 - A_1x_0^*).\end{aligned}$$

The term $(b_1 - A_1x_0^*)$ is called the *innovation term*. Thus we have derived the following recursive formula:

$$\begin{aligned}G_{k+1} &= G_k + A_{k+1}^TA_{k+1}\\ x_{k+1}^* &= x_k^* + G_{k+1}^{-1}A_{k+1}^T(b_{k+1} - A_{k+1}x_k^*).\end{aligned}$$

In the above formula, note that we must invert the matrix G_{k+1} at each step. We can derive a simpler formula that updates the inverse of G_k using the so-called *Sherman–Morrison formula* [30]

$$(A + UV)^{-1} = A^{-1} - A^{-1}U(I + VA^{-1}U)^{-1}VA^{-1}.$$

Applying this formula to G_{k+1} gives

$$(G_k + A_{k+1}^TA_{k+1})^{-1} = G_k^{-1} - G_k^{-1}A_{k+1}^T(I + A_{k+1}G_k^{-1}A_{k+1}^T)^{-1}A_{k+1}G_{k+1}^{-1}.$$

Writing $P_k = G_k^{-1}$ for simplicity, the final recursive least-squares algorithm becomes

$$\begin{aligned}P_{k+1} &= P_k - P_kA_{k+1}^T(I + A_{k+1}P_kA_{k+1}^T)^{-1}A_{k+1}P_k\\ x_{k+1}^* &= x_k^* + P_{k+1}A_{k+1}^T(b_{k+1} - A_{k+1}x_k^*).\end{aligned}$$

An advantage of this formulation of the problem is that if we receive a single new row of data, then A_{k+1} is just a row and the matrix we need to invert,

$$(I + A_{k+1}P_kA_{k+1}^T)^{-1},$$

is a scalar and is therefore easy to compute.

Example 9.6 ***(Linear Regression with New Data Arriving Daily)***
Consider the data

$$A_0 = \begin{bmatrix}1 & 0\\ 0 & 1\\ 1 & 1\end{bmatrix},\ b_0 = \begin{bmatrix}1\\ 1\\ 1\end{bmatrix},$$

$$A_1 = [2{,}1],\ b_1 = [3],$$

$$A_2 = [3{,}1],\ b_2 = [3].$$

Then on the first day we have

$$P_0 = (A_0^TA_0)^{-1} = \begin{bmatrix}2/3 & -1/3\\ -1/3 & 2/3\end{bmatrix} \Rightarrow x_0^* = P_0A_0^Tb_0 = \begin{bmatrix}2/3\\ 2/3\end{bmatrix}.$$

On the second day we have $1 + A_1 P_0 A_1^T = [3]$ and so

$$P_1 = P_0 - \frac{P_0 A_1^T A_1 P_0}{1 + A_1 P_0 A_1^T} = \begin{bmatrix} 1/3 & -1/3 \\ -1/3 & 2/3 \end{bmatrix}$$

and therefore

$$x_1^* = x_0^* + P_1 A_1^T (b_1 - A_1 x_0^*) = \begin{bmatrix} 1 \\ 2/3 \end{bmatrix}.$$

On the third day we have $1 + A_2 P_1 A_2^T = [8/3]$ and so

$$P_2 = P_1 - \frac{P_1 A_2^T A_2 P_1}{1 + A_2 P_1 A_2^T} = \begin{bmatrix} 1/6 & -1/4 \\ -1/4 & 5/8 \end{bmatrix}.$$

Therefore, we have

$$x_2^* = x_1^* + P_2 A_2^T (b_2 - A_2 x_1^*) = \begin{bmatrix} 5/6 \\ 3/4 \end{bmatrix}.$$

Nonlinear Regression

We have already discussed regression using higher-order polynomials, which is a special case of *nonlinear regression.* It turns out that the key feature enabling the extension to more general nonlinear problems is the property of *linearity in the parameters.* Consider the input/output system

$$y = a_1 \phi_1(x) + a_2 \phi_2(x) + \cdots + a_m \phi_m(x) = \phi(x)^T a.$$

The function $\phi(x)$ can be any suitably smooth known function of x. Since the output y is a linear combination of the nonlinear functions ϕ_i, the least-squares solution for the parameter vector proceeds in an identical fashion to the case of linear regression. As before, we take N measurements and define the system

$$\begin{aligned} y(1) &= \phi(x(1))^T a \\ &\vdots \\ y(N) &= \phi(x(N))^T a, \end{aligned}$$

which we write as

$$Y = \Phi(x) a.$$

The least-squares solution is therefore

$$a^* = (\Phi^T \Phi)^{-1} \Phi^T Y,$$

provided that $\Phi^T \Phi$ is invertible. The matrix function Φ is called the *regressor matrix*. The condition that $(\Phi^T \Phi)^{-1}$ is invertible is called the *excitation condition.* Since Φ is an $n \times m$ matrix, it is necessary (but not sufficient) that the number of measurements N be larger than m for this condition to hold. In the polynomial case, it is also sufficient.

Example 9.7 Consider the system

$$I\ddot{\theta} + mg\ell \sin(\theta) = u,$$

which is a simple model of a one-link robot arm, with θ the joint angle and u a torque input generated by a motor, and suppose that the inertia, I, and mass, m, are not known. This may happen, for example, when the robot picks up an unknown load.

From our earlier studies of the simple pendulum we know that the total energy E is given by

$$E = \frac{1}{2}I\dot{\theta}^2 + mg\ell(1 - \cos(\theta)) \tag{9.3}$$

and that the derivative of the energy is

$$\dot{E} = \dot{\theta}u. \tag{9.4}$$

If we define $a_1 = I$ and $a_2 = m$, then we can write Equation (9.3) as

$$E = \phi^T a,$$

where the regressor is $\phi = [\frac{1}{2}\dot{\theta}, g\ell(1 - \cos(\theta))]^T$ and $a = [a_1, a_2]^T$. If we integrate Equation (9.4) from time 0 to time t, with $E(0) = 0$, we have

$$E(t) = \int_0^t \dot{\theta}(s)u(s)ds.$$

Therefore, applying a known input function $u(t)$, which generates an output trajectory $\theta(t), \dot{\theta}(t)$, we can divide the interval $[0, t]$ into a sequence of times $t_1, \ldots, t_n$, and write

$$E(t_i) = \phi(t_i)a, \quad i = 1, \ldots, n.$$

We can then determine the parameter vector a as the least-squares solution of

$$Y = \Phi a,$$

where

$$Y = \begin{bmatrix} E(t_1) \\ \vdots \\ E(t_n) \end{bmatrix}, \quad \Phi = \begin{bmatrix} \phi^T(t_1) \\ \vdots \\ \phi^T(t_n) \end{bmatrix},$$

provided that Φ has full rank (see [26]).

9.5 Logistic Regression

We have seen that the logistic equation is important in several applications, such as modeling limited growth in population dynamics and product adoption in business models. In this section we show an important use of the logistic equation in regression. *Logistic regression*, also called *logit regression*, is used to model the probability of events that take on only two values, such as pass/fail, win/lose, or healty/sick. Logistic regression is widely used in machine learning, for example to determine whether a given email is legitimate or spam.

We will illustrate logistic regression with the following example. Consider a bank or other lending institution deciding whether or not to grant approval for a loan application. One criterion used to make such decisions is the applicant's credit score, for example, the Fair Isaac Corporation (FICO) score, which is based on factors such as outstanding debts, payment history and others. The standard FICO score ranges between 300 and 850. Since other decisions besides the FICO are generally involved in the decision model, one cannot determine approval or denial with certainty based solely on the FICO score. With logistic regression we can determine a probability of acceptance or denial given historical data of past applications.

The table below shows a hypothetical data set of 12 credit scores and whether or not they were approved or denied, with 1 representing approval and 0 representing denial. The data set is plotted as the black dots on the figure shown. Since the data take on values of either 0 or 1, linear regression is not useful to fit a line to these data. Logistic regression attempts to fit the best S-curve function to the data, from which the probability of any credit score can be estimated. The data set is therefore the vector of credit scores X and the binary output vector Y.

FICO Score X	Approval Y
320	0
355	0
420	0
480	0
492	1
520	0
550	0
575	1
633	1
649	1
675	1
720	1

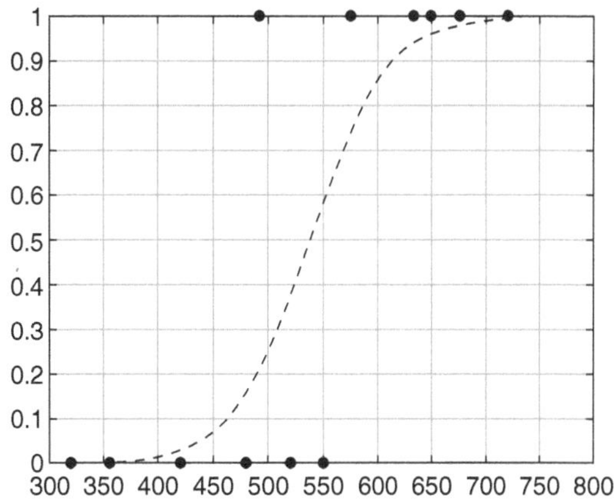

Definition 9.2 ***(Logit Function)*** Let Y be a Bernoulli random process, so that Y takes values 0 or 1, and let $p = Prob(Y = 1)$. We define the *odds* of an event as $p/(1-p)$, which is the probability of success divided by the probability of failure. The *logit* function, ℓ, is defined as the natural logarithm of the odds as

$$\ell = \ln\left(\frac{p}{1-p}\right).$$

The logit function stretches the probability interval [0,1] to $[-\infty, \infty]$.

It is easy to show (Problem 9.7) that the inverse of the logit function is given by

$$p = \frac{e^{\ell}}{1+e^{\ell}} = \frac{1}{1+e^{-\ell}}, \tag{9.5}$$

which is again the logistic function. Figure 9.5 shows the logit function and its inverse.

The procedure of logistic regression is now to find $\beta_1 + \beta_2 x$ as a solution to

$$\ln\left(\frac{p}{1-p}\right) = \beta_1 + \beta_2 x, \tag{9.6}$$

which leads to the estimated probability

$$p = \frac{1}{1+e^{-\beta_1-\beta_2 x}}. \tag{9.7}$$

Estimation of the Coefficients

The coefficients β_1 and β_2 in the above expression can estimated by so-called *maximum likelihood estimation*. We will briefly discuss the main idea behind the maximum likelihood estimate. Given a linear expression

$$y = \beta_1 + \beta_2 x,$$

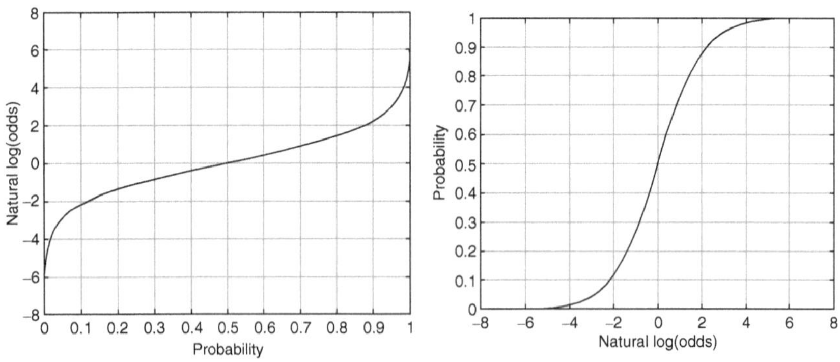

Figure 9.5 The logit function (left) and the inverse logit function (right). The inverse logit function gives the familiar S-curve.

we say that the *likelihood* of the event $x(i) \in X$ is

$$p_i = \frac{1}{1 + e^{-\beta_1 - \beta_2 x(i)}} \tag{9.8}$$

if the event $x(i)$ results in a success $Y(i) = 1$, and

$$p_i = 1 - \frac{1}{1 + e^{-\beta_1 - \beta_2 x(i)}} \tag{9.9}$$

if the event $x(i)$ results in a failure $Y(i) = 0$. Then the total likelihood is the product

$$L(\beta_1, \beta_2) = \prod_{i=1}^{N} p_i.$$

The coefficients β_1 and β_2 are then chosen to maximize the total likelihood function $L(\beta_1, \beta_2)$. The Matlab command

$$B = \texttt{mnrfit}(X, Y),$$

where `mnrfit` is called the ***multinomial logistic regression*** function, will return the optimal values of β_1 and β_2 given the data set X and Y.

Example 9.8 For the previous credit score data, the Matlab script below implements the logistic regression procedure to determine the best-fit logistic function. The resulting function is shown in Figure 9.6.

Logistic Regression Example

```
x = [320 355 420 480 492 520 550 575 633 649 675 720]';
y = [0 0 0 0 1 0 0 1 1 1 1 1]';
y2=y+1; % mnrfit() does not accept 0 data so add 1 to y
plot(x,y,'o','MarkerEdgeColor','k','MarkerFaceColor','k');
hold on;
beta = mnrfit(x,y2); % optimum coefficients beta1 and beta2
t=linspace(300,800,100);
phat=1./(1+exp(beta(1)+beta(2)*t));
plot(t,phat,'k');
grid on
%
f=@(z)1/(1+exp(beta(1)+beta(2)*z));
% f(z) returns the probability of approval for FICO score z
```

Running this script returns $\beta_1 = 15.6276$, $\beta_2 = -0.0291$. The anonymous function f can then calculate the probability of approval given any other credit score. For example, if your own credit score is 580, then $f(580) = 0.7745$ and the odds of approval are 3.4346.

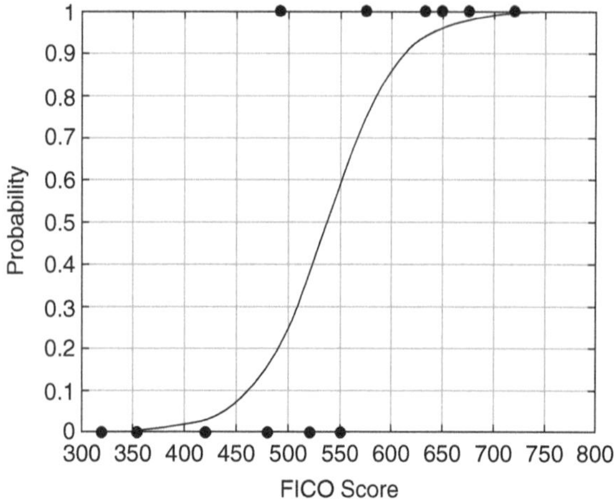

Figure 9.6 Best-fit logistic curve for the given credit score data.

9.6 Neural Networks

In this section we give a brief introduction to *neural networks*. Neural networks, or *artificial neural networks* are modeled after biological neurons and are useful in many applications, such as:

- speech recognition;
- image analysis;
- classification of data;
- function approximation;
- feedback control.

We limit the treatment to so-called *perceptrons*, which are the simplest examples of neural networks. Perceptron networks are primarily used for classifying binary data sets, i.e., sets containing two types of data. For example, a yes/no decision on a credit application, a decision on whether a given image is a dog or a cat, and so on.

Biological Neurons

The main structure of a biological neuron is shown in Figure 9.7.

- *Dendrites* bring signals from other neurons into the cell body, called the *soma*.
- In the soma, cell capacitance "integrates" the signals. Once the composite signal exceeds a *threshold*, another signal, the *action potential*, is transmitted through the *axon*. The axon connects through *synapses* and dendrites to other neurons.
- The synapses operate through the discharge of neurotransmitter chemicals (ions) across intercellular gaps and can either be *excitatory* (tending to increase the likelihood that the next neuron fires) or *inhibitory* (tending to prevent firing of the next neuron).
- *learning* takes place as the interconnections among the neurons are strengthened or weakened as a result.

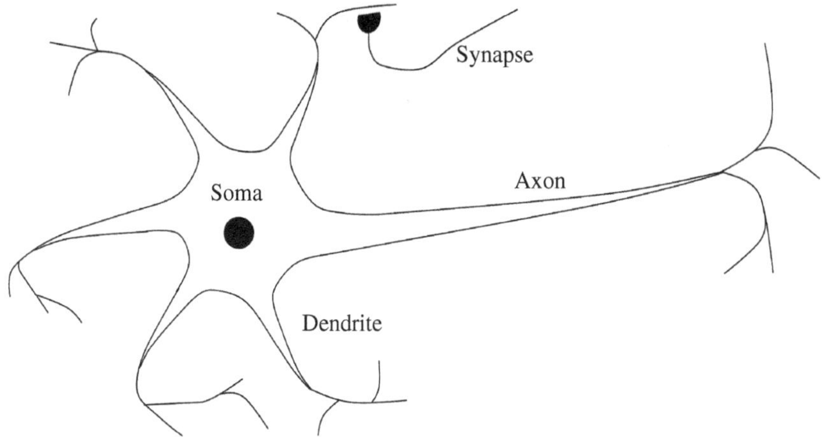

Figure 9.7 A brain neuron

Artificial Neuron

An artificial neuron is shown in Figure 9.8. The incoming edges are analogous to the dendrites and the outgoing edge is analogous to the axon. The signals $x_1, x_2, \ldots, x_n$ represent input signals that enter the neuron body through each (weighted) edge. The parameters $w_1, w_2, \ldots, w_n$ are the edge weights and w_0 is a bias term. The output signal is then given by the formula

$$y(t) = \sigma\left(\sum_{j=1}^{n} w_j x_j(t) + w_0\right).$$

The cell function, $\sigma(\cdot)$ is known as the *activation function*. Two common examples of activation functions are shown in Figure 9.9. The weight w_0 represents a threshold value.

Positive weights $w_j > 0$ represent excitatory synapses and negative weights $w_j < 0$ represent inhibitory synapses. This model is known as a *perceptron*.

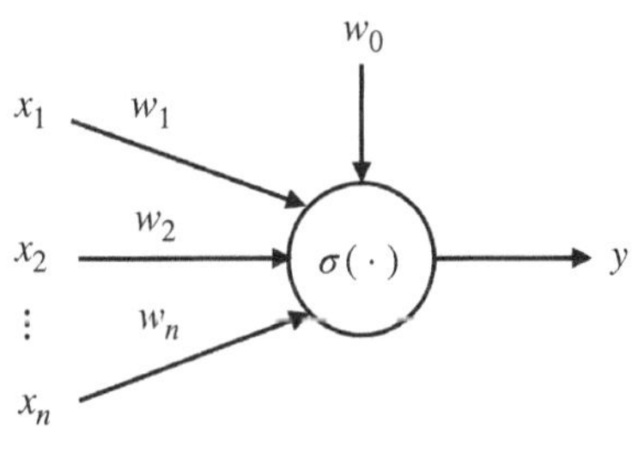

Figure 9.8 An artificial neuron model

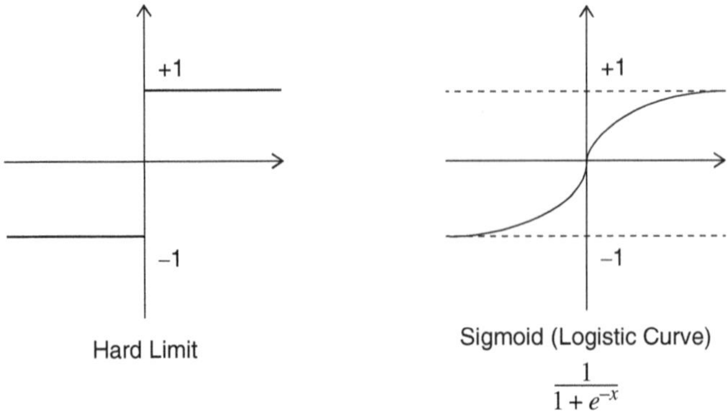

Figure 9.9 Examples of activation functions

- The basic idea is that there is no output below a certain value of the argument of $\sigma(\cdot)$ and a specific output magnitude above that value of the argument.
- The bias value w_0 shifts the activation function left or right.
- Using vector notation $x = [x_1, x_2, \dots, x_n]^T$, $w = [w_1, w_2, \dots, w_n]^T$, we can write the output equation as

$$y = \sigma(w^T x + w_0).$$

A *neural network* is then composed of an interconnection of neurons, an example of which is shown in Figure 9.10.

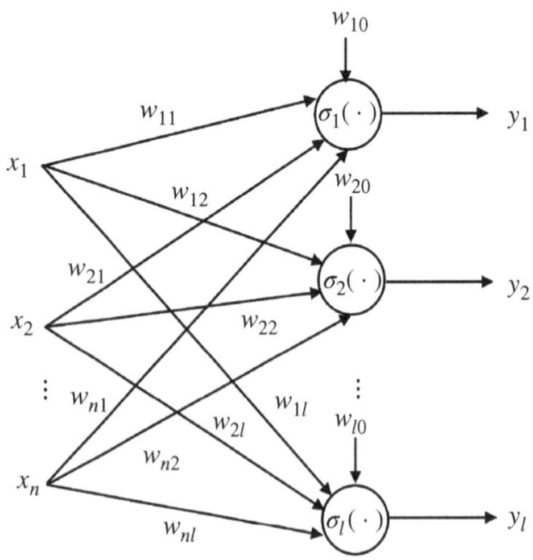

Figure 9.10 A one-layer neural network

The network in Figure 9.10 is called a *one-layer neural network*. In the one-layer neural network, there are ℓ cells, n inputs, and ℓ outputs. The output equations are written as

$$y_k = \sigma\left(\sum_{j=1}^{n} w_{jk}x_j + w_{k0}\right), \quad k = 1, 2, \ldots, \ell,$$

which we can write as

$$y = \overline{\sigma}(W^T x + w_0),$$

where $y = [y_1, y_2, \ldots, y_\ell]^T$, $x = [x_1, x_2, \ldots, x_n]^T$, $w_0 = [w_{10}, w_{20}, \ldots, w_{\ell 0}]^T$ and

$$W = \begin{bmatrix} w_{11} & w_{12} & \cdots & w_{1\ell} \\ w_{21} & w_{22} & \cdots & w_{2\ell} \\ \vdots & \vdots & & \vdots \\ w_{n1} & w_{n2} & \cdots & w_{n\ell} \end{bmatrix}.$$

The vector activation function is defined for a vector $v = [v_1, v_2, \ldots v_\ell]^T$ as

$$\overline{\sigma}(\cdot) = [\sigma(v_1), \sigma(v_2), \ldots, \sigma(v_\ell)]^T.$$

Suppose that we have a set of (noisy) data that we want to classify into two regions A and B. For example, in a camera image, the problem is to determine if a given pixel is part of the background or part of an object of interest.

A one layer neural network with two inputs x_1 and x_2 and one output is given by

$$y = \sigma(w_1 x_1 + w_2 x_2 + w_0).$$

If we use the hard-limit function for σ then y takes on values ± 1. When $y = 0$ we have

$$0 = w_1 x_1 + w_2 x_2 + w_0$$

or, in other words,

$$x_2 = -\frac{w_1}{w_2}x_1 - \frac{w_0}{w_2},$$

which defines a line in the (x_1, x_2) plane, below which $y = -1$ and above which $y = +1$ (Figure 9.11).

Therefore, each time a measurement is made, the class to which it belongs can be determined from the value of the output y. The trick is to determine the best values for the weights and threshold so that the classifier performs as expected.

There are several methods for determining the weights. This is called *training*. The basic idea is to take a sequence of inputs for which one knows the classification and use that to determine a best guess for the weights. The more "training examples" one uses, the better the confidence one has in the weights. Once the

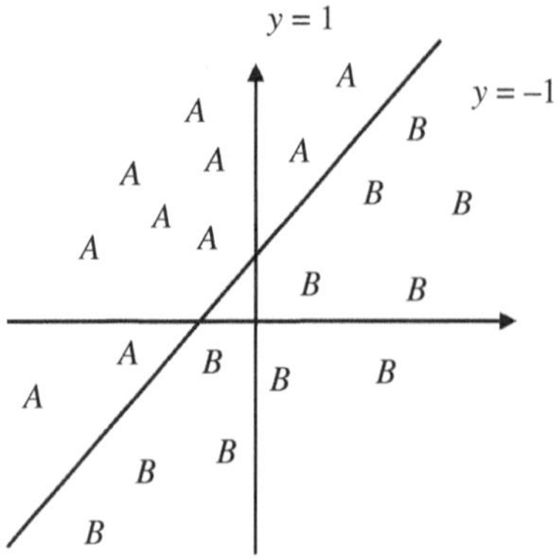

Figure 9.11 The decision boundary separating the two data sets.

weights are determined, then future (unknown) inputs can be classified with a certain percent confidence.

In order for this one-layer perceptron classifier to work, the data set must be *linearly separable*, which means that there is a hyperplane separating the data into two sets, with all data on each side having the same classification.

A simple training algorithm for the perceptron can be described as follows. Let $((x_{i1}, x_{i2}), y_i)$, $i = 1, \dots, N$ be a data set with known output y_i for each input vector (x_{i1}, x_{i2}). Suppose

$$\hat{y}_i = \sigma(w_1 x_{i1} + w_2 x_{i2} + w_0)$$

is the output corresponding to a weight vector $w = (w_0, w_1, w_2)$. Then we choose w to minimize the cost function

$$f(w) = \sum_{i=1}^{N} \frac{1}{2}(y_i - \hat{y}_i(w))^2.$$

Finding w so that $\hat{y}_i(w) = y_i$ for all i means that each element of the training set is correctly classified. We can use a gradient-descent algorithm

$$w_i^{k+1} = w_i^k - \eta \nabla f(w_i^k),$$

where the step size η is called the *learning rate*, the subscript i corresponds to a training sample, and k is a given iteration. Computing the components of the gradient, we have that

$$\frac{\partial f(w_i^k)}{\partial w_i^k} = -[y_i - \hat{y}_i^k(w_i^k)]x_{ik}$$

and therefore the gradient descent algorithm becomes

$$w_i^{k+1} = w_i^k + \eta[y_i - \hat{y}_i^k(w_i^k)]x_{ik}.$$

Example 9.9 Suppose that we have a training set consisting of eight samples

bias	x_1	x_2	y
1	2	1	1
1	3	3	0
1	3.3	1.5	1
1	4	2	1
1	4.5	4.2	0
1	6.2	5	0
1	5.5	3.8	0
1	5.8	3	1

The Matlab script below implements the above training algorithm to estimate the neural network weights.

Neural Network Training

```
x= [ [1 2 1]; [1 3 3]; [1 3.3 1.5];[1 4 2];...
    [1 4.5 4.2]; [1 6.2 5]; [ 1 5.5 3.8]; [1 5.8 3]];
y=[1 0 1 1 0 0 0 1]; % the data set
w=[0;0;0]; %initial weights
n=0;
num_iterations=5;
z=0;      % set the threshold
eta=0.1; % set the learning rate
while n<num_iterations
for i=1:8
    f=w'*x(i,:)';
    if f>z
        yhat=1;
    else
        yhat=0;
    end
for j=1:3  % update rule for the weights
    w(j)=w(j) + eta*(y(i) -yhat)*x(i,j);
end
```

Running the Matlab script gives the values for the weights as

$$w_0 = 0.1, \quad w_1 = 0.67, \quad w_2 = -1.06.$$

With these values for the neural network weights, the line separating the training data is shown in Figure 9.12.

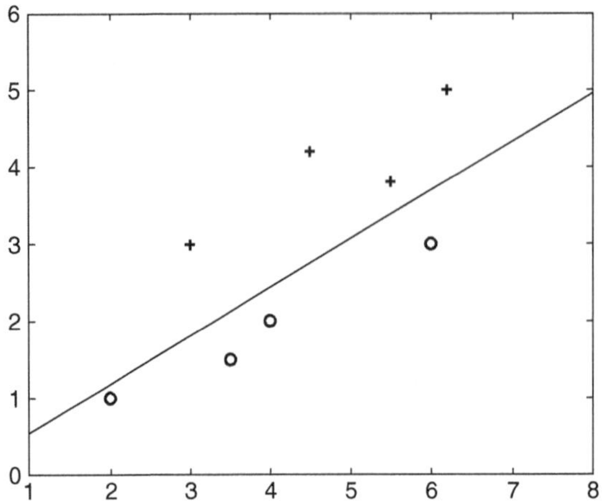

Figure 9.12 Results of training the perceptron on the data set

Problems

9.1 Compute the range space and null space of the matrix

$$A = \begin{bmatrix} 1 & 1 \\ 0 & 0 \\ -1 & -1 \end{bmatrix}.$$

9.2 The United States constitution requires that a census of the population be conducted every 10 years. The data below show the US population (in millions) from 1900 to 2010. In this problem you will use Matlab to predict the 2020 population using these data.

Year	Population
1900	75.995
1910	91.972
1920	105.711
1930	123.203
1940	131.669
1950	150.697
1960	179.323
1970	203.212
1980	226.505
1990	249.633
2000	281.422
2010	308.746

Let $y(k)$ be the population at year $u(k)$. Fit the following models to the data using the method of least squares:

- *Model 1*: $y = a_0 + a_1 u$ (linear model)
- *Model 2*: $y = a_0 + a_1 u + a_2 u^2$ (quadratic model)

You may use the Matlab functions `polyfit` and `polyval` to do this.
For each model:

(a) Plot the results from the models and compare to the true output $y(k)$.
(b) Compute the predicted 2020 population from each model.
(c) Compute the residuals.
(d) Which model would you select and why? Explain.

9.3 Consider the recursive least-squares problem in Example 9.6. Write a Matlab script to implement the recursive least-squares algorithm and verify the solution obtained in the example. Add two more days of data as follows:

$$A_3 = [4, 2],\ b_1 = [4.5],$$
$$A_4 = [5, 2],\ b_2 = [5].$$

9.4 Verify that the quadratic polynomial model in Example 9.5 is a better fit than the linear model by computing the respective mean-squared errors.

9.5 The velocity of a rocket, measured at half-second intervals in meters per second, is

0, 14.70, 29.40, 44.10, 58.80, 73.50, 88.20, 102.90, 117.60, 132.30, 147.00.

With $g = 9.8\ \text{m/s}^2$, what g-force is an astronaut in the rocket experiencing?

9.6 Professor Gauss has discovered a planet orbiting a star in the constellation Orion. He makes 11 measurements of the position of the planet relative to the star and would like to determine the parameters of its elliptical orbit about the star. Assuming that the position measurements are:

$$x = [0.9874, 0.5055, -0.4309, -0.9755, -0.6104,$$
$$0.2430, 0.9840, 0.7094, -0.1198, -0.9148, -0.8846],$$
$$y = [-0.0126, 0.8067, 0.8945, 0.1556, -0.7136,$$
$$-0.9996, -0.2555, 0.6125, 1.0151, 0.4085, -0.5895],$$

as shown in the figure below, fit an ellipse of the form

$$\frac{x^2}{\alpha^2} + \frac{y^2}{\beta^2} = 1$$

to the measured data, i.e., estimate the values of α and β using the method of least squares.

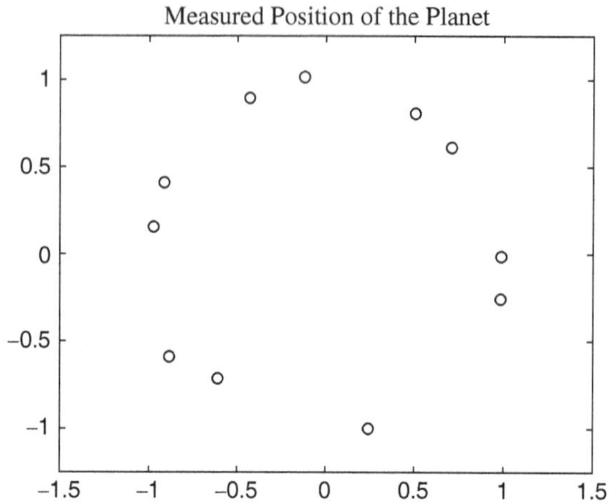

9.7 Show by direct calculation that the inverse of the logit function $\ell = \ln\left(\frac{p}{1-p}\right)$ is equal to $p = \frac{e^{\ell}}{1+e^{\ell}} = \frac{1}{1+e^{-\ell}}$.

9.8 Consider the perceptron network in Example 9.9 and suppose that new data points are received as $(x, y) = \{(1.0, 1.5), (5,3), (7, 3.4)\}$, which are known to be classified as 1, and $(x, y) = \{(4,5), (5,5), (6.5, 4)\}$, which are known to be classified as 0.

(a) Which points are correctly classified and which are misclassified?

(b) Is the complete data set still linearly separable?

(c) Retrain the perceptron network so that it correctly classifies both the old and new data.

10

Stochastic Systems

In this chapter we turn to a discussion of stochastic systems, which are systems that are governed by random, rather than deterministic, dynamics. The word "stochastic" comes from the Greek, meaning "able to guess or conjecture." A deterministic system is repeatable in the sense that repeated trials with the same input and initial conditions will produce the same output. A stochastic system will generally produce different outputs with each trial that can only be predicted in a probabilistic sense.

We first give an introductory treatment of discrete Markov chains. We discuss *regular*, *ergodic*, and *absorbing* Markov chains and give some examples and applications.

We then discuss Monte Carlo methods, which use repeated sampling from a given probability distribution to solve problems that may be deterministic. These methods are useful for systems for which deterministic methods are impractical due to the size or complexity of the problem. We will discuss Monte Carlo methods for three applications, namely, simulation, numerical integration, and optimization. Before continuing in this chapter, the reader should review Appendix C, which contains the necessary background on probability theory that is used in the remainder of the chapter.

10.1 Markov Chains

Recall the definition of a random variable from Appendix C. By a (discrete-time) *random process* we mean a sequence of random variables $\{X_n\}_{n=0}^{\infty}$, indexed by integers n. In this section we introduce the concept of a *Markov chain*, which is a particular class of random processes with numerous applications in areas such as physics, chemistry, biology, finance, statistics, games, and genetics. Recall, as described in Appendix C, that in a Bernoulli process like flipping a coin, events are independent, i.e., the past does not influence the future. There are

Introduction to Modeling and Simulation: A Systems Approach, First Edition. Mark W. Spong.

Companion Website: www.wiley.com/go/spong/engineeringmodeling

many phenomena where this is not the case. For example, the probability of rain tomorrow will generally depend on whether or not it is raining today. Likewise, the probability that the price of a stock rises or falls on a given day likely depends on whether the stock price rose or fell on the previous day. Such phenomena are captured by the concept of a Markov process. A Markov process is a random process where the probability of an event at step n depends on the state obtained at the previous step $n-1$. We can formalize this notion in terms of conditional probabilities as follows.

Definition 10.1 ***(Markov Chain)*** Let $\{X_n\}_{n=0}^{\infty}$ be a discrete-time random process. We say that $\{X_n\}$ is a *Markov chain* if it satisfies the Markov property,

$$\begin{aligned} &P(X_n = i_n \mid X_{n-1} = i_{n-1}, X_{n-2} = i_{n-2}, \ldots, X_0 = i_0) \\ &\quad = P(X_n = i_n \mid X_{n-1} = i_{n-1}), \end{aligned}$$

where $i_0, i_1, \ldots, i_n$ are the states and $0,1,\ldots,n$ are the time steps.

Thus, the probability that $X_n = i_n$ depends only on X_{n-1} and not on $X_0, \ldots, X_n, X_{n-2}$. In other words, the next step depends on where you are but not how you got there. Let

$$p_i(n) = P(X_n = i)$$

be the probability that the process is in state i at time n, and let p_{ij} be the conditional probability

$$p_{ij}(n) = P(X_{n+1} = j \mid X_n = i).$$

In other words, $p_{ij}(n)$ is the probability that the process will be in state j at time $n+1$, given that it is in state i at time n. We assume that the given Markov chain is homogeneous, which means that

$$p_{ij} = P(X_{n+1} = j \mid X_n = i) = P(X_1 = j \mid X_0 = i).$$

In other words, the probability of moving from state i to state j is independent of n. We can represent a homogeneous Markov chain by a so-called *transition diagram*, such as shown in Figure 10.1. Note that, since the p_{ij} are probabilities, the outgoing probabilities at each node must sum to 1, for example, $p_{11} + p_{12} + p_{1i} = 1$ in Figure 10.1, which simply says that the process must either stay in its current state or move to another state.

Example 10.1 Suppose Company A makes widgets and currently has a 10% market share. Their monthly data suggest that there is a 20% chance that a current customer will switch to a competitor Company B each month. At the same time, there is a 40% chance that a customer of Company B will switch to Company A.

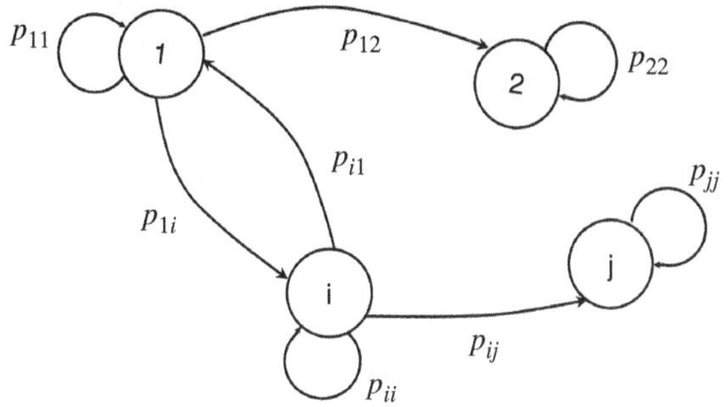

Figure 10.1 Transition diagram for a Markov chain. The numbered circles represents the states and the weights p_{ij} represent the probability that the state will change from i to j in one step.

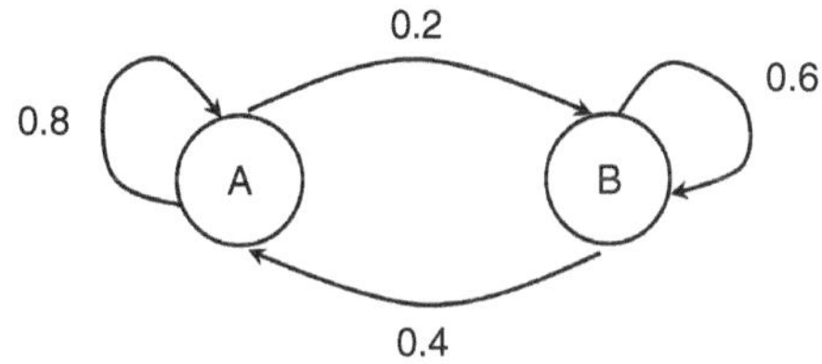

Figure 10.2 Transition diagram for Example 10.1. Each month a fraction of customers will switch from company A to B or from B to A according to the given probabilities.

We can represent this as a Markov chain with transition diagram as shown in Figure 10.2.

We can represent the respective outgoing probabilities for each state as a row of a matrix, called the *transition matrix*.

Definition 10.2 ***(Transition Matrix)*** The transition matrix for a Markov chain $\{X_n\}$ with N states is an $N \times N$ matrix defined as

$$P = \begin{bmatrix} p_{11} & p_{12} & \cdots \\ p_{21} & p_{22} & \cdots \\ \vdots & \vdots & \ddots \end{bmatrix}.$$

Other terms for the transition matrix P are “probability matrix” and “Markov matrix”.

The transition matrix in the above example is therefore,

$$\begin{bmatrix} 0.8 & 0.2 \\ 0.4 & 0.6 \end{bmatrix},$$

where the first row represents company A and the second row represents company B.

Now, with the given probabilities of customers switching allegiance, we can ask what is the projected market share of each company from month to month.

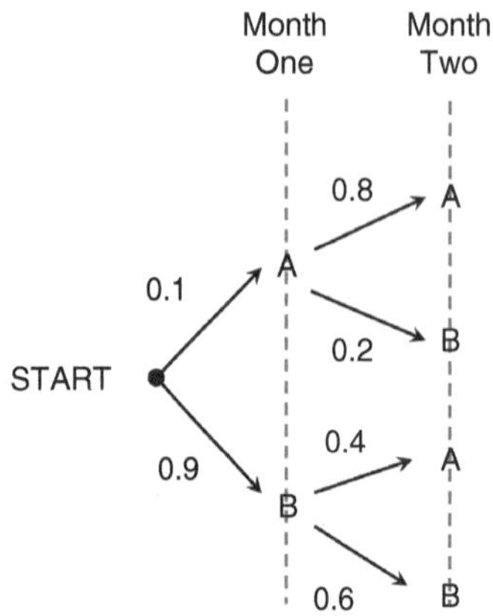

Figure 10.3 Computing the transition probabilities for months 1 and 2.

To answer this question, consider the diagram in Figure 10.3. Let $S_0 = [0.1, 0.9]$ be the initial market share percentage of A and B at month 1. We can think of S_0 as representing the probability that a given customer purchases from A (0.1) and from B (0.9). Then we can see from the diagram that the probability that a customer buys from A at month 2 is

$$0.1 \times 0.8 + 0.9 \times 0.4 = 0.44.$$

Likewise, the probability that a customer buys from B at month 2 is

$$0.1 \times 0.2 + 0.9 \times 0.6 = 0.56.$$

We can represent the above calculation using the transition matrix as

$$[0.44, 0.56] = [0.1, 0.9] \begin{bmatrix} 0.8 & 0.2 \\ 0.4 & 0.6 \end{bmatrix}.$$

Thus, with $S_1 = [0.44, 0.56]$, we can write this as

$$S_1 = S_0 P,$$

where P is the state transition matrix. We see that the projected market share for A has increased from 10% to 44% while the projected market share for B has decreased from 90% to 56%.

Assuming that the probabilities do not change from month to month, so that the transition matrix P is constant, we can compute the projected market shares in subsequent months as

$$S_2 = S_1 P = S_0 P^2 = [0.5760, 0.4240]$$
$$S_3 = S_2 P = S_0 P^3 = [0.6304, 0.3696].$$

In general, we see that the projected market share at month k is given by $S_k = S_0 P^k$. A crucial question is to determine if there is a limiting value,

$S_* = \lim_k S_k = S_0 \lim_k P^k$. In this example, we see that $S_* = [0.667, 0.333]$ satisfies

$$S_* P = S_*.$$

The vector S_* is called a *stationary matrix* or *steady state* and is a left eigenvector of the transition matrix P corresponding to the eigenvalue 1. Therefore, we expect that, in the long term, the market share for Company A will approach 75%. Note also that P^k converges to

$$P^* = \begin{bmatrix} 0.667 & 0.333 \\ 0.667 & 0.333 \end{bmatrix},$$

from which we can read off the long-term probabilities, in this case the long term market share. The long-term market share is thus independent of the initial market share, i.e., the choice of S_0, and depends only on the transition probabilities.

We should note that not every left eigenvector of P corresponding to the eigenvalue 1 is a stationary matrix since the elements of S_* must sum to 1. For example, the code

Compute a Stationary Matrix for *P*

```
P=[0.8 0.2; 0.4 0.6];
[v,D,W]=eig(P);
S=W(:,1);
Sstar = S/sum(S);
```

returns $S = [0.8944, 0.4472]$ as a unit left eigenvector of P, corresponding to the eigenvalue $\lambda = 1$, but S is not a stationary matrix. By adding the constraint that the elements of the left eigenvector should sum to 1, we obtain `Sstar` $= [0.667, 0.333]$.

Stochastic Matrices

The transition matrix for a Markov chain is an example of a so-called stochastic matrix.

Definition 10.3 ***(Stochastic Matrix)*** An $n \times n$ matrix $P = (p_{ij})$ is a *stochastic matrix* if

$$p_{ij} \geq 0 \text{ and } p_{i1} + p_{i2} + \cdots + p_{in} = 1, \quad \text{for all } i,$$

which means that the sum of the elements in each row of P is equal to 1.

We note that a stochastic matrix is a special case of a nonnegative matrix, which is one whose elements p_{ij} are nonnegative. A positive matrix is a matrix P with $p_{ij} > 0$. This should not be confused with a positive-definite matrix. A stochastic matrix P has the following properties.

- P has an eigenvalue equal to $+1$.
- The absolute value of any eigenvalue of P is less than or equal to 1.

The first property follows from the fact that the row elements add up to 1, which we can write as

$$\begin{bmatrix} p_{11} & p_{12} & \cdots & p_{1n} \\ p_{21} & p_{22} & \cdots & p_{2n} \\ \vdots & \vdots & \vdots & \vdots \\ p_{n1} & p_{n2} & \cdots & p_{nn} \end{bmatrix} \begin{bmatrix} 1 \\ 1 \\ \vdots \\ 1 \end{bmatrix} = 1 \begin{bmatrix} 1 \\ 1 \\ \vdots \\ 1 \end{bmatrix}$$

showing that 1 is an eigenvalue with eigenvector $[1,1,\dots,1]^T$. For the second property, let

$$Pv = \lambda v,$$

where λ is an eigenvalue of P and v is an associated eigenvector. Comparing the ith row of both sides, we have

$$p_{i1}v_1 + \cdots + p_{in}v_n = \lambda v_i, \quad \text{for } i = 1, \dots, n.$$

Let $|v_k| = \max\{|v_1|, \dots, |v_n|\}$. Then

$$\begin{aligned} |\lambda||v_k| &= |p_{k1}v_1 + \cdots + p_{kn}v_n| \\ &\leq p_{k1}|v_1| + \cdots + p_{kn}|v_n| \\ &\leq p_{k1}|v_k| + \cdots + p_{kn}|v_k| \\ &\leq (p_{k1} + \cdots + p_{kn})|v_k| = |v_k|, \end{aligned}$$

which implies that $|\lambda| \leq 1$.

Example 10.2 ***(The Markov Mouse)***
As a second example, suppose a mouse is released into a maze with four cells (representing states) as illustrated in Figure 10.4.

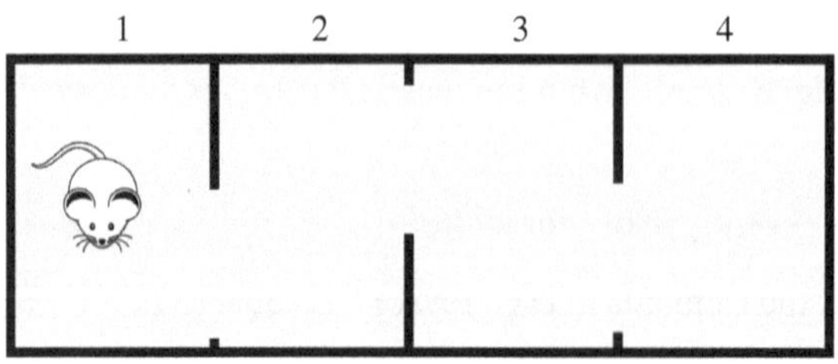

Figure 10.4 The Markov mouse.

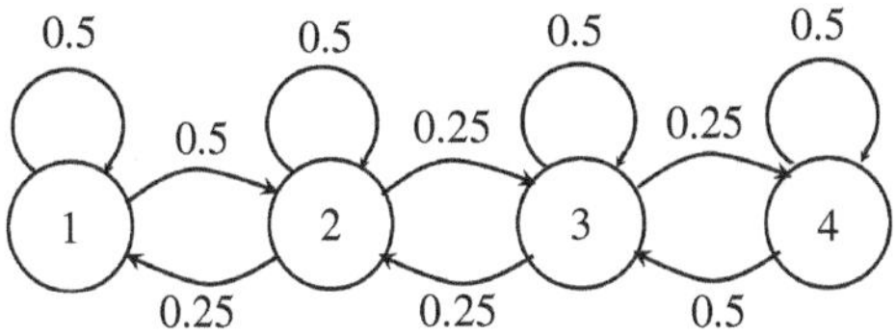

Figure 10.5 Transition diagram for the Markov mouse. The probabilities are calculated assuming that the mouse either stays or moves with equal probability 0.5. The probability of being in an adjacent cell at the next step depends on the number of cells adjacent to the current cell.

We assume that the mouse is a "Markov mouse," which means that it moves randomly from cell to cell and the probability of moving into a given cell depends only on where the mouse currently is and not how it reached its current cell. At each discrete time step the mouse may remain in its current cell or move to an adjacent cell with equal probability. If X_n is the location of the mouse (cells 1,2,3,4) at each discrete time step, then X_n is a random variable and the location of the mouse at time $n + 1$ depends only on its location at time n. Therefore, $\{X_n\}$ is a Markov chain.

Suppose that at time 0 the mouse is in state 1. Then, at time 2 the mouse could be either in state 1 with probability $p_{11} = 0.5$ or in state 2 with probability $p_{12} = 0.5$. The mouse cannot be in state 3 or 4, as it requires more than one time step to transition from state 1. Thus, $p_{13} = p_{14} = 0$.

Now, suppose the mouse is in state 2. It is easy to see that the state transition probabilities are $p_{22} = 0.5$, $p_{21} = p_{23} = 0.25$ and $p_{24} = 0$. The remaining probabilities are determined likewise. Thus, we can draw the transition diagram in Figure 10.5 to represent the movement of the mouse.

From the transition diagram, we can construct the associated transition matrix, P, as

$$P = \begin{bmatrix} 0.5 & 0.5 & 0 & 0 \\ 0.25 & 0.5 & 0.25 & 0 \\ 0 & 0.25 & 0.5 & 0.25 \\ 0 & 0 & 0.5 & 0.5 \end{bmatrix}.$$

We can follow the behavior of the mouse over time starting in a particular cell using the powers of the transition matrix. Suppose the mouse starts in cell 1 and set $S_0 = [1,0,0,0]$. Then the probability that the mouse is in cell i at step n is given by the ith entry of $S_n = S_0P^n$. For example, the first few steps are

$$S_1 = S_0 * P = [0.5, 0.5, 0,0]$$
$$S_2 = S_0 * P^2 = [0.375, 0.5, 0.125, 0]$$
$$S_3 = S_0 * P^3 = [0.3125, 0.4688, 0.1875, 0.0313]$$
$$S_4 = S_0 * P^4 = [0.2734, 0.4375, 0.2266, 0.0625].$$

The limiting value P^* of P^n is given by

$$P^* = \begin{bmatrix} 0.1667 & 0.3333 & 0.3333 & 0.1667 \\ 0.1667 & 0.3333 & 0.3333 & 0.1667 \\ 0.1667 & 0.3333 & 0.3333 & 0.1667 \\ 0.1667 & 0.3333 & 0.3333 & 0.1667 \end{bmatrix}.$$

Thus, after a long time, on average the mouse will spend 1/3 of its time in cells 2 and 3 and 1/6 of its time in cells 1 and 4. We can also check this result by computing the left eigenvector of P corresponding to the eigenvalue $\lambda = 1$.

The Markov mouse example

```
P=[1/2 1/2 0 0; 1/4 1/2 1/4 0; 0 1/4 1/2 1/4; 0 0 1/2 1/2];
[v,D,W]=eig(P);
S=W(:,1);
Sstar = S/sum(S);
```

Executing this script gives

$$\texttt{Sstar} = [0.1667,\ 0.3333,\ 0.3333,\ 0.1667].$$

10.1.1 Regular and Ergodic Markov Chains

The previous examples are examples of *regular Markov chains*, in which the transition matrices are so-called *primitive matrices.*

Definition 10.4 ***(Primitive Matrix)*** An $n \times n$ matrix P is a *primitive matrix* if P^k is a positive matrix for some $k > 0$.

Definition 10.5 ***(Regular Markov Chain)*** A Markov chain is *regular* if its transition matrix P is primitive, i.e., if there exists $k > 0$ such that P^k is a *positive matrix.*

Example 10.3 Consider the matrix

$$P_1 = \begin{bmatrix} 0.20 & 0.80 \\ 1 & 0 \end{bmatrix}.$$

It is easy to see that P_1 is primitive since

$$P_1^2 = \begin{bmatrix} 0.84 & 0.16 \\ 0.20 & 0.80 \end{bmatrix}.$$

Now consider the matrix

$$P_2 = \begin{bmatrix} 0 & 1 \\ 1 & 0 \end{bmatrix}.$$

Since $P_2^2 = I,\ P_2^3 = P_2, \ldots$ and so on, there is no power of P_2 that has all positive entries. Therefore, P_2 is not a primitive matrix.

The implication of the fact that the transition matrix P for a regular Markov chain has all positive elements for some k is that, over time, the Markov process will visit every state in the chain with a nonzero probability.

Theorem 10.1 *Let P be the transition matrix for a regular Markov chain. Then there is a unique stationary matrix S_* such that $S_* P = S_*$. Moreover, $P^* = \lim_k P^k$ is such that every row of P^* is equal to S_*. The element $S_j^* = \lim_{n\to\infty} p^n(i,j)$ is the long-run probability of going from state i to state j.*

We can get an idea of why this is true from the following argument. Note that the property that every row of P^* is equal to S_* implies that every column of P^* is a constant vector. Now, if y is any column of P, then Py replaces y by the average (mean) of its components. Thus, the components of Py are more similar (to each other) than those of y. As this process continues, $P^n y$ converges to a constant vector.

So, if $P^n \to P^*$, then $P^{n+1} = P^n P \to P^* P$. But $P^{n+1} \to P^*$ and so $P^* = P^* P$, which means that any row w of P^* satisfies $w = wP$. As a result, there is a unique probability vector satisfying the above and the convergence is independent of the initial state.

Definition 10.6 ***(Ergodic Markov Chain)*** A Markov chain is *ergodic*, also called *irreducible*, if it is possible to go from every state to every other state (not necessarily in one step).

It is easy to see that every regular Markov chain is ergodic, since all entries of its transition matrix are eventually positive, but an ergodic Markov chain may not be regular.

Example 10.4 In the Markov mouse example, suppose that the mouse will not stay in its current cell but will always move to an adjacent cell with equal probability. Then the transition diagram is as shown in Figure 10.6.

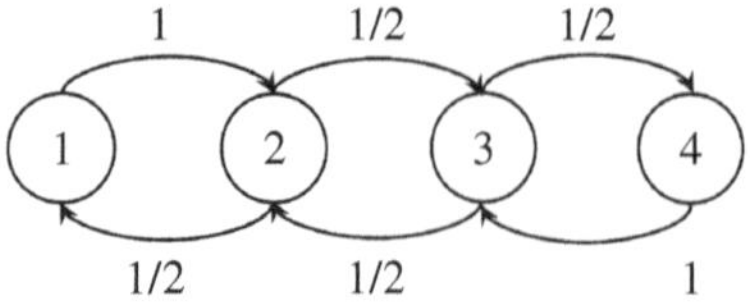

Figure 10.6 Transition diagram for the Markov mouse assuming that the mouse never stays in its current cell.

Since it is possible for the mouse to go from any cell to any other cell, the Markov chain is ergodic. In this case, the transition matrix is

$$P = \begin{bmatrix} 0 & 1 & 0 & 0 \\ 1/2 & 0 & 1/2 & 0 \\ 0 & 1/2 & 0 & 1/2 \\ 0 & 0 & 1 & 0 \end{bmatrix}.$$

Computing powers of P, we see that odd powers of P converge to

$$P^{\text{odd}} = \begin{bmatrix} 0 & 2/3 & 0 & 1/3 \\ 1/3 & 0 & 2/3 & 0 \\ 0 & 2/3 & 0 & 1/3 \\ 1/3 & 0 & 2/3 & 0 \end{bmatrix}$$

and even powers of P converge to

$$P^{\text{even}} = \begin{bmatrix} 1/3 & 0 & 2/3 & 0 \\ 0 & 2/3 & 0 & 1/3 \\ 1/3 & 0 & 2/3 & 0 \\ 0 & 2/3 & 0 & 1/3 \end{bmatrix}.$$

Hence the Markov chain is not regular. In this case, if n is odd, it is not possible for the mouse to go from cell 1 to cells 1 and 3 in n steps and, if n is even, it is not possible for the mouse to go from cell 1 to cells 2 and 4 in n steps.

As in the case of a regular Markov chain, an ergodic Markov chain always has a unique stationary matrix $S^* = S^*P$. However, unlike for regular Markov chains, the entries of the stationary matrix S^* do not necessarily give us the long-run probabilities for being in the various states.

Example 10.5 ***(Ehrenfest Model of Gaseous Diffusion)***
As another example, let us consider the Ehrenfest Markov chain model for gaseous diffusion [21]. Suppose we have a box consisting of two compartments, labeled 1 and 2, filled with a number of gas molecules and an opening where molecules can pass between the compartments (Figure 10.7).

Figure 10.7 Two compartments containing a number of gas molecules.

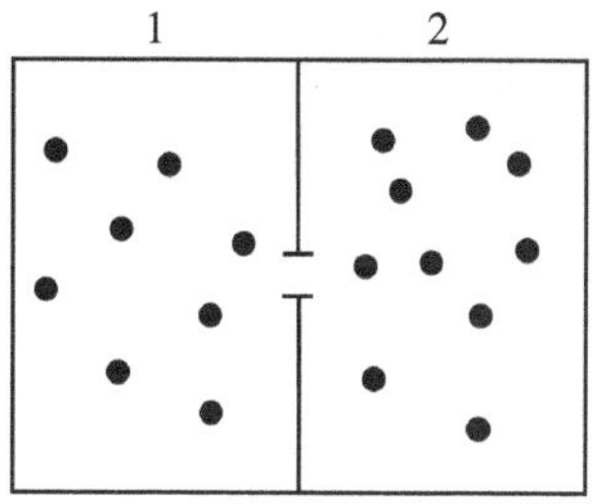

Figure 10.8 Transition diagram for the Ehrenfest model.

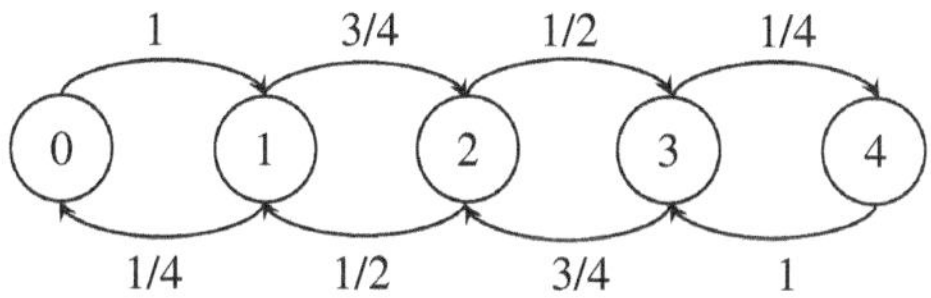

If we start with one compartment empty, we expect that over time the compartments will reach equilibrium with roughly half the number of molecules in each compartment. We can model this process as a Markov chain as follows. For purposes of illustration, let us assume that there are four molecules. We choose as states $S = (0, 1, 2, 3, 4)$ representing the number of molecules at time i in compartment 1. At each time step, we then choose one of the four molecules at random and move it from its current compartment to the other compartment. The transition diagram for this process is shown in Figure 10.8, which has transition probability matrix

$$P = \begin{bmatrix} 0 & 1 & 0 & 0 & 0 \\ 1/4 & 0 & 3/4 & 0 & 0 \\ 0 & 1/2 & 0 & 1/2 & 0 \\ 0 & 0 & 3/4 & 0 & 1/4 \\ 0 & 0 & 0 & 1 & 0 \end{bmatrix}.$$

This is an ergodic Markov chain and the stationary matrix is computed as $S^* = [0.0625, 0.2500, 0.3750, 0.2500, 0.0625]$ (Problem 10.2).

The interpretation of S^* as that, at any time step in the long term, there will be a 6.25% chance that compartment 1 will be empty, a 25% chance that compartment 1 will contain a single molecule, a 37.5% chance that compartment 1 will contain 2 molecules and so on.

Figure 10.9 plots the number of molecules in compartment 1 over time, starting with 0 molecules in compartment 1 and 1000 molecules in compartment 2. We see that the number of molecules in compartment 1 rises quickly toward an equilibrium value of 500 and fluctuates randomly about this value.

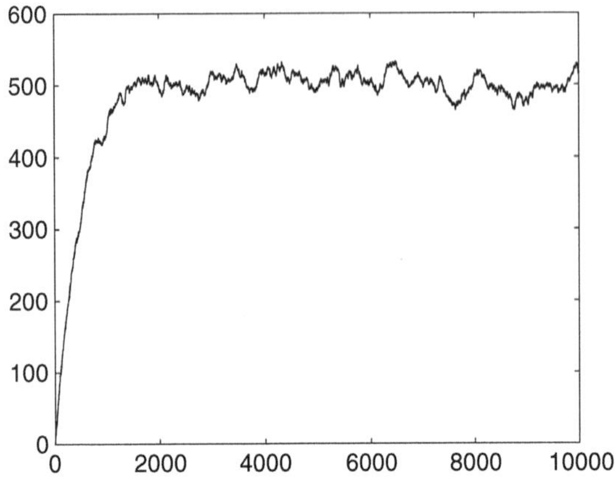

Figure 10.9 Number of gas molecules in compartment 1 over time. The number fluctuates randomly about the equilibrium value.

Gaseous Diffusion Example

```
NUM_MOLECULES = 1000;          % number of gas molecules
NUM_STEPS = 10*NUM_MOLECULES; % number of time steps
COMPARTMENT = zeros(1,NUM_MOLECULES);
% Initialize the number of molecules in compartment 1
NUM1=[sum(COMPARTMENT)];
for j=1:NUM_STEPS
i=ceil(NUM_MOLECULES*rand);% choose a molecule at random
                           % switch molecule i to the other
                           % compartment
COMPARTMENT(i)=rem(COMPARTMENT(i)+1,2);
NUM1=[NUM1 sum(COMPARTMENT)]; % number of molecules in
                              % compartment 1 at step j
end
plot(0:NUM_STEPS,NUM1)
```

10.1.2 Absorbing Markov Chains

We next discuss *absorbing states* and *absorbing Markov chains.* A state S is called *absorbing* if once entered, it cannot be left.

Definition 10.7 ***(Absorbing Markov Chain)*** A Markov Chain is *absorbing* if:

1. there exists at least one absorbing state; and
2. it is possible to go from each nonabsorbing state to at least one absorbing state in finitely many steps.

As an example, the Markov chain in Figure 10.10 is absorbing. State A is absorbing and it is possible to reach A from each state B, C, and D.

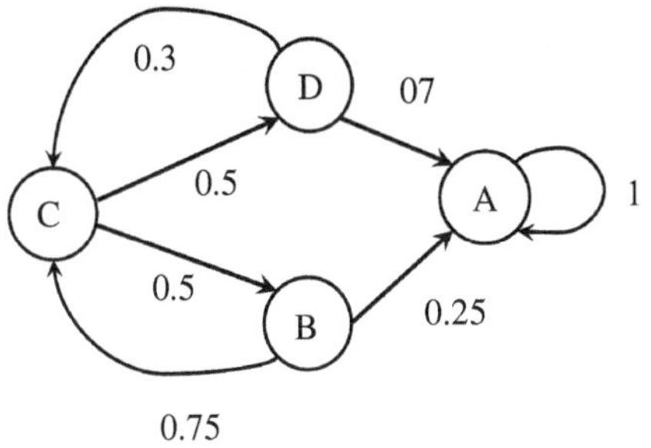

Figure 10.10 An example of an absorbing Markov chain. Once the state reaches A, it can never leave it.

Isomorphic Chains and Standard Forms

Consider the two Markov chains in Figure 10.11 with transition matrices

$$P_1 = \begin{bmatrix} 0.8 & 0.2 & 0 \\ 0 & 1 & 0 \\ 0.1 & 0 & 0.9 \end{bmatrix}, \quad P_2 = \begin{bmatrix} 1 & 0 & 0 \\ 0.2 & 0.8 & 0 \\ 0 & 0.1 & 0.9 \end{bmatrix}.$$

Clearly, these two transition diagrams represent the same Markov chain – we have just labeled the states differently. An important question in general is, therefore, to determine when two transition diagrams represent the same Markov chain. Such transition diagrams are called *isomorphic*. We will say more about graph isomorphism in the next chapter.

Note that in the above example we can consider the diagram labels on the left (D, A, C) as a *permutation* of the labels (A, B, C) on the right.

Definition 10.8 ***(Permutation Matrix)*** A *permutation matrix* is a matrix formed from the $n \times n$ identity matrix by permuting the rows.

Example 10.6 Suppose $\pi : \{A, B, C\} \to \{A, B, C\}$ is a permutation given by

$$\pi(A, B, C) = (B, A, C).$$

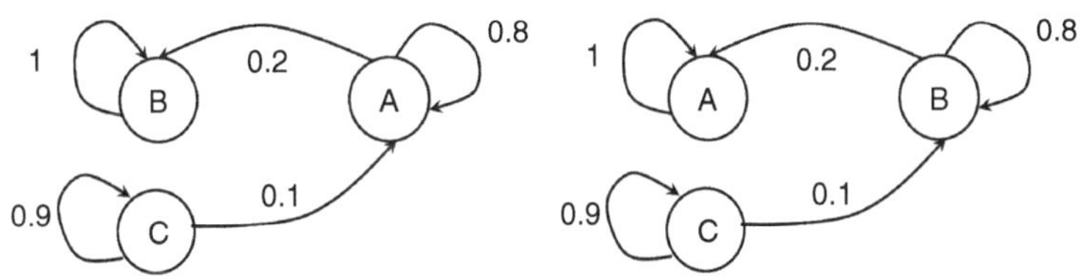

Figure 10.11 An example of isomorphic Markov chains.

We see that the permutation switches A and B and leaves C fixed. We can represent this permutation by a matrix obtained from the 3×3 identity matrix by switching the first two rows and leaving the third row fixed, resulting in the permutation matrix

$$\Pi = \begin{bmatrix} 0 & 1 & 0 \\ 1 & 0 & 0 \\ 0 & 0 & 1 \end{bmatrix}.$$

We note that permutation matrices are orthogonal, i.e., $\Pi^{-1} = \Pi^T$.

Theorem 10.2 *Suppose P_1 and P_2 are transition matrices associated with given Markov chains. Then the Markov chains are equivalent (said to be* isomorphic*) if and only if there exists a permutation matrix Π such that*

$$\Pi P_1 \Pi^T = P_2.$$

Returning to the previous example, it is straightforward now to see that

$$\begin{bmatrix} 0 & 1 & 0 \\ 1 & 0 & 0 \\ 0 & 0 & 1 \end{bmatrix} \begin{bmatrix} 0.8 & 0.2 & 0 \\ 0 & 1 & 0 \\ 0.1 & 0 & 0.9 \end{bmatrix} \begin{bmatrix} 0 & 1 & 0 \\ 1 & 0 & 0 \\ 0 & 0 & 1 \end{bmatrix} = \begin{bmatrix} 1 & 0 & 0 \\ 0.2 & 0.8 & 0 \\ 0 & 0.1 & 0.9 \end{bmatrix}.$$

Next, we state without proof that any transition matrix for an absorbing Markov chain can be put into a particular form, called a *standard form*, by a suitable permutation transformation. Let P be the $N \times N$ transition matrix for an absorbing Markov chain. Then there is an $N \times N$ permutation matrix Π such that

$$\Pi P \Pi^T = \left[\begin{array}{c|c} I & 0 \\ \hline R & Q \end{array}\right].$$

The rows of the block submatrix I correspond to all absorbing states; the rows of R correspond to the nonabsorbing states. For the proof, see [71].

Example 10.7 Suppose that P is given by

$$P = \begin{array}{c} A \\ B \\ C \\ D \end{array} \begin{bmatrix} 0 & 0.3 & 0.3 & 0.4 \\ 0 & 1 & 0 & 0 \\ 0 & 0 & 1 & 0 \\ 0.8 & 0.1 & 0.1 & 0 \end{bmatrix}.$$

Then B, C are absorbing states and A, D are nonabsorbing states. Thus, the appropriate permutation of states is $(A, B, C, D) \to (B, C, A, D)$. Therefore, with permutation matrix

$$\Pi = \begin{bmatrix} 0 & 1 & 0 & 0 \\ 0 & 0 & 1 & 0 \\ 1 & 0 & 0 & 0 \\ 0 & 0 & 0 & 1 \end{bmatrix},$$

we get

$$\Pi P \Pi^T = \begin{bmatrix} 1 & 0 & 0 & 0 \\ 0 & 1 & 0 & 0 \\ 0.3 & 0.3 & 0 & 0.4 \\ 0.1 & 0.1 & 0.8 & 0 \end{bmatrix}$$

which is in standard form with

$$R = \begin{bmatrix} 0.3 & 0.3 \\ 0.1 & 0.1 \end{bmatrix} \quad \text{and} \quad Q = \begin{bmatrix} 0 & 0.4 \\ 0.8 & 0 \end{bmatrix}.$$

Using the standard form for the transition matrix P, we can now compute the limiting matrix $\overline{P} = \lim_k P^k$. Using properties of block matrix multiplication, it is easy to show that

$$P^k = \left[\begin{array}{c|c} I & 0 \\ \hline (I + Q + \cdots + Q^k)R & Q^k \end{array}\right].$$

Since P is a stochastic matrix, it follows that the eigenvalues of Q are strictly less than 1in absolute value, and, therefore $Q^k \to 0$ as $k \to \infty$.

Also the series $I + Q + \cdots + Q^k$ converges to $(I - Q)^{-1}$ as $k \to \infty$. Therefore, we have

$$\overline{P} = \lim_{k \to \infty} P^k = \left[\begin{array}{c|c} I & 0 \\ \hline FR & 0 \end{array}\right],$$

where $F = (I - Q)^{-1}$. The matrix $F = (I - Q)^{-1}$ is called the *fundamental matrix*. Some properties of the limiting matrix $\overline{P}$ are as follows.

1. The element $\overline{p}_{ij}$ of $\overline{P}$ is the long-run probability of going from state i to state j.
2. The sum of the entries in each row of the fundamental matrix F is the average number of trials it will take to go from each nonabsorbing state to some absorbing state.

Example 10.8 Suppose that a bank classifies loans into four categories:

- F, paid in full and removed from the books;
- G, in good standing;
- A, in arrears;
- B, a bad debt and turned over to a collection agency.

Past records indicate that each month 10% of accounts in good standing will be paid in full; 80% will remain in good standing; and 10% will become in arrears. For accounts in arrears, 10% are paid in full, 40% remain in arrears, and 10% are classified as bad debts. The states F and B are clearly absorbing states, since once an account is put into one of these categories no further action is necessary and the account remains in the respective state. The transition diagram for this problem is therefore shown in Figure 10.12.

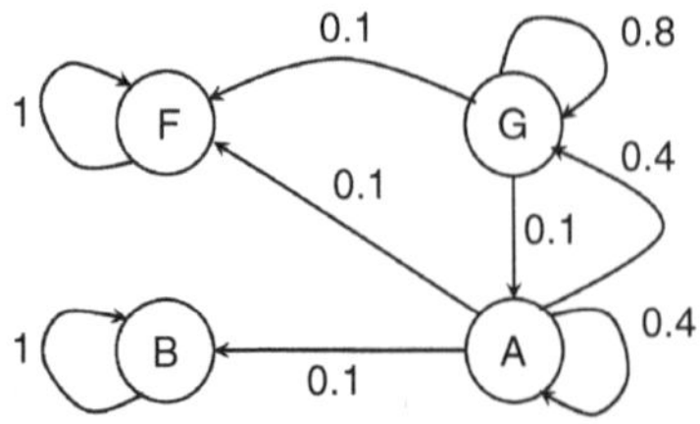

Figure 10.12 Transition diagram for Example 10.8.

We can see that it is possible to go from any nonabsorbing state to an absorbing state in finitely many steps, so that the Markov chain is absorbing. The transition matrix for this example in standard form is

$$\begin{array}{c} \\ F \\ B \\ G \\ A \end{array}\begin{array}{c} \begin{array}{cccc} F & B & G & A \end{array} \\ \left[\begin{array}{cccc} 1 & 0 & 0 & 0 \\ 0 & 1 & 0 & 0 \\ \hline 0.1 & 0 & 0.8 & 0.1 \\ 0.1 & 0.1 & 0.4 & 0.4 \end{array}\right] \end{array}$$

and thus

$$R = \begin{bmatrix} 0.1 & 0 \\ 0.1 & 0.1 \end{bmatrix}, \quad Q = \begin{bmatrix} 0.8 & 0.1 \\ 0.4 & 0.4 \end{bmatrix}.$$

Therefore, we can compute F and FR as

$$F = (I - Q)^{-1} = \begin{bmatrix} 7.5 & 1.25 \\ 5 & 2.5 \end{bmatrix}, \quad FR = \begin{bmatrix} 0.875 & 0.125 \\ 0.75 & 0.25 \end{bmatrix}.$$

From the entries of FR we conclude that 75% of the accounts in arrears eventually pay in full and 25% of the accounts in arrears become classified as bad debt. Likewise, 12.5% of accounts in good standing become classified as bad debt.

By examining F we see that accounts remain in arrears on average for 7.5 months and accounts remain in good standing on average for 8.75 months.

Example 10.9 Rafael Nadal is one of the top tennis players in the world. Analysis of past tournaments reveals that in games where the score is deuce, he wins the next point 60% of the time, and when he has the advantage, he wins the game 70% of the time. In addition, when his opponent has the advantage, Nadal loses the game 25% of the time. With these data we can create a Markov chain model with the transition diagram shown in Figure 10.13 and transition matrix

$$P = \left[\begin{array}{cc|ccc} 1 & 0 & 0 & 0 & 0 \\ 0 & 1 & 0 & 0 & 0 \\ \hline 0 & 0 & 0 & 0.6 & 0.4 \\ 0.7 & 0 & 0.3 & 0 & 0 \\ 0 & 0.25 & 0.75 & 0 & 0 \end{array}\right]$$

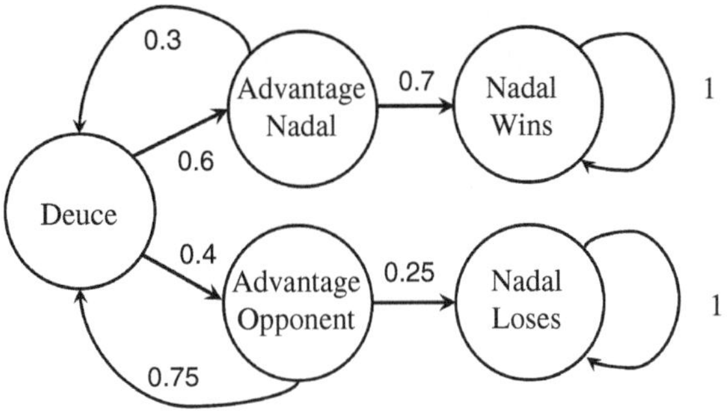

Figure 10.13 Transition diagram for the tennis example. The data are fictitious and for example only.

Therefore, with R and Q given by

$$R = \begin{bmatrix} 0 & 0 \\ 0.7 & 0 \\ 0 & 0.25 \end{bmatrix}, \quad Q = \begin{bmatrix} 0 & 0.6 & 0.4 \\ 0.3 & 0 & 0 \\ 0.75 & 0 & 0 \end{bmatrix},$$

we can compute F and FR as

$$F = (I - Q)^{-1} = \begin{bmatrix} 1.9231 & 1.1538 & 0.7692 \\ 0.5769 & 1.3462 & 0.2308 \\ 1.4423 & 0.8654 & 1.5769 \end{bmatrix}, \quad FR = \begin{bmatrix} 0.8077 & 0.1923 \\ 0.9423 & 0.0577 \\ 0.6058 & 0.3942 \end{bmatrix}.$$

It is left as an exercise (Problem 10.4) to give an interpretation of the entries of F and FR.

10.2 Monte Carlo Methods

Monte Carlo methods refer to computational algorithms that use repeated random sampling to obtain numerical results. They are often used in problems where exact deterministic solutions are difficult or impossible to obtain. Monte Carlo methods were first developed by the mathematician Stanslaw Ulam for use in nuclear weapons development. The name comes from the Monte Carlo casino in Monaco.

Applications

Monte Carlo methods are mainly used in three application areas: optimization, numerical integration, and simulation, and have numerous applications in many areas of science, engineering, mathematics, and business.

In physical sciences. Monte Carlo methods are important in computational biology, computational physics, physical chemistry, and related applied fields. In statistical physics Monte Carlo molecular modeling is an alternative to computational molecular dynamics.

In computer graphics. Path tracing, occasionally referred to as Monte Carlo ray tracing, renders a three-dimensional scene by randomly tracing samples of possible light paths. Repeated sampling of any given pixel will eventually cause the average of the samples to converge to the correct solution of the rendering equation, making it one of the most physically accurate three-dimensional graphics rendering methods available.

In finance. Monte Carlo simulation is commonly used to evaluate the risk and uncertainty that would affect the outcome of different decision options.

10.2.1 Random Number Generation

The development of Monte Carlo methods stimulated great interest in ways to generate sequences of random numbers. Computer-based random number generators are generally pseudo-random number generators (PRNGs), which means that they are generated by deterministic algorithms but can seem random based on standard statistical tests for randomness. A sequence of numbers generated by a PRNG will eventually repeat. The sequence of values generated is determined by a fixed number called a *seed*. If the seed is known, then the sequence of values can be recomputed exactly, which is useful in certain applications.

An early PRNG is the *linear congruential generator*, which uses a recurrence relation

$$X_{n+1} = (aX_n + b) \bmod m$$

to generate the sequence $\{X_1, X_2 \dots\}$, where a, b, and m are large integers. The maximum number of numbers the formula can produce is $m - 1$, one less than the modulus. The performance of this generator is highly sensitive to the choice of parameters a and m. When $b \neq 0$, correctly chosen parameters allow a period equal to m for all seed values. This will occur if and only if:

1. m and b are relatively prime;
2. $a - 1$ is divisible by all prime factors of m;
3. $a - 1$ is divisible by 4 if m is divisible by 4.

These three requirements are referred to as the Hull-Dobell theorem. The Matlab script below will generate a sequence of N random numbers between 0 and 1 starting with a given seed.

Generate a sequence of *N* random numbers

```
a=1664525; b=1013904223; m=2^32;
N = input('Enter the length of the sequence: ');
seed = input('Enter the value of the seed: ');
x = zeros(N+1,1);
x(1) = seed;
for i=1:N
    x(i+1)=mod((a*x(i)+b),m);
end
seq = x/1.0e+09;
seq(1) = [];
seq = seq-floor(seq);
```

Running this script with $N = 8$ gives the sequences shown in Table 10.1.

Several other pseudo-random number generators have been developed, such as the Mersenne twister, so-called because it has a period defined by a Mersenne prime $M_n = 2^n - 1$.

Matlab's `rand()` pseudo-random number generator is, in fact, based on the Mersenne twister algorithm. The basic idea is to define a series x_i through a recurrence relation, and then output numbers of the form $x_i T$, where T is an invertible binary matrix, meaning the elements of T are 0s and 1s, called a tempering matrix.

The Matlab function `rand()` generates random numbers from the uniform distribution on [0,1] with

$$p(x) = \begin{cases} 1, & \text{if } 0 \leq x \leq 1 \\ 0, & \text{otherwise.} \end{cases}$$

To generate a sequence of random numbers from a different distribution p we can employ the following procedure.

1. Compute the cumulative distribution function $y = F(x) = \int_{-\infty}^{x} \overline{p}(x)dx$.
2. Determine the inverse function $F^{-1}(y)$.

Table 10.1 Pseudo-random sequences of numbers between 0 and 1, with different seeds.

Seed	Pseudo-random sequence of numbers
1.2	0.0159, 0.6639, 0.0359, 0.2003, 0.2255, 0.3704, 0.9799, 0.0348
3.4	0.0196, 0.5210, 0.4927, 0.6910, 0.0740, 0.2815, 0.0397, 0.6655
7.5	0.0264, 0.9707, 0.9535, 0.9134, 0.7906, 0.9863, 0.3277, 0.1482

3. Generate a uniform random number y in the interval [0,1].
4. Then $z = F^{-1}(y)$ is distributed according to $\overline{p}$.

Example 10.10 Suppose we want to generate a sequence of random numbers that is uniformly distributed on the interval $[a, b]$. In this case

$$\overline{p}(x) = \begin{cases} 1, & \text{if } a \leq x \leq b \\ 0, & \text{otherwise.} \end{cases}$$

Thus the cumulative distribution is

$$F(x) = \frac{x - a}{x - b}.$$

Computing F^{-1}, we get $z = a + (b - a)y$, where y is uniformly distributed on [0,1].

As another example consider the exponential distribution,

$$\overline{p}(x) = \begin{cases} \lambda \exp(-\lambda x), & x \geq 0 \\ 0, & x < 0 \end{cases}$$

The cumulative distribution function is given by

$$F(x) = 1 - \exp(-x/\lambda),$$

and by inverting $F(x)$ we obtain $z = -\lambda \ln(1 - y)$. $1 - y$ is uniformly distributed since y is uniformly distributed. Therefore, we can take $z = -\lambda \ln(y)$, which has mean λ.

Example 10.11 A simple example of the function `rand()` is the following script that counts the number of heads in N flips of a fair coin.

Coin flip example

```
N=20;                    % number of flips
r=2.*rand(N,1);          % generate N random numbers
                         % uniformly distributed between 0 and 2
HT=floor(r);             % changes the random numbers to a
                         % sequence of zeros and ones,
                         % where 1 corresponds to heads
num_heads=0;             % initialize the counter
for i=1:size(HT)
num_heads=num_heads+HT(i);   % count the number of heads
end
```

Example 10.12 As another example, consider a quarter circle inscribed in a unit square. Since the area of the quarter circle is $\frac{\pi}{4}$, we can estimate the value of π with the following experiment.

1. Distribute points in the unit square at random.
2. Count the fraction of points inside the quarter circle.
3. Multiply this fraction by 4.

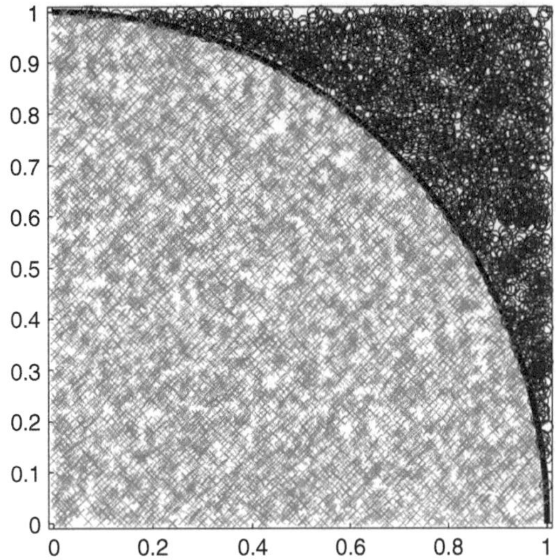

Figure 10.14 Counting the fraction of points inside the quarter circle gives an estimate of π.

The approximation gets closer to the value of π as the number of points increases (Figure 10.14).

Monte Carlo Approximation of Pi

```
N=1000; % number of samples
xval=rand(N,1);
yval=rand(N,1);
counter=0;
for i=1:N
    norm_val = xval(i)^2+yval(i)^2;
 if norm_val<1.0
    counter=counter+1;
    plot(xval(i),yval(i),'Marker','x','Color','#A3A3A3')
 else
    plot(xval(i),yval(i),'Marker','o','Color','#303030')
 end
end
pi_estimate=4*counter/N;
```

10.2.2 Monte Carlo Integration

An important application of Monte Carlo methods is in numerical integration. Deterministic techniques for numerical integration, such as the trapezoid approximation, work well for functions $f(x)$ of a single variable. However, for higher-dimensional problems, these deterministic methods suffer from the *curse of dimensionality*, meaning that the convergence rate becomes exponentially worse as the dimension of the integral increases, requiring N^n samples for an n-dimensional integral of the form

$$\int \cdots \int f(x_1, \ldots, x_n) dx_1 \cdots dx_n.$$

The convergence rate of Monte Carlo integration, on the other hand, is independent of the dimension, which makes it the only practical integration method for very large problems, such as those arising in computer graphics.

To understand how Monte Carlo integration works, consider the one-dimensional problem

$$F = \int_0^1 f(x)dx, \quad \text{where } f : \mathbb{R} \to \mathbb{R}.$$

Let $X = \text{rand}(N, 1)$ be a sequence of random numbers chosen from the uniform distribution on [0,1], and define

$$F_N = \frac{1}{N} \sum_{i=1}^{N} f(X_i).$$

Then the expected value of the random variable F_N is given by

$$\begin{aligned} E[F_N] &= E\left[\frac{1}{N}\sum_{i=1}^{N} f(X_i)\right] = \frac{1}{N}\sum_{i=1}^{N} E[f(X_i)] \\ &= \frac{1}{N}\sum_{i=1}^{N} \int_0^1 f(x)P(x)dx = \frac{1}{N}\sum_{i=1}^{N} \int_0^1 f(x)dx \\ &= \int_0^1 f(x)dx = F. \end{aligned}$$

Moreover, F_N converges to F with probability equal to 1 as $N \to \infty$. Also the standard deviation σ is proportional to $\frac{1}{\sqrt{N}}$, which means that to reduce the error by one half requires four times as many points N.

Example 10.13 Consider the problem

$$F = \int_0^1 \int_0^1 \int_0^1 \frac{xy^2}{z+1} dxdydz,$$

which is equal to $\ln(2)/6 = 0.1155$. The Matlab script below approximates this integral.

Monte Carlo Integration

```
f=@(x,y,z) x.*y.^2./(z+1);
N=1000;
x=rand(N,1);
```

```
y=rand(N,1);
z=rand(N,1);
I = sum(f(x,y,z))/N;
```

10.2.3 Monte Carlo Optimization

There are several optimization methods that are based on the use of randomized samples, such as simulated annealing and genetic algorithms. A study of these methods is beyond the scope of this text. We can give a simple example of a one-dimensional optimization problem using Monte Carlo methods. Consider the one-dimensional problem

$$\text{minimize } f(x)$$
$$\text{subject to } a \le x \le b.$$

1. Generate N random numbers uniformly in $[a, b]$.
2. Evaluate $f(x)$ at these random numbers.
3. Pick the smallest value of $f(x)$ from this set as the estimated minimum.

Example 10.14 Consider the problem

$$\begin{aligned} &\text{minimize} && f(x) = (x-1)^2 + 4 \\ &\text{subject to} && 0 \le x \le 2. \end{aligned}$$

This problem has a global minimum at $x^* = 1$ with $f(x^*) = 4$. The Matlab script below generates 1000 random numbers between 0 and 2 and uses them to estimate the minimum value of f.

Monte Carlo Optimization

```
f=@(x) (x-1).^2 +4;
x=2*rand(1000,1);
[y,i]=min(f(x));
% The optimum is at
x_star = x(i);f_star = f(x(i));
```

10.2.4 Monte Carlo Simulation

Monte Carlo simulation is useful to study uncertainty in systems [35]. A wide-spread use of Monte Carlo simulation is in stock portfolio analysis to predict the future returns of various groups of stocks.

Example 10.15 ***(Stock Price Prediction)***
Let $P(i)$ be the price of a certain stock on day i. The uncertainty in daily stock price can be modeled by taking P as a random variable with model

$$P(i+1) = P(i)e^{\sigma r}$$

where σ is a standard deviation determined from historical daily averages of the stock, and r is a random number. A Monte Carlo prediction of the future price of the stock is generated by simulation this model over a number of days and averaging the results. The Matlab script below simulates this model and produces the response in Figure 10.15.

Monte Carlo Stock Price Simulation

```
NUM_ITERATIONS=50; % number of simulation runs
NUM_DAYS=100;      % number of days
AVERAGE_PRICE=0;
hold on
for i=1:NUM_ITERATIONS
PRICE = 1;         % initial price
DAILY_PRICE=PRICE;
for i=1:NUM_DAYS
a= rand(1);
sig=0.02;
PRICE=PRICE*exp(sig*rand);
DAILY_PRICE=[DAILY_PRICE,PRICE];
end
plot(DAILY_PRICE,':','Color','k')
AVERAGE_PRICE=AVERAGE_PRICE+DAILY_PRICE;
end
AVERAGE_PRICE=AVERAGE_PRICE/50;
plot(AVERAGE_PRICE,'Linewidth',1,'Color','k')
```

Example 10.16 As a second example, consider the SIR epidemic model

$$\dot{S} = -\frac{\beta}{K}S + bI$$
$$\dot{I} = \frac{\beta}{K}S + bI$$
$$\dot{R} = bI,$$

and suppose that we do not know the parameters β and b exactly. If we assume that β and b are normally distributed, then we can generate random samples from the normal distribution and average multiple simulation runs to predict the

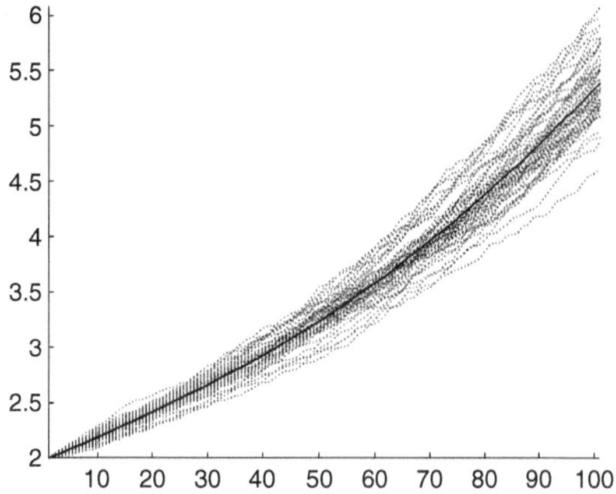

Figure 10.15 Monte Carlo simulation to predict stock prices. Fifty random simulations over 100 days are averaged to produce a predicted price trend.

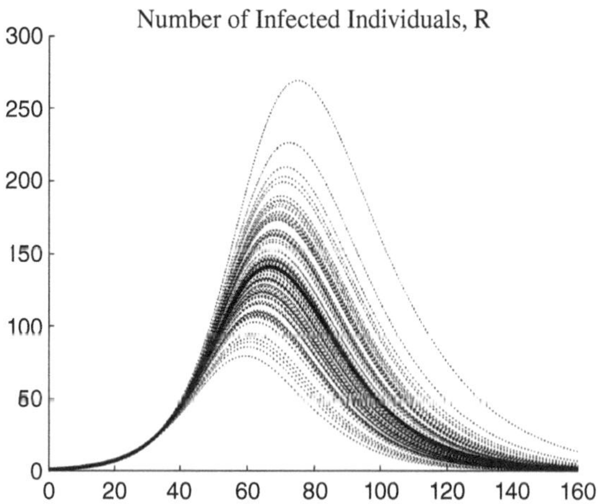

Figure 10.16 Monte Carlo simulation of the SIR model showing the estimated number of infected individuals, assuming that the parameters β and b are unknown but normally distributed relative to nominal values.

values. Figure 10.16 shows the number of infected individuals, I, averaged over 10 simulation runs.

Remark 10.1 Matlab can be used to sample from a given probability distribution using the commands `makedist` and `random`. For example:

Sampling from a Probability Distribution

```
pd = makedist('Normal','mu',0.25,'sigma',0.5);
rng('default');
r=random(pd,100,1);
```

generates 100 samples from a normal distribution with mean `mu` = 0.25 and standard deviation `sigma` = 0.5. Likewise,

Sampling from a Probability Distribution

```
pd = makedist('Poisson','lambda',4);
rng('default');
r=random(pd,100,1);
```

generates 100 samples from a Poisson distribution with mean `lambda` = `4`. Type `list = makedist` to see a list of distributions that are supported in this way.

Problems

10.1 Draw the transition diagram for the Markov mouse in the case where the mouse will not stay in the same cell but always move to an adjacent cell with equal probability.

10.2 Verify that the steady-state matrix S^* in the Ehrenfest Markov chain model is

$$S^* = [0.0625, 0.2500, 0.3750, 0.2500, 0.0625]$$

10.3 Suppose the matrices

$$P_1 = \begin{bmatrix} 1 & 0 & 0 \\ 0 & 1/2 & 1/2 \\ 1/3 & 1/3 & 1/3 \end{bmatrix}, \; P_2 = \begin{bmatrix} 0 & 1/2 & 1/2 \\ 1/3 & 1/3 & 1/3 \\ 1/2 & 1/2 & 0 \end{bmatrix},$$

$$P_3 = \begin{bmatrix} 1/3 & 1/3 & 1/3 \\ 1/4 & 0 & 3/4 \\ 0 & 1 & 0 \end{bmatrix}, \; P_4 = \begin{bmatrix} 1/3 & 1/3 & 1/3 \\ 1/4 & 0 & 3/4 \\ 0 & 0 & 1 \end{bmatrix},$$

$$P_5 = \begin{bmatrix} 0 & 1 & 0 \\ 1 & 0 & 0 \\ 1/4 & 1/2 & 1/4 \end{bmatrix}$$

represent transition probability matrices for Markov chains. Draw the transition diagrams corresponding to each matrix. Classify each Markov chain as either regular, ergodic, or absorbing. For each of the matrices, compute the limiting case $\lim_{k\to\infty} P^k$ and interpret the results.

10.4 Give an interpretation of the matrices F and FR in the Rafael Nadal example. For example, if the score is deuce, how many points on average does he play before he eventually wins (or loses).

10.5 The summer weather in Texas can be hot, mild, or rainy. The rules are:

- There are never two mild days in a row.
- On a mild day, there is an equal probability of heat or rain the next day.
- If it is hot or rainy, there is an equal chance of the same thing the next day.
- If there is a change from hot or rainy, only half the time is it a change to a mild day.

1. With states $\{h, m, r\}$, draw a transition diagram and compute the transition matrix.
2. Is this a regular Markov chain? Explain why or why not.
3. Compute the stationary matrix S^*. What does this tell you about the long-term weather of Texas.

10.6 In the Texas weather problem, change the row of the transition matrix corresponding to h to [1,0, 0]. This makes h an absorbing state.

1. Write the transition matrix in standard form.
2. Computer the fundamental matrix F and also compute FR.
3. Interpret the results.

10.7 Professor Johnson tells student A that the final exam will be closed book. Then student A relays the information to student B, who passes it to student C, ... There is a 10% chance that a student will change "closed book" to "open book" and a 10% chance that a student will change "open book" to "closed book" when communicating the information to a fellow student.

1. Define states, draw a transition diagram, and compute the transition matrix.
2. After the information has passed to 50 students, what is the probability that student number 51 will get the correct information?

10.8 Professor Weiner is famous for being absent minded. Suppose he divides his time among his office (O), his lab (L), the library (B), and the cafeteria (C). Suppose that the cafeteria is in the library so one must pass through the library to get to the cafeteria. Assume the following transition probabilities:

- between his office and lab, $p_{OL} = 1/3, p_{LO} = 1$
- between his office and the library, $p_{OB} = 2/3, p_{BO} = 2/3$
- between the library and cafeteria, $p_{BC} = 1/3$, and $p_{CB} = 1$.

Note that, since Professor Weiner is absent minded, after he returns from the lunch at the cafeteria, there is a $1/3$ chance that he will forget and go back to the cafeteria from the library. Note also that, from his lab, he always returns to his office.

1. Draw the transition diagram and write the transition probability matrix P.
2. In the long run, how much of his time does he spend in each building?

10.9 Now suppose that once Professor Weiner reaches the cafeteria, he remains there the rest of the day.

1. Redraw the transition diagram and recompute the transition probability matrix P.
2. Put the transition probability matrix P in standard form and compute the long-run probability matrix $\overline{P} = \lim_{k\to\infty} P^k$.
3. Interpret the results. What does $\overline{P}$ tell you about where you can find him?

10.10 Suppose you are tracking the stock price of a certain company. You know that if the stock price rises today there is a 70% chance that it will rise tomorrow and a 30% chance that it will fall tomorrow. If the stock price falls today there is a 40% chance that it will fall tomorrow and a 60% chance that it will rise tomorrow.

(a) Define states and represent this as a Markov chain, i.e., draw the transition diagram and compute the transition probability matrix.
(b) Compute by hand the steady-state matrix S^*. Do not use Matlab. Show the details of your calculation.

10.11 We note that

$$\int_0^1 \frac{4}{1+x^2} dx = \pi.$$

Using Monte Carlo integration estimate the value of π to two decimal places.

10.12 Solve the following optimization problem using Monte Carlo simulation:

$$\begin{aligned} \text{minimize} \quad & (x_1^3 - 5)^2 + (x_2 - 3)^2 \\ \text{subject to} \quad & 0 \le x_1 \le 2 \\ & 0 \le x_2 \le 1. \end{aligned}$$

Verify the solution using `fmincon`.

11

Feedback Systems

11.1 Introduction

In this chapter we discuss feedback and control systems [9, 10]. A history of engineering by Frederick Kilgour [45] describes the development of automatic control in the twentieth century as "[one] of the great events of history which totally changed the ways of human life." Indeed, our present age is the age of automatic control. Factory automation, robots, autonomous vehicles, drones, medical devices, and a host of other products, processes, and applications of sensor-based feedback control are profoundly changing the ways we live, work, play, and interact.

Feedback is also one of the most fundamental properties in natural systems in addition to engineered systems. In biological systems, feedback is present in regulating many processes, from body temperature, to cell metabolism, gene expression, hormone production, as well as balance and locomotion. In weather and climate dynamics, feedback is an important mechanism affecting global temperatures, weather patterns, ocean currents, and so on.

We recall our definition of a (closed-loop) feedback system from Chapter 1 as "a system where the input is influenced by the output" (Figure 11.1)

Identifying when feedback occurs in a particular system may not be easy at first glance. For example, a thermostat that measures room temperature and turns the heating, ventilation, and air conditioning system on and off to maintain a constant temperature is an obvious use of feedback. On the other hand, it may be less obvious that feedback is naturally occurring in the predatory-prey system, where a change in the population of one species affects the change in the population of another species. The logistic model of population growth is an example of *positive* feedback initially as the growth of a population causes further growth, followed by *negative* feedback as saturation occurs and growth is inhibited.

Introduction to Modeling and Simulation: A Systems Approach, First Edition. Mark W. Spong.

Companion Website: www.wiley.com/go/spong/engineeringmodeling

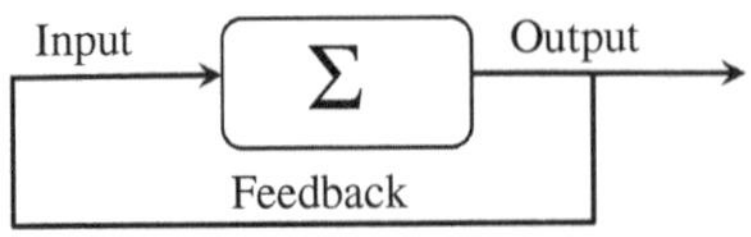

Figure 11.1 A closed-loop feedback system where the output affects future inputs.

The Effects of Feedback

Consider a simple example of a static system defined by $y = 0.1u$ and suppose we want the output y to equal a reference input r. By careful calibration, we can take $u = 10r$ so that $y = r$ as desired. Now, suppose that there is an unknown disturbance $-d$ acting on the system, so that $\bar{y} = 0.1(10r - d) = r - 0.1d$. The percent error is given by

$$\% \text{ error} = \frac{y - \bar{y}}{y} \times 100 = \frac{r - (r - 0.1d)}{r} \times 100 = 10\frac{d}{r}$$

so, for example, if $r = 50$ and $d = 10$, the percent error is 2%.

Now, suppose that we employ feedback by taking $u = ke$, where $e = r - y$ is the error between the reference input r and the output y. Now we have

$$\bar{y} = 0.1(ke - d) = 0.1k(r - \bar{y}) - 0.1d = 0.1kr - 0.1k\bar{y} - 0.1d.$$

Solving for $\bar{y}$ gives us

$$\bar{y} = \frac{0.1k}{1 + 0.1k}r + \frac{0.1}{1 + 0.1k}d.$$

Thus, the percent error between $y = \frac{0.1k}{1+0.1k}r$ and $\bar{y}$ due to the disturbance becomes

$$\% \text{ error} = \frac{\bar{y} - y}{y} \times 100 = \frac{10}{0.1k}\frac{d}{r}$$

and so the percent error is reduced by a factor of $0.1k$, which is called the *loop gain*. Now, if $d = 10$, $r = 50$, and $k = 100$, the percent error is reduced by a factor of 10 or 0.2%. By increasing the gain k further, the error can be reduced by any amount desired.

Thus, by introducing negative feedback, the effect of the disturbance is greatly reduced. Figure 11.2 is the block diagram representation of these open and closed loop systems.

Another (equally important) property of feedback is to reduce uncertainty in modeling errors. For example, if the actual plant model is $y = 0.11u$ instead of $y = 0.1u$, then the control $u = 10r$ will result in an error of 10% in the open-loop

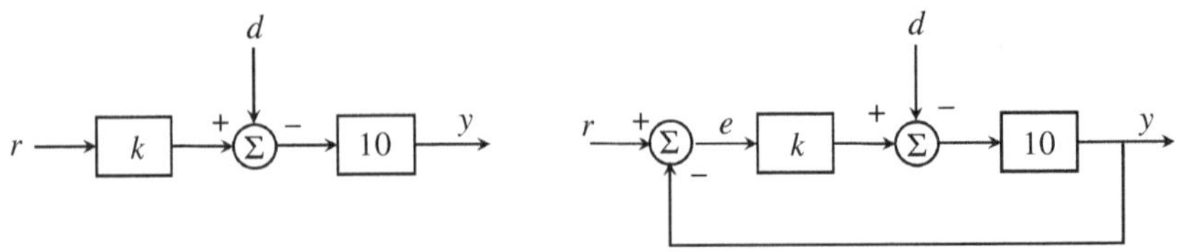

Figure 11.2 Open-loop system (left) and closed-loop system (right).

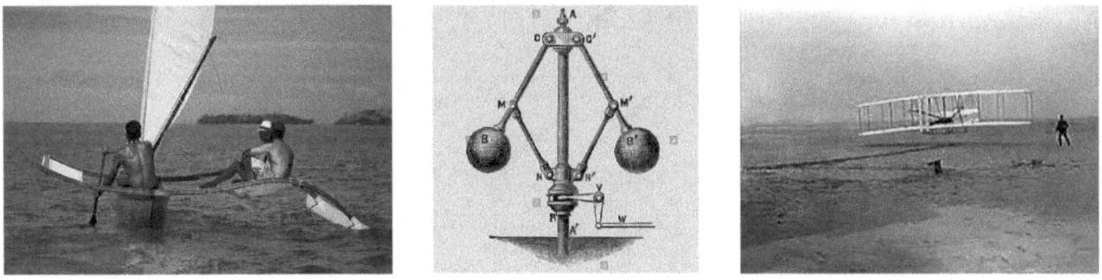

Figure 11.3 Examples of feedback systems: the outrigger canoe, flyball governor and the Wright flyer.

system. In the closed-loop system, the error $r - y$ can be made arbitrarily small for large k, even if the plant model is unknown (Problem 11.2).

Feedback in History

One of the earliest examples of human ingenuity in exploiting feedback is the outrigger canoe, developed by Polynesian society in prehistoric times (Figure 11.3, left). The outrigger greatly improves the roll stability of the canoe [1]. Another important example of the use of feedback was the application of the flyball governor by James Watt to regulate the speed of steam engines (Figure 11.3, middle). Watt's application spurred the Industrial Revolution, beginning in the eighteenth century. The 1868 paper by James Clerk Maxwell [51] analyzed the stability of such governors and is generally recognized as the first scientific article in control theory.

In the first decade of the twentieth century, the Wright brothers solved the problem of flight by developing a three-axis control system that enabled the pilot to steer the aircraft and maintain stability (Figure 11.3, right). In the 1920s, the invention of the negative feedback amplifier by Harold Black at Bell Laboratories was the breakthrough that enabled long distance telephone communication. The space program of the 1960s spurred the development of state-space methods in control. The invention of the microprocessor in the 1970s has resulted in feedback control permeating all aspects of society to the point that we now talk about the "Internet of Things" in which sensors, control, and wireless communication are ubiquitous in household appliances, traffic signals, medical devices, security systems, industrial and personal robots, and more.

11.2 Transfer Functions

In this section we introduce the transfer function representation of continuous-time linear systems. Consider the autonomous, nonhomogeneous linear system

$$\frac{d^n y}{dt^n} + a_{n-1}\frac{d^{n-1}y}{dt^{n-1}} + \cdots + a_0 y = b_m \frac{d^m u}{dt^m} + b_{m-1}\frac{d^{m-1}}{dt^{m-1}} + \cdots + b_0 u, \qquad (11.1)$$

where u is the input, y is the output and the coefficients a_i and b_i are constants. If we associate a variable $s^i Y(s)$ with the derivative $\frac{d^i y}{dt^i}$ and likewise $s^i U(s)$ with $\frac{d^i u}{dt^i}$, for each i, then we can represent Equation (11.1) as an algebraic equation

$$(s^n + a_{n-1}s^{n-1} + \cdots + a_0)Y(s) = (b_m s^m + \cdots + b_0)U(s). \tag{11.2}$$

Definition 11.1 ***(Transfer Function)*** The *transfer function* $H(s)$ for an autonomous linear system of the form Equation (11.1) is defined in terms of Equation (11.2) as

$$H(s) = \frac{Y(s)}{U(s)} = \frac{b_m s^m + b_{m-1}s^{m-1} + \cdots + b_1 s + b_0}{s^n + a_{n-1}s^{n-1} + \cdots + a_1 s + a_0}.$$

The functions $Y(s)$ and $U(s)$ are the Laplace transforms of $y(t)$ and $u(t)$, respectively, but we will not need to consider detailed properties of the Laplace transform in what follows. It suffices to note that, with zero initial conditions, the differential Equation (11.1) is completely determined by the coefficients a_i and b_i and so the algebraic representation captures essential information about the system in a compact and useful form. We can write the transfer function as

$$H(s) = \frac{q(s)}{p(s)},$$

where

$$q(s) = b_m s^m + b_{m-1}s^{m-1} + \cdots + b_1 s + b_0,$$
$$p(s) = s^n + a_{n-1}s^{n-1} + \cdots + a_1 s + a_0$$

are the numerator and denominator polynomials, respectively, of $H(s)$. The block diagram in Figure 11.4 shows the transfer function of the system that transforms the input U to output Y. We treat s as a complex variable since roots of polynomials are, in general, complex numbers. The complex s-plane is referred to as the *Laplace domain.*

Definition 11.2 Consider a system with transfer function $H(s) = \frac{q(s)}{p(s)}$. We assume that $q(s)$ and $p(s)$ are relatively prime, which means that they have no common roots.

- $H(s)$ is called a *proper* rational function if the degree m of $q(s)$ is less than or equal to the degree n of $p(s)$.
- $H(s)$ is *strictly proper* if m is strictly less than n.

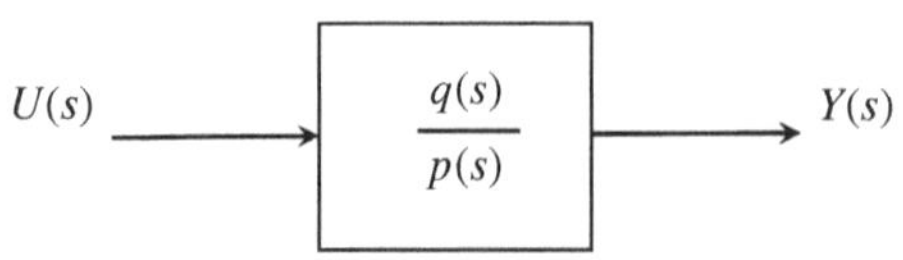

Figure 11.4 A transfer function representation of a linear system as a ratio of polynomials in the complex variable s.

- The polynomial $p(s)$ is called the *characteristic polynomial* of the system.
- The roots of $p(s)$ and $q(s)$ are, respectively, the *poles* and *zeros* of the system.
- The difference $r = n - m$ between the degrees of the numerator and denominator polynomials is called the *relative degree* of the system.

Example 11.1 Consider the system

$$\frac{d^3y}{dt^3} + 2\frac{d^2y}{dt^2} + 3\frac{dy}{dt} + 6 = \frac{du}{dt} - 2u.$$

Then the transfer function is

$$H(s) = \frac{Y(s)}{U(s)} = \frac{s-2}{s^3 + 2s^2 + 3s + 6},$$

which is strictly proper and has relative degree 2. The poles of $H(s)$ are $s = -2, \pm 1.732j$ and $H(s)$ has one zero at $s = +2$.

As an aside, we note that

$$y(t) = \dot{u} \Rightarrow Y(s) = sU(s).$$

In other words, multiplication by s represents a pure differentiation. Likewise,

$$y(t) = \int_0^t u(\sigma)d\sigma \Rightarrow Y(s) = \frac{1}{s}U(s),$$

and hence multiplication by $\frac{1}{s}$ represents a pure integration. This is, in fact, the reason that Simulink uses the notation $\frac{1}{s}$ to represent an integrator.

System Response

The roots of the characteristic polynomial $p(s)$ determine the form of the system time response (refer to our discussion of the eigenvalues in the system response in Chapter 2). Suppose that the system is stable so that all roots have negative real part. Then those roots closest to the imaginary axis (i.e., with largest real part) have a greater influence on the long-term behavior of the system, since the exponential decay is slower for these poles. These are called the *dominant poles* (Figure 11.5), and give further justification for studying first-order and second-order systems. The astute reader will recognize the relation between this discussion and our treatment of singularly perturbed systems in Chapter 3.

Example 11.2 Consider the second-order system, called the *damped oscillator*,

$$\begin{aligned}
\dot{x}_1 &= x_2 \\
\dot{x}_2 &= -\omega^2 x_1 - 2\zeta\omega x_2 + \omega^2 u \\
y &= x_1.
\end{aligned}$$

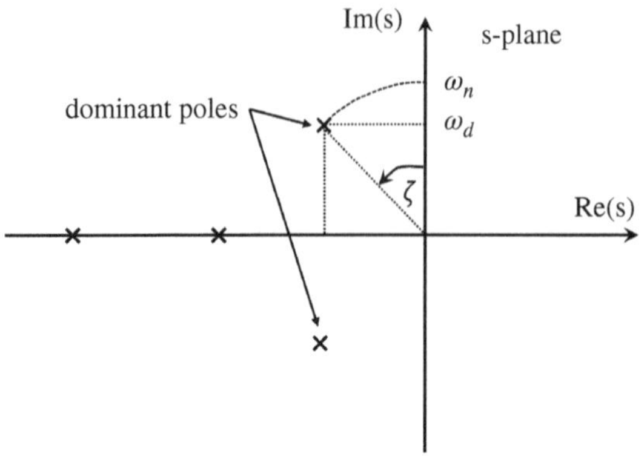

Figure 11.5 Dominant second-order pole location. Second-order systems are an important class of systems to study in detail.

In this system ω is the *(undamped) natural frequency* and ζ is called the *damping ratio*. Note that, with $\zeta = 0$, the system is the harmonic oscillator. A straightforward calculation shows that the transfer function from u to y is

$$\frac{Y(s)}{U(s)} = \frac{\omega^2}{s^2 + 2\zeta\omega s + \omega^2}.$$

If the damping ratio $0 < \zeta < 1$, the response is called *underdamped*. If $\zeta > 1$, the response is *overdamped* and, for $\zeta = 1$, the response is *critically damped*. Figure 11.6 shows three step responses for $\zeta = 0.5$, 1.0, and 1.5. The accompanying Matlab script generates these step responses using the commands `tf` and `step` from Matlab's Control Systems Toolbox.

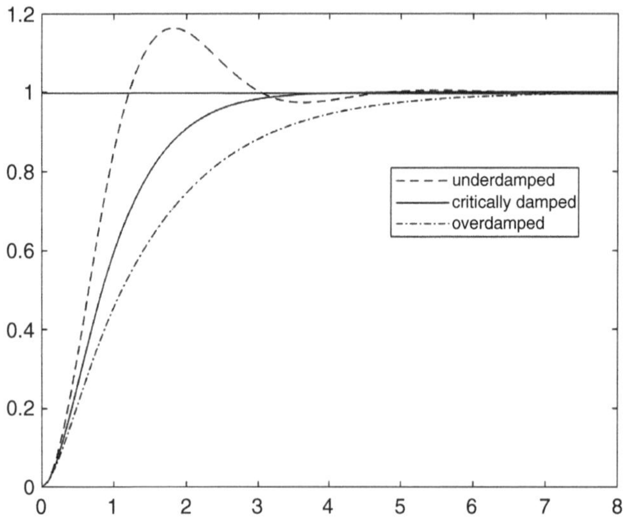

Figure 11.6 Underdamped, critically damped, and overdamped second-order step responses. $\zeta = 0.5, 1.0$, and 1.5, respectively.

Step Responses for a Second-Order System

```
% plots of underdamped, critically damped,
% and overdamped responses
omega = 2;              % undamped natural frequency
t=(0:0.1:8)';
for i=1:3
    zetai = i*0.5;    % damping ratio = 0.5,1,1.5
num = [omega^2];                            % numerator
den = [1 2*zetai*omega omega^2];  % denominator
sys = tf(num,den);    % defines the transfer function
y = step(sys,t);      % computes the step response
plot(t,y,'k')
hold on
end
```

As a rule of thumb, control systems are often designed to produce a slightly underdamped response so that the response is rapid without too much overshoot. Design specifications can be stated in terms of *percent overshoot*, *rise time* and *settling time* as illustrated in Figure 11.7. The rise time is typically defined as the time for the response to go from 10% to 90% of its final value (sometimes 25% and 75%

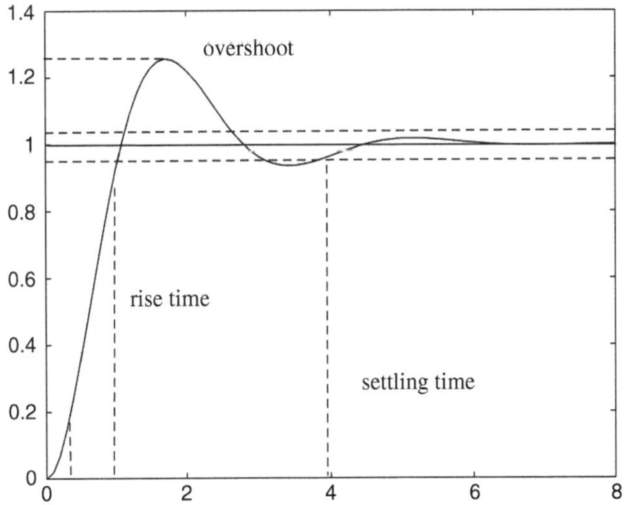

Figure 11.7 A second-order system step response showing overshoot, rise time, and settling time.

are used). The settling time is typically defined as the time it takes the system to settle to with 2% to 5% of its final value.

We note that the roots of the characteristic polynomial $s^2 + 2\zeta\omega s + \omega^2$ are $s = -\zeta\omega \pm j\omega\sqrt{1-\zeta^2}$ in the underdamped case. It can be shown with a little calculation (Problem 11.3) that the analytical solution of the step response is

$$y(t) = 1 - \frac{1}{\sqrt{1-\zeta^2}} e^{-\zeta\omega t} \sin(\omega_d t + \phi), \tag{11.3}$$

where $\omega_d = \omega(\sqrt{1-\zeta^2})$ is the damped natural frequency and $\phi = \arccos(\zeta)$ is the phase shift. The effect of the damping ζ is to reduce the frequency of oscillation.

Steady-State Response

Given a transfer function

$$H(s) = \frac{s^m + b_{m-1}s^{m-1} + \cdots + b_0}{s^n + a_{n-1}s^{n-1} + \cdots + a_0},$$

the constant $H(0) = \frac{b_0}{a_0}$ is called the *DC gain* and is equal to the steady-state value of the step response provided that the system is stable.

Example 11.3 Consider the system

$$H(s) = \frac{s^2 + 2}{s^3 + 4s^2 + 4s + 4}.$$

The DC gain is equal to 0.5 and the step response, displayed in Figure 11.8, shows that the response converges to 0.5 as $t \to \infty$.

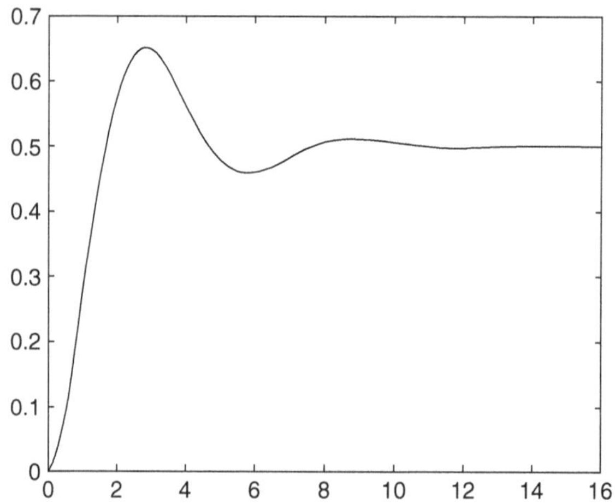

Figure 11.8 Step response of the third-order system with DC gain equal to 0.5.

11.3 Feedback Control

Figure 11.9 shows a block diagram of a typical feedback control system. In this figure $P(s)$ is the *plant* transfer function, which is a model of the system to be controlled, and $C(s)$ is the transfer function for the *compensator*, which is to be designed to achieve a desired output response. The goal is typically to have the output y match or *track* the reference signal r. If $r(t)$ is a constant (called a *set point*), the problem is called the *regulation problem*, whereas if $r(t)$ is a function of time, the problem is called the *tracking problem*. At the same time, the compensator should guarantee good performance despite the presence of the disturbance input d. This is known as *disturbance rejection*. Given that there is generally uncertain knowledge of the exact plant model and disturbances acting on the system, not only is good performance difficult to achieve, but also a poorly designed compensator can actually result in an unstable system. For this reason, control engineers must pay close attention to issues of modeling, identification, and stability.

Closed-Loop Transfer Functions

The block diagram in Figure 11.9 represents the equations

$$
\begin{aligned}
y &= P(s)(u + d)\\
u &= C(s)(r - y).
\end{aligned}
$$

Setting $d = 0$, we can solve for the transfer function from the reference r to y as

$$y = \frac{P(s)C(s)}{1 + P(s)C(s)} r.$$

Setting $r = 0$, we can solve for the transfer function from the disturbance d to y as

$$y = \frac{1}{1 + P(s)C(s)} d.$$

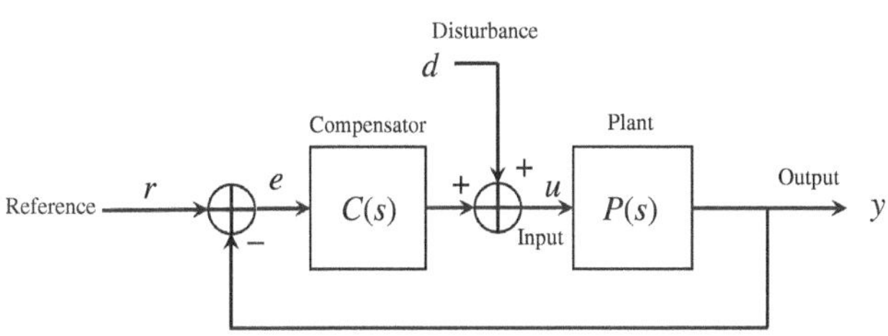

Figure 11.9 Block diagram showing the basic components of a control system.

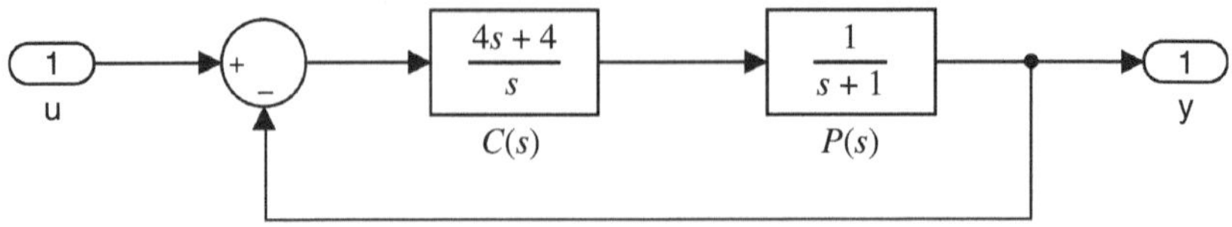

Figure 11.10 First-order system $P(s)$ with compensator $C(s)$.

By superposition, the response of the system to both the reference and disturbance is given by

$$y = \frac{P(s)C(s)}{1 + P(s)C(s)} r + \frac{1}{1 + P(s)C(s)} d.$$

Example 11.4 Consider a first-order system with transfer function $P(s)$ and compensator $C(s)$:

$$P(s) = \frac{1}{s+1}, \quad C(s) = \frac{4s+4}{s}.$$

A Simulink model for this system is shown in Figure 11.10.

The closed-loop transfer function is given by

$$\frac{Y(s)}{U(s)} = \frac{P(s)C(s)}{1 + P(s)C(s)} = \frac{4s+1}{s^2+5s+1}.$$

The compensator $C(s)$ is known as a PI compensator. We will explain this and other compensators in the next section.

PID Control

In this subsection we discuss the design of compensators using proportional–integral–derivative (PID) control. This is the most widely used method of control. It is estimated that well over 90% of industrial controllers in use are PID controllers. In the time domain, a PID control is of the form

$$u = K_p e + K_i \int_0^t e(\sigma) d\sigma + K_d \dot{e}.$$

In this equation, u is the control input, $e = r - y$ is the error signal, and K_p, K_i, and K_d are the proportional, integral and derivative gains, respectively. The transfer function for a PID compensator is thus

$$C(s) = K_p + K_i \frac{1}{s} + K_d s = \frac{K_d s^2 + K_p s + K_i}{s}.$$

Roughly speaking, the proportional term in the PID controller produces a fast response but can lead to overshoot. The derivative term is viewed as anticipatory and adds damping to reduce overshoot. The integral term accumulates the past error and is generally designed to eliminate the steady-state error. An often used procedure is first to design the proportional and derivative gains to achieve desirable transient properties, such as fast rise time and low overshoot, and then to add integral control to eliminate any steady-state error. The design of the gains in the PID compensator is a problem of great interest and numerous articles and books are devoted to the subject (see [6]). Special cases include PI control and PD control, depending on the particular application.

Example 11.5 ***(PD control of satellite orientation)***
Consider the problem of controlling the rotation of a satellite with thrusters producing a force F as shown in Figure 11.11.

A simple model of the satellite dynamics is

$$J\ddot{\theta} = F,$$

where J is the rotational inertia about the axis of rotation. Let θ_d be a reference angle and choose a PD control law

$$F = k_p(\theta_d - \theta) - k_d\dot{\theta},$$

where k_p and k_d are the proportional and derivative gains, respectively. Then the closed-loop system is given by

$$J\ddot{\theta} + k_d\dot{\theta} + k_p\theta_d = k_p\theta_d.$$

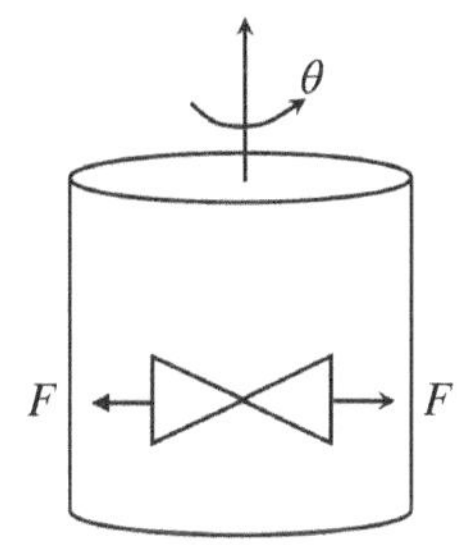

Figure 11.11 Satellite attitude control. The satellite is equipped with thruster that can produce a force F to cause the satellite to rotate about its principal axis.

In theory we can choose the gains k_p and k_d to achieve an arbitrary step response, i.e., an arbitrary natural frequency and damping ratio so that the angle θ will converge to the desired reference angle θ_d arbitrarily fast. However, in practice we must deal with realistic constraints. The first is the problem of measuring the position and velocity in order to implement the control. The second is the fact that the thrusters cannot produce arbitrarily large force, so we must consider realistic limits on the control magnitude.

For the first problem, suppose that the satellite is equipped with an accelerometer that can measure angular acceleration. We can then obtain the velocity and position by integrating the acceleration. For the second problem, we can use the Simulink saturation block to model the limits on the magnitude of the control input.

Figure 11.12 is a Simulink model for this system and Figure 11.13 shows the output response, where we have chosen gains, based on the saturation level, to achieve a 20-degree rotation in under 10 seconds.

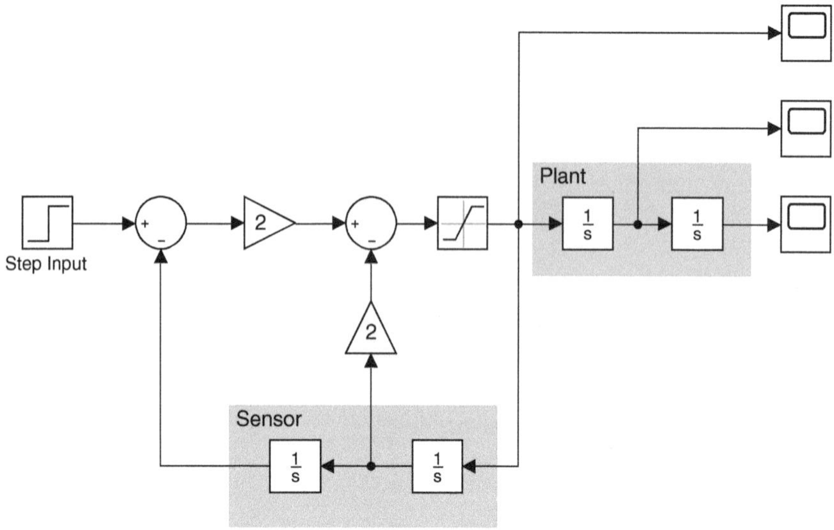

Figure 11.12 Simulink model of the satellite control. The sensor block integrates the measured acceleration once to estimate velocity and a second time to estimate position. The control magnitude F is limited to $|F| \leq 6$.

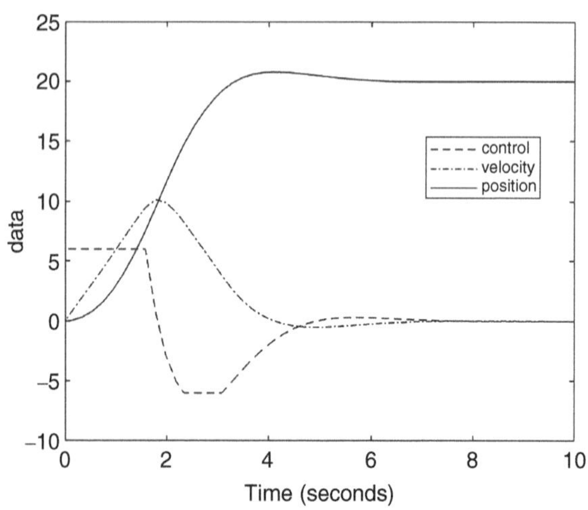

Figure 11.13 Position, velocity, and control input torque for the satellite. During the time that the acceleration is saturated, the velocity is linear, as can be seen in the plot.

Example 11.6 ***(PI Control)***

PI control is often used in regulating speed in electric motors, automobile cruise control, and other systems governed by first-order models. Consider the system

$$\dot{v} = -av + u + d,$$

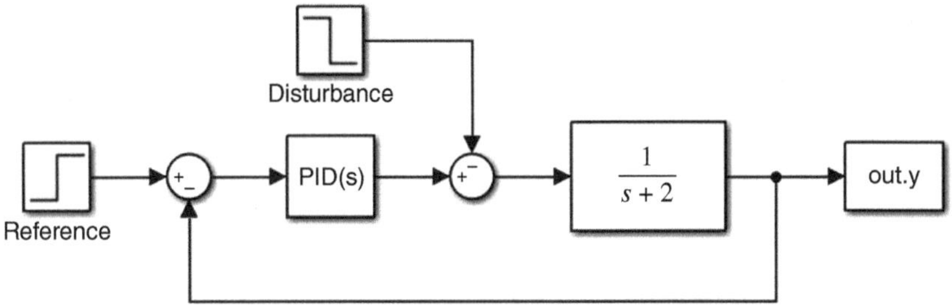

Figure 11.14 Simulink model of a PI speed control system.

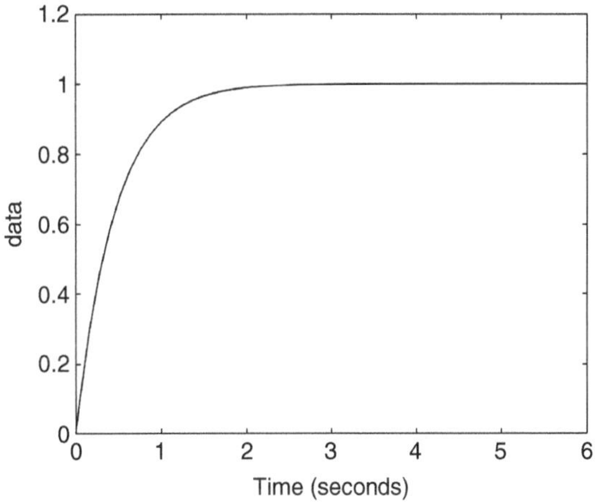

Figure 11.15 Responses of the PI control gains, $K_p = 2$, $K_I = 4$, $K_D = 0$. The disturbance input is $d = -0.2$.

where v is the velocity of the system under consideration, u is the control input, and d is a disturbance. A PI control is of the form

$$u = k_p e + k_i \int_0^t e(\sigma)d\sigma, \quad \text{where } e = v_r - v$$

and v_r is a reference speed. A Simulink model for this system is shown in Figure 11.14 using Matlab's PID compensator block.

Figure 11.15 shows the step response. The PID gains in the PID block were chosen as $K_p = 2, K_I = 4, K_D = 0$.

11.4 State-Space Models

In this section we first discuss the relation between transfer function models and state-space models for single-input/single-output systems. If we have a single-input/single output linear system

$$\dot{x} = Ax + bu \tag{11.4}$$

$$y = c^T x, \tag{11.5}$$

with $x \in \mathbb{R}^n$, $u \in \mathbb{R}$, and $y \in \mathbb{R}$, we can write this in terms of the Laplace variable s as

$$\begin{aligned} sX(s) &= AX(s) + bU(s) \\ Y(s) &= c^T X(s). \end{aligned}$$

Eliminating $X(s)$ and solving these equations for $Y(s)$ in terms of $U(s)$ results in

$$Y(s) = c^T (sI - A)^{-1} bU(s).$$

It follows that the transfer function associated with the linear system (11.4)–(11.5) is given by

$$H(s) = c^T (sI - A)^{-1} b. \tag{11.6}$$

Example 11.7 Suppose that we have the linear system (11.4)–(11.5) with

$$c^T = \begin{bmatrix} 1 & 0 \end{bmatrix}, \quad A = \begin{bmatrix} 0 & 1 \\ -1 & -2 \end{bmatrix}, \quad b = \begin{bmatrix} 0 \\ 1 \end{bmatrix}.$$

Then a direct calculation (Problem 11.11) shows that

$$c^T (sI - A)^{-1} b = \begin{bmatrix} 1 & 0 \end{bmatrix} \begin{bmatrix} s & -1 \\ 1 & s+2 \end{bmatrix}^{-1} \begin{bmatrix} 0 \\ 1 \end{bmatrix} = \frac{1}{s^2 + 2s + 1}. \tag{11.7}$$

We note that the eigenvalues of A coincide with the poles of the transfer function.

11.4.1 Minimal Realizations

Turning the above argument around, suppose that we are given a strictly proper transfer function of the form

$$H(s) = \frac{q(s)}{p(s)},$$

where the polynomials $p(s)$ and $q(s)$ are relatively prime, and suppose that we can find matrices A, b, c satisfying Equation (11.6). Then the linear system (11.4)–(11.5) is called a *realization* of $H(s)$.

A *minimal realization* is one where the dimension of the state x is equal to n, the degree of the denominator polynomial $p(s)$. In that case, it can be shown that

$$p(s) = \det (sI - A),$$

and consequently the poles of the transfer function $H(s)$ are precisely the eigenvalues of the matrix A.

Remark 11.1 Minimal realizations of a given $H(s)$ are unique up to a linear change of variables. Suppose that

$$\dot{x} = Ax + bu$$
$$y = c^T x$$

is a minimal realization of a system with transfer function $H(s)$. Let $x = Tz$ be a linear change of coordinates in state space, so that

$$\dot{z} = T^{-1}ATz + T^{-1}bu$$
$$y = c^T Tz.$$

Then a straightforward calculation shows that

$$c^T T(sI - T^{-1}AT)^{-1}T^{-1}b = c^T(sI - A)^{-1}b = H(s).$$

Thus there are infinitely many (equivalent) state-space systems of dimension n that realize the same transfer function.

Canonical Forms

There are several special state-space realizations, called *canonical forms*, that are particularly useful for feedback control and estimation. We present three of them here, called the *controller canonical form*, the *reachable canonical form*, and *observer canonical form*. For simplicity, we illustrate these canonical forms for a third-order system,

$$\frac{d^3y}{dt^3} + a_2\ddot{y} + a_1\dot{y} + a_0 y = b_2\ddot{u} + b_1\dot{u} + b_0 u, \tag{11.8}$$

which, in the Laplace domain, is

$$(s^3 + a_2 s^2 + a_1 s + a_0)Y(s) = (b_2 s^2 + b_1 s + b_0)U(s). \tag{11.9}$$

The extension to the general nth-order case follows the same pattern and is straightforward.

Controller Canonical Form

Given the system (11.8), we introduce the variable z and write the system as

$$\frac{d^3z}{dt^3} + a_2\ddot{z} + a_1\dot{z} + a_0 z = u$$
$$y = b_3\ddot{z} + b_2\dot{z} + b_1 z.$$

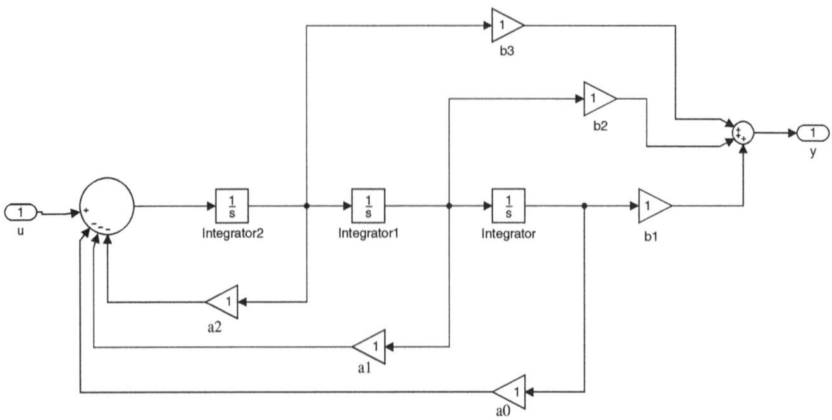

Figure 11.16 Controller canonical form. The z variables are the outputs, x_i, of the integrators and y is therefore a linear combination of the z-variables.

It is easy to see that these two systems are equivalent by expressing them in the Laplace domain.

$$p(s)Y(s) = p(s)q(s)Z(s) = q(s)p(s)Z(s) = q(s)U(s).$$

The Simulink model in Figure 11.16 is a realization of this system, called the *controller canonical form.*

From the Simulink block diagram, we can write the state equations by inspection

$$\begin{aligned}
\dot{x}_1 &= x_2 \\
\dot{x}_2 &= x_3 \\
\dot{x}_3 &= -a_0 x_1 - a_1 x_2 - a_2 x_3 + u \\
y &= b_0 x_1 + b_1 x_2 + b_2 x_3,
\end{aligned}$$

which is of the form

$$\begin{aligned}
\dot{x} &= Ax + bu \\
y &= c^T x,
\end{aligned}$$

where

$$A = \begin{bmatrix} 0 & 1 & 0 \\ 0 & 0 & 1 \\ -a_0 & -a_1 & -a_2 \end{bmatrix}, \quad b = \begin{bmatrix} 0 \\ 0 \\ 1 \end{bmatrix}, \quad c^T = [b_0, b_1, b_2].$$

The Reachable Canonical Form

Matlab has the function `tf2ss` that will compute a state-space realization from a transfer function description of a given system. Note, however, that Matlab gives the realization in a different form, called the *reachable canonical form* of the system (11.8), which is given by

$$\bar{A} = \begin{bmatrix} -a_2 & -a_1 & -a_0 \\ 1 & 0 & 0 \\ 0 & 1 & 0 \end{bmatrix}, \quad \bar{b} = \begin{bmatrix} 1 \\ 0 \\ 0 \end{bmatrix}, \quad \bar{c}^T = [b_2, b_1, b_0].$$

The reachable canonical form and the controller canonical form are related by

$$A = T^{-1}\bar{A}T, \quad b = T^{-1}\bar{b}, \quad c^T = \bar{c}^T T,$$

where

$$T = \begin{bmatrix} 0 & 0 & 1 \\ 0 & 1 & 0 \\ 1 & 0 & 0 \end{bmatrix}.$$

This amounts to renumbering the state variables as $x_1 \leftrightarrow x_3, x_2 \leftrightarrow x_2, x_3 \leftrightarrow x_1$. The Simulink diagram for the reachable canonical form is identical to Figure 11.16 by relabeling the outputs of the integrators as explained above.

Example 11.8 Consider the system with transfer function $H(s) = \dfrac{2s+1}{s^2+5s+6}$. The Matlab code below computes a state-space realization of $H(s)$ in reachable canonical form.

Matlab State-Space Realization

```
% state-space realization of H(s) = (2s+1)/(s^2+5s+6)
clear
num=[2 1]; den = [1 5 6];     % numerator and denominator
                              % of the transfer function H(s)
sys = tf(num,den);            % define the transfer function
[Ab,bb,cb,~] = tf2ss(num,den); % reachable canonical form
```

Running this script produces the output

$$\text{Ab} = \begin{bmatrix} -5 & -6 \\ 1 & 0 \end{bmatrix}, \quad \text{bb} = \begin{bmatrix} 1 \\ 0 \end{bmatrix}, \quad \text{cb} = [2, 1].$$

We can recover the controller canonical form as follows.

Matlab State-Space Realization

```
%conversion to controller canonical form (A,b,c)
T=[0 1;1 0];
A = T*Ab*T; b = T*bb; c=cb*T;
```

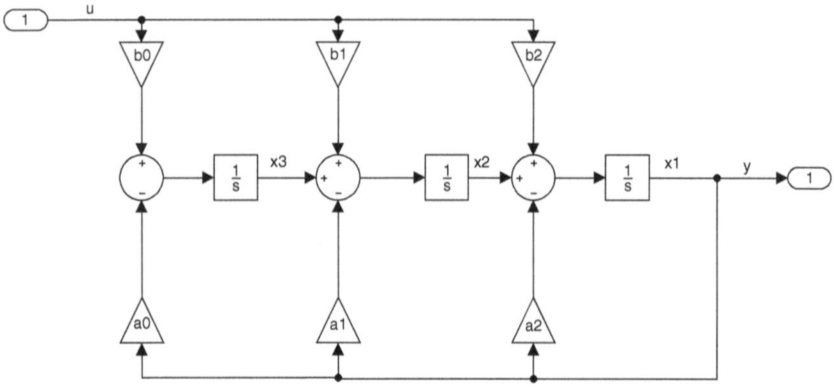

Figure 11.17 Observer canonical form

This results in the system in controller canonical form

$$\mathtt{A} = \begin{bmatrix} 0 & 1 \\ -6 & -5 \end{bmatrix}, \quad \mathtt{b} = \begin{bmatrix} 0 \\ 1 \end{bmatrix}, \quad \mathtt{cb} = [1, 2].$$

These two canonical forms are particularly useful for the problem of pole placement, which we will discuss below in Section 11.4.2.

The Observer Canonical Form

The observer canonical form is useful for the problem of state estimation, as we shall see below in Section 11.4.3. The *observer canonical form* of the system (11.8) is

$$A = \begin{bmatrix} -a_2 & 1 & 0 \\ -a_1 & 0 & 1 \\ -a_0 & 0 & 0 \end{bmatrix}, \quad b = \begin{bmatrix} b_2 \\ b_1 \\ b_0 \end{bmatrix}, \quad c^T = [1\ 0\ 0]\,.$$

Figure 11.17 shows a Simulink block diagram of the observer canonical form.

Controllability

The notion of *controllability* is a fundamental property of linear systems that determines the ability to control the system with state feedback. The notion of controllability was introduced by Kalman [41] (see also [27]). Roughly speaking, a system is controllable if there exists a control input to move the system state between any two given states.

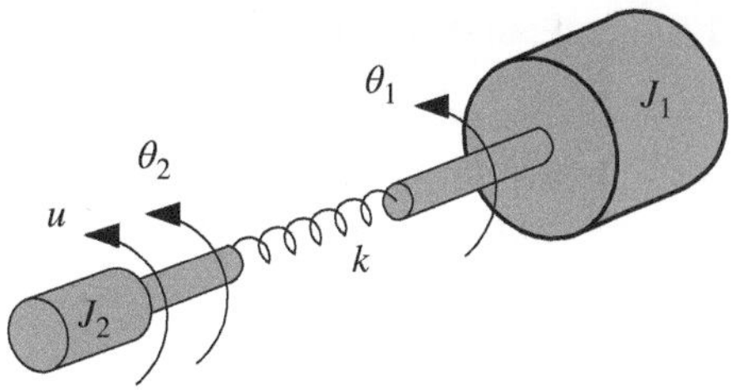

Figure 11.18 Inertias coupled by a torsional spring.

Ignoring the output equation for now, let us consider the linear system

$$\dot{x} = Ax + bu, \tag{11.10}$$

which we refer to as the system (A, b). The following proposition gives a constructive way to determine controllability of the system (11.10) from the system matrices.

Proposition 11.1 A single-input linear system (A, b) is controllable if and only if

$$\det[b, Ab, A^2b, A^3b, \dots, A^{n-1}b] = n.$$

Example 11.9 Consider the system in Figure 11.18 consisting of inertias J_1 and J_2 connected by a flexible coupling with spring constant K. This system models, for example, the torsional flexibility of the shaft connecting a motor to a load. Assume that the control input u acts on the inertia J_2. Then, the equations describing this system are

$$J_1\ddot{\theta}_1 + K(\theta_1 - \theta_2) = 0 \tag{11.11}$$

$$J_2\ddot{\theta}_2 + K(\theta_2 - \theta_1) = u. \tag{11.12}$$

In state space with $x_1 = \theta_1, x_2 = \dot{\theta}_1, x_3 = \theta_2, x_4 = \dot{\theta}_2$, we can write the system as

$$\begin{bmatrix} \dot{x}_1 \\ \dot{x}_2 \\ \dot{x}_3 \\ \dot{x}_4 \end{bmatrix} = \begin{bmatrix} 0 & 1 & 0 & 0 \\ -\frac{K}{J_1} & 0 & \frac{K}{J_1} & 0 \\ 0 & 0 & 0 & 1 \\ \frac{K}{J_2} & 0 & -\frac{K}{J_2} & 0 \end{bmatrix} \begin{bmatrix} x_1 \\ x_2 \\ x_3 \\ x_4 \end{bmatrix} + \begin{bmatrix} 0 \\ 0 \\ 0 \\ \frac{1}{J_2} \end{bmatrix} u. \tag{11.13}$$

Thus, the dimension of the state space is $n = 4$. It is left as an exercise (Problem 11.4) to show that

$$\det[b, Ab, A^2b, A^3b] = \frac{K^2}{J_1^2 J_2^4} \tag{11.14}$$

and therefore the system is controllable.

11.4.2 Pole Placement

Given a linear system (A, b) with input u, a linear state feedback controller is one of the form $u = -k_1x_1 - k_2x_2 - \cdots - k_nx_n + k_rr = -k^Tx + k_rr$. The vector k is called the *gain matrix* and r is a reference input. With this control input, the system is now

$$\dot{x} = (A - bk^T)x + bk_rr. \tag{11.15}$$

This means that the linear state feedback changes the open-loop poles, which are the eigenvalues of A, to the closed-loop poles, which are the eigenvalues of $A - bk^T$. An important consequence of controllability is the following,

Theorem 11.1 ***(Pole Assignment Theorem)*** *Let $p(s)$ be an arbitrary polynomial of degree n with real coefficients. If the linear system (A, b) is controllable, then there exists a vector $k \in \mathbb{R}^n$ such that*

$$\det(sI - A + bk^T) = p(s).$$

Suppose that $p(s) = s^n + p_{n-1} + \cdots + p_0$ is the desired closed-loop characteristic polynomial and $p_{o\ell}(s) = s^n + a_{n-1}s^{n-1} + \cdots + a_0$ is the characteristic polynomial of the open-loop system matrix A.

If the system (A, b) is in controller canonical form, the gain k is very easy to determine as $k = [p_0 - a_0, \cdots, p_{n-1} - a_{n-1}]$.

Example 11.10 Suppose that

$$A = \begin{bmatrix} 0 & 1 & 0 \\ 0 & 0 & 1 \\ -2 & -3 & -5 \end{bmatrix}, \quad b = \begin{bmatrix} 0 \\ 0 \\ 1 \end{bmatrix}.$$

Then we have that

$$A - bk^T = \begin{bmatrix} 0 & 1 & 0 \\ 0 & 0 & 1 \\ -2 & -3 & -5 \end{bmatrix} - \begin{bmatrix} 0 \\ 0 \\ 1 \end{bmatrix} \begin{bmatrix} k_1, & k_2, & k_3 \end{bmatrix}$$

$$= \begin{bmatrix} 0 & 1 & 0 \\ 0 & 0 & 1 \\ -2 - k_1 & -3 - k_2 & -5 - k_3 \end{bmatrix}.$$

Thus the choice $k = \left[p_0 - 2, p_1 - 3, p_2 - 5\right]$ produces the correct closed-loop polynomial $p(s) = s^3 + p_2s^2 + p_1s + p_0$.

If (A, b) is not in controller canonical form, then we can find a change of coordinates matrix T so that $(T^{-1}AT, T^{-1}b)$ is in controller canonical form, after which we can apply the above procedure. This leads to the following formula for the required gain matrix k [5]:

$$k^T = [p_0 - a_0, p_1 - a_1, \cdots, p_{n-1} - a_{n-1}]W^{-1},$$

$$W = [b, Ab, A^b, \ldots, A^{n-1}b] \begin{bmatrix} a_1 & a_2 & \cdots & a_{n-1} & 1 \\ a_2 & \cdots & a_{n-1} & 1 & 0 \\ \vdots & \vdots & 1 & 0 & 0 \\ a_{n-1} & \vdots & 0 & \cdots & 0 \\ 1 & 0 & \cdots & 0 & 0 \end{bmatrix}.$$

Example 11.11 Suppose that (A, b) are given by

$$A = \begin{bmatrix} 1 & -2 \\ 2 & 0 \end{bmatrix}, \quad b = \begin{bmatrix} -1 \\ 1 \end{bmatrix}.$$

Then the characteristic polynomial of A is $s^2 + a_1 s + a_0 = s^2 - s - 4$ and so the open-loop system is unstable. Suppose we want to assign closed-loop poles at $s = -1, -1$, which implies that $p(s) = s^2 + p_1 s + p_0 = s^2 + 2s + 1$. One can check by direct calculation that (A, b) is a controllable pair and W is given by

$$W = \begin{bmatrix} -2 & -1 \\ -3 & 1 \end{bmatrix}.$$

Therefore, we have

$$k^T = [-3, -3]W^{-1} = [-1.2,\ 1.8]$$

and

$$A - bk^T = \begin{bmatrix} -0.2 & -0.2 \\ 3.2 & -1.8 \end{bmatrix},$$

which has eigenvalues at $-1, -1$ as desired.

Example 11.12 ***(Pole Placement in Matlab)***
Matlab's Control System Toolbox has the commands `acker` and `place` for placing the poles of a linear control system. The command `place` is better conditioned numerically for larger problems. However, with `place` it is not possible to choose both poles with the same value (i.e., with multiplicity 2), whereas there is no such restriction using `acker`, which is based on Ackermann's formula [2].

Pole Assignment: Comparison with Ackermann's Formula

```
clear
A = [1 -2;2 0];   % define A
b = [-1;1];       % define b
C = [b A*b];      % controllability matrix
p=[1 2];          % coefficients of s^2+p_1s + p_0
al = [4 -1];      % coefficients of s^2+a_1s + a_0
V = [-1 1;1 0];
W=C*V;
k = (p-al)*inv(W);        % calculation of gain
Acl = A-b*k
eig(Acl)                  % check of closed-loop eigenvalues
kt=acker(A,b,[-1 -1]); % comparison with Ackermann's formula
```

Running this script computes the gains k and kt as

$$\begin{aligned} \mathtt{k} &= [-1.2000,\ 1.8000] \\ \mathtt{kt} &= [-1.2000,\ 1.8000]. \end{aligned}$$

Choosing the Gain k_r

To compute the step response for the system (11.15) we need to know how to choose the gain k_r that multiplies the reference input r. It turns out the k_r should be chosen as the inverse of the DC gain of the system. For a state-space system (A, b, c^T) it can be shown that the DC gain is equal to $-c^T A^{-1} b$.

Example 11.13 The step response for the system in Example 11.12 with output $y = [1, 0]x$ is plotted in Figure 11.19, based on the following Matlab script.

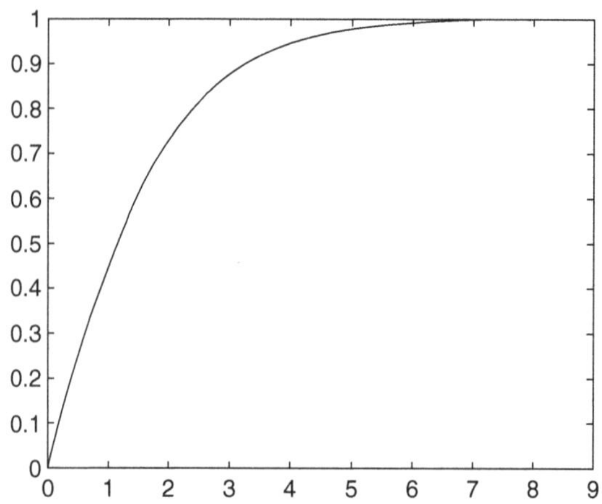

Figure 11.19 Step response of the closed-loop system.

Step Response of the Closed-Loop System

```
clear
A = [1 -2;2 0];
b = [-1;1];
c = [1 0];
kt=acker(A,b,[-1 -1]);
Acl = A-b*kt;                          % closed-loop system
syscl = ss(Acl,b,c,0);
kr=1/dcgain(syscl);                    % compute kr
syscl_scaled = ss(Acl,b*kr,c,0);  % scale reference by kr
t=0:0.1:9;
y=step(syscl_scaled,t);
plot(t,y,'k')                  % plot the scaled step response
```

11.4.3 State Estimation

The state feedback control $u = -k^T x$ requires knowledge of all the state variables, which may be unreasonable in practice. To be practical, a control algorithm should be based only on the measured output(s). This motivates the problem of *state estimation*, which is the problem of reconstructing the entire state from the measured output.

Definition 11.3 ***(Observer)*** An *observer* for a linear system (c, A, b) is a system in the form

$$\dot{\hat{x}} = A\hat{x} + bu + \ell(y - c^T\hat{x}),$$

where $\hat{x}$ is a estimate of the state x. The vector ℓ is the *observer gain*.

Thus, an observer is a simulation model of the system together with an added term that is proportional to the error between the measured output y and an estimated output $c^T\hat{x}$.

Define $e = x - \hat{x}$ as the error between the state $x(t)$ and the estimated state $\hat{x}(t)$. Then the problem is to choose the observer gain, ℓ, so that $e(t) \to 0$ for any initial conditions $e(0) = x(0) - \hat{x}(0)$. To this end, we compute

$$\dot{e} = \dot{x} - \dot{\hat{x}} = Ax + bu - (A\hat{x} + bu + \ell(y - c^T\hat{x})). \tag{11.16}$$

Using the fact that $y = c^T x$, Equation (11.16) reduces to

$$\dot{e} = (A - \ell c^T)e. \tag{11.17}$$

Thus, if the observer gain, ℓ, can be found so that $A - \ell c^T$ is a Hurwitz matrix, Equation (11.17) says that the estimation error will converge exponentially to zero for any initial error $e(0)$, which is to say that the estimate $\hat{x}(t)$ will converge exponentially to the true state $x(t)$. The condition under which such a gain matrix

ℓ exists is called *observability*, and is dual, in a precise mathematical sense, to controllability.

Definition 11.4 ***(Observability)*** Consider the single-output linear system (11.4)–(11.5). The pair (c, A) is *observable* if and only if

$$\det \begin{bmatrix} c^T \\ c^T A \\ \vdots \\ c^T A^{n-1} \end{bmatrix} \neq 0.$$

Exactly as in the case of controllability, it turns out that if (c, A) is observable, then there exists an observer gain matrix ℓ so that $\det(A - \ell c^T) = p_o(s)$, where $p_o(s)$ is any desired polynomial of degree n. Therefore, choosing $p_o(s)$ so that $A - \ell c$ is Hurwitz, the estimation error e converges exponentially to zero for all initial conditions.

Example 11.14 Consider the system (11.4)–(11.5) with

$$A = \begin{bmatrix} -2 & 1 \\ -1 & 0 \end{bmatrix}, \quad b = \begin{bmatrix} b_1 \\ b_2 \end{bmatrix}, \quad c^T = [1\ 0].$$

We will choose the input u to be zero, so the definition of the input matrix b is not important in this example.

Checking observability of (c, A) we have

$$\begin{bmatrix} c^T \\ c^T A \end{bmatrix} = \begin{bmatrix} 1 & 0 \\ -2 & 1 \end{bmatrix},$$

which is seen to be nonsingular and hence the system is observable.

With the observer gain given as $\ell = \left[\ell_1, \ell_2\right]^T$ we have

$$A - \ell c^T = \begin{bmatrix} -2 - \ell_1 & 1 \\ -1 - \ell_2 & 0 \end{bmatrix}$$

and so $\det\,(sI - A + \ell c^T) = s^2 + (2 + \ell_1)s + 1 + \ell_2$. If we choose the desired observer poles as, say, $s_1 = s_2 = -4$, we get $\ell_1 = 2$ and $\ell_2 = 3$.

Figures 11.20 and 11.21 show the Simulink system to compute the observer and a plot of the error $e(t) = x(t) - \hat{x}(t)$, respectively. The observer error converges to zero independent of the initial conditions.

Remark 11.2 In order to design the observer gain matrix ℓ we can use any algorithm, such as `place` or `acker`, by replacing A by A^T and b by c^T.

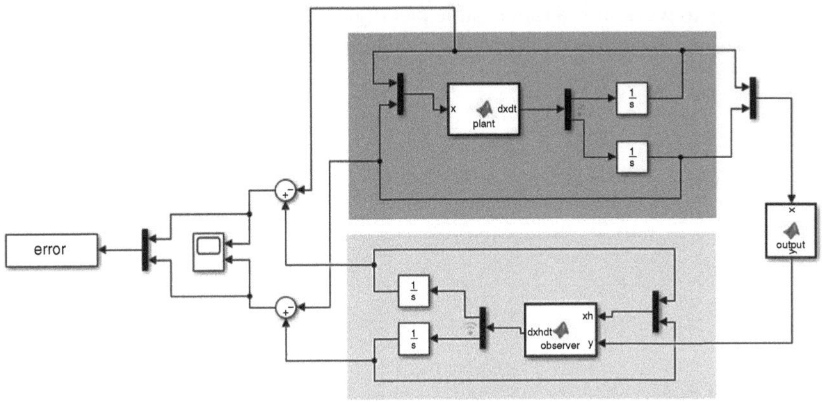

Figure 11.20 Simulink system for the observer

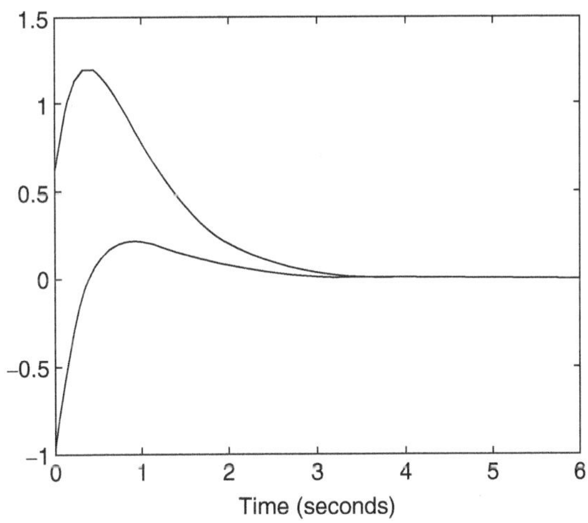

Figure 11.21 Estimation errors for the observer. The initial conditions were chosen as $x(0) = (1, -0.6)$ and $\hat{x}(0) = (0, 0)$.

11.4.4 The Separation Principle

We now have a controller $u = -k^T x$, defined in terms of the system state, x, that can place the poles of the closed-loop system in any desired locations. We also have a state estimator $\hat{x}$ that can estimate the true state x. This suggests that, to implement the control, we can try to substitute the estimated state $\hat{x}$ into the control law (i.e., $u = -k^T \hat{x}$). Intuitively, if $\hat{x}(t)$ converges sufficiently fast to $x(t)$, we might expect this strategy to work. It is a remarkable fact that this strategy does indeed work for linear systems. The analysis proceeds as follows. With $u = -k^T\hat{x}$, we have

$$\begin{aligned}\dot{x} &= Ax - bk^T\hat{x}\\ &= Ax - bk^Tx - bk^T\hat{x} + bk^Tx\\ &= (A - bk^T)x + bk^Te.\end{aligned}$$

Combining this with the error equation

$$\dot{e} = (A - \ell c^T)e$$

yields

$$\begin{bmatrix}\dot{x}\\ \dot{e}\end{bmatrix} = \begin{bmatrix}A - bk^T & bk^T\\ 0 & A - \ell c^T\end{bmatrix}\begin{bmatrix}x\\ e\end{bmatrix}.$$

It follows that the set of closed-loop poles is the union of the controller poles, which are the eigenvalues of $A - bk^T$, and observer poles, which are the eigenvalues of $A - \ell c^T$. This result is known as the *separation principle.*

Example 11.15 As a simple example, let us consider control of a double integrator model

$$\begin{aligned}\dot{x}_1 &= x_2\\ \dot{x}_2 &= u\\ y &= x_1.\end{aligned}$$

To design the observer/controller, we need to choose the observer gain $\ell = (\ell_1, \ell_2)^T$ and controller gain $k^T = (k_1, k_2)$. As a rule of thumb, it is desirable to set the real part of the observer poles more negative than the closed-loop plant poles so that the estimation error decays rapidly. For example, we can set the observer poles at $s = (-10, -10)$ and the closed-loop plant poles at $s = (-4, -4)$. A straightforward calculation then gives

$$\ell = (20, 100)^T, \quad \text{and} \quad k^T = (4, 2).$$

Thus, to follow a step reference r, the control law is

$$u = -4(\hat{x}_1 - r) - 2\hat{x}_2,$$

where $\hat{x}_1$ and $\hat{x}_2$ are the observer states. Figure 11.22 is a Simulink model for this system and Figure 11.23 is the output response y.

So, why not just make the observer and controller gains as large as we want in order to achieve fast convergence? Clearly, there must be physical constraints that will limit these gains. Since all measurements are subject to noise, large gains in

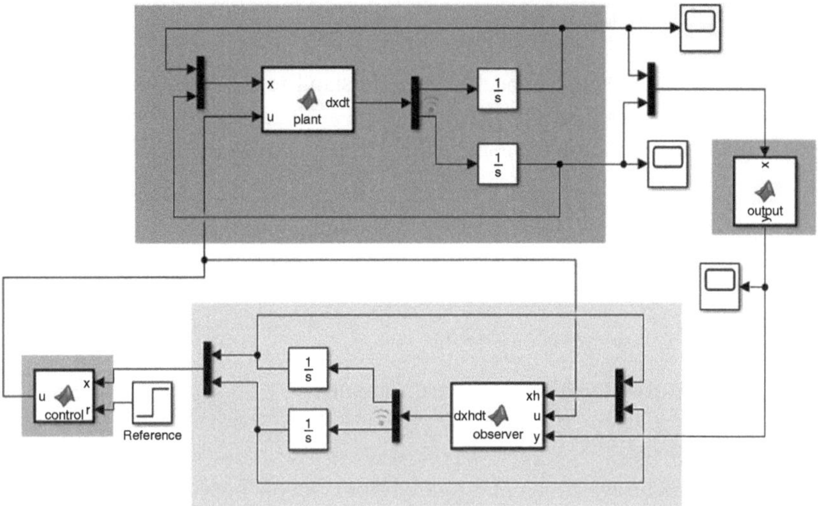

Figure 11.22 Simulink model of the observer/state feedback control system.

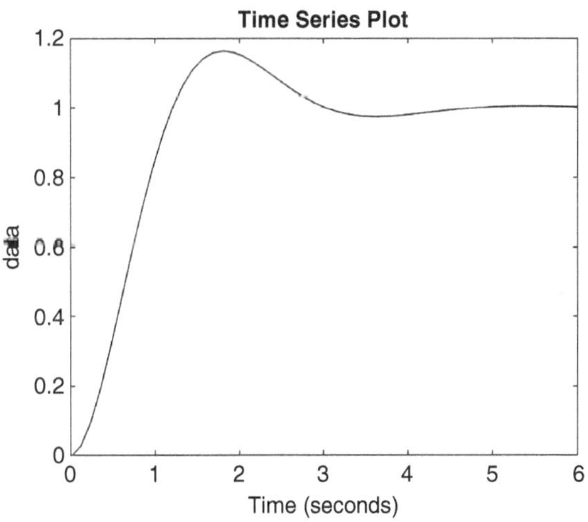

Figure 11.23 Output response of the observer/state feedback control system.

the observer will tend to amplify the noise. Also, since all physical actuators are limited in the amount of "force" they can produce, large control gains will tend to saturate the input. Large control gains may also excite unmodeled dynamics in the plant. The design of the observer/controller gains must, therefore, take these realistic constraints into consideration.

11.5 Optimal Control

On way to design a state-feedback control to stabilize a linear system is to use an optimization procedure. For example, given a single-input controllable linear system (A, b), let us choose the control input u to minimize the cost function

$$J = \int_0^\infty (x^T Q x + r u^2) dt,$$

where Q is positive definite and $r > 0$. A fundamental result, due to Kalman [40], says that the optimal control input u is given by

$$u = -k^T x, \quad \text{where } k^T = \frac{1}{r} b^T P,$$

and where P satisfies the *algebraic Riccati equation*

$$A^T P + PA - \frac{1}{r} P b b^T P + Q = 0.$$

Thus the control design problem is shifted from one of choosing the closed-loop poles to one of choosing the weights Q and r in the cost function J.

Example 11.16 The Matlab function `icare` solves the algebraic Riccati equation. Let us consider the problem of stabilizing the inverted pendulum. The linear approximation about the inverted equilibrium is given by

$$\begin{aligned} \dot{x}_1 &= x_2 \\ \dot{x}_2 &= x_1 + u, \end{aligned}$$

where $x_1 = \theta - \pi$, $x_2 = \dot{\theta}$ are the pendulum angle and velocity, respectively, and u is the control input torque. The open-loop equilibrium is a saddle point, as we know. With Q equal the 2×2 identity matrix and $r = 1$, the Matlab script below computes the optimal state-feedback gains $k = (2.4142, 2.4142)$. The closed-loop eigenvalues are at $(-1, -1.4124)$. The response of the system with initial conditions $(0.1, 0)$ is plotted in Figure 11.24, showing that the equilibrium has been stabilized, meaning that the pendulum will balance about the inverted position.

stabilizing the inverted pendulum

```
clear
A = [0 1;1 0]; b = [0;1]; c = [1 0];d=0;
r=1;
G=-b*b'/r;Q=eye(2);
P=icare(A,[],Q,[],[],[],G);
k=P*b/r;
Abar = A-b*k';
```

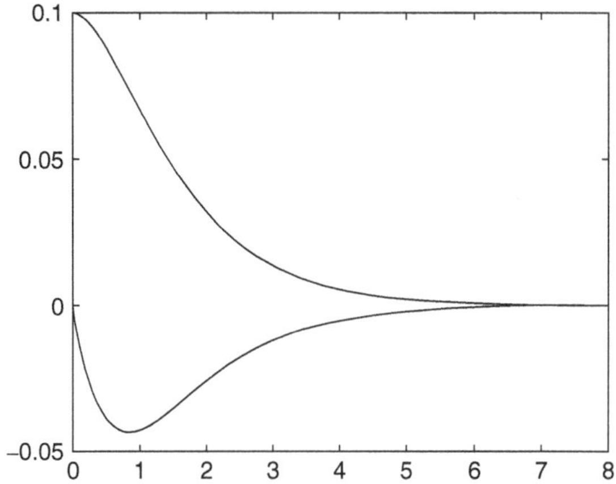

Figure 11.24 Pendulum response.

11.6 Control of Nonlinear Systems

If our system is nonlinear of the form

$$\dot{x} = f(x) + g(x)u$$
$$y = h(x),$$

we can design a controller for a linear approximation of the nonlinear system about an equilibrium point. A successfully designed controller will keep the state of the system near the equilibrium point, which provides further justification for the use of linear control theory for nonlinear systems. Suppose that (x_e, u_e) is an equilibrium point for the nonlinear system. Then the linear approximation about (x_e, u_e) is

$$\dot{\overline{x}} = A\overline{x} + b\overline{u}$$
$$\overline{y} = c^T\overline{x},$$

where $\overline{x} = x - x_e$, $\overline{u} = u - u_e$, $\overline{y} = y$, and A, b, c are Jacobians

$$A = \frac{\partial f}{\partial x}(x_e), \quad b = \frac{\partial g}{\partial x}(x_e), \quad c = \frac{\partial h}{\partial x}(x_e).$$

Example 11.17 ***(Reaction-Wheel Pendulum)***
The reaction-wheel pendulum [12] is a simple pendulum with a rotating disk at the bob. The acceleration of the disk produces a torque at the pendulum pivot that can be used to control the pendulum. A simplified model of the reaction-wheel pendulum is given by

$$\ddot{\theta}_1 - \sin(\theta_1) = -u$$
$$\ddot{\theta}_2 = u,$$

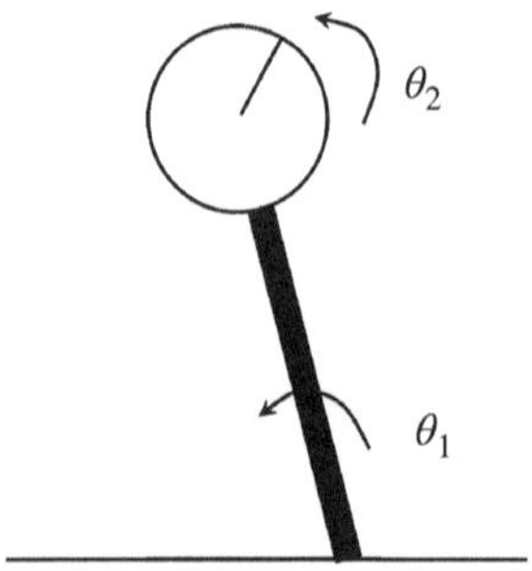

Figure 11.25 The reaction-wheel pendulum.

where θ_1 is the angle of the pendulum, θ_2 is the angle of the disk, and u is the torque produced by a motor attached to the disk. This is, therefore, a fourth-order nonlinear system.

We leave it as a case study to write this system as

$$\dot{x} = f(x) + g(x)u$$

and design a controller to stabilize the system about the inverted position (Problem 11.15).

Example 11.18 ***(Adaptive Control Example)***
Consider the first-order system

$$\dot{x} = ax + u, \quad a > 0,$$

which is unstable for $u = 0$. Clearly, the control $u = -kx$ stabilizes this system provided that $k > a$. If the coefficient a is unknown, then we do not know how large to choose the gain k to stabilize the system. The field of adaptive control treats problems like this by attempting to estimate unknown parameters, like a, which is called *indirect adaptive control*, or estimating k, which is called *direct adaptive control*. Let us consider the problem of estimating the gain k in order to stabilize the system. One way to do this is to make k a function of time and adjust the value of k online based on the measured value of x. Let us choose k as the solution of the differential equation

$$\dot{k} = \gamma x^2, \quad \gamma > 0. \tag{11.18}$$

Equation (11.18) is called the *MIT rule*. The rationale for this choice of k is that k will increase faster than x and so eventually k will get large enough to stabilize the system. With the control $u = -kx$, the system now becomes

$$\dot{x} = (a - k)x \tag{11.19}$$

$$\dot{k} = \gamma x^2. \tag{11.20}$$

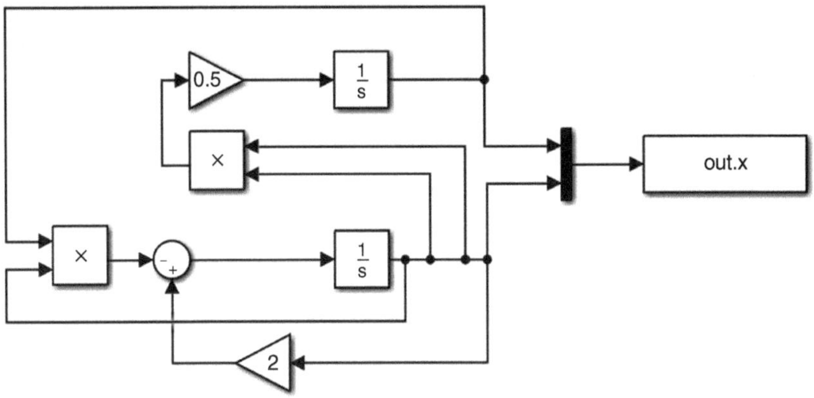

Figure 11.26 Simulink model of adaptive control using the MIT rule.

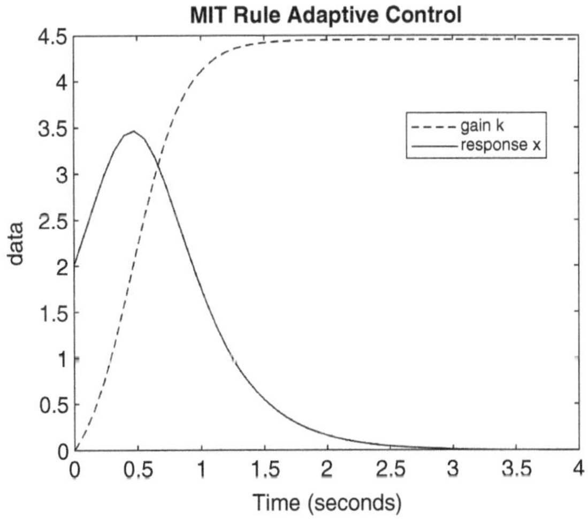

Figure 11.27 Response of the adaptive control system. The trajectory $x(t)$ initially rises since the equilibrium is unstable for small k. Once the gain k reaches a large enough value, the state $x(t)$ converges to zero and k converges to a constant value.

Note that now $k(t)$ is no longer a parameter, but rather a state variable and the system is a two-dimensional nonlinear system. Figure 11.26 is a Simulink model and Figure 11.27 shows the response with $a = 2$ and $\gamma = 0.5$. The final value of k depends on the gains and initial conditions.

Remark 11.3 Notice that if we define state variables $x_1 = y$ and $x_2 = k$, then the system (11.19)–(11.20) is the system previously considered in Example 5.10, which was analyzed using Lyapunov theory and LaSalle's invariance principle. Thus, Lyapunov theory is important to understand the responses of nonlinear control systems, where the concepts of poles zeros and transfer functions do not apply.

Problems

11.1 Examples of open-loop systems are a washing machine, a camera, and an automobile braking system. Describe how you could turn them into closed-loop systems: what sensors you would add and what variables you would control. What are the potential advantages to making these systems closed loop? What, if any, are the pitfalls?

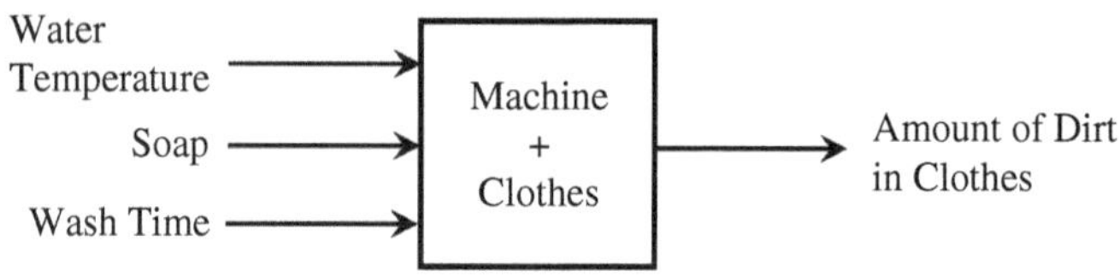

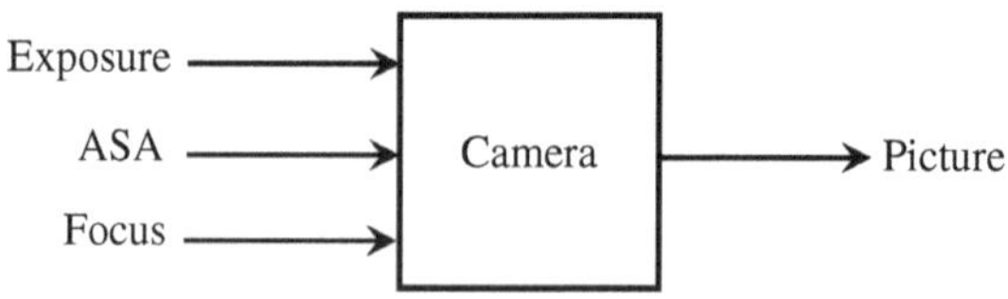

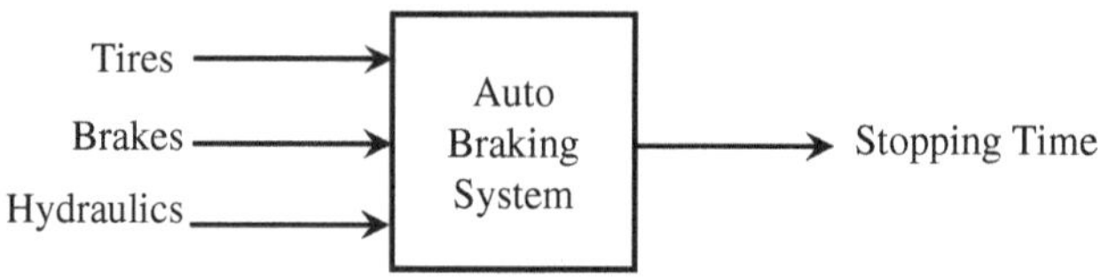

11.2 For the system below, show that $y \to r$ as $k \to \infty$, in other words, the error $r - y$ can be made arbitrarily small independent of the plant parameter a.

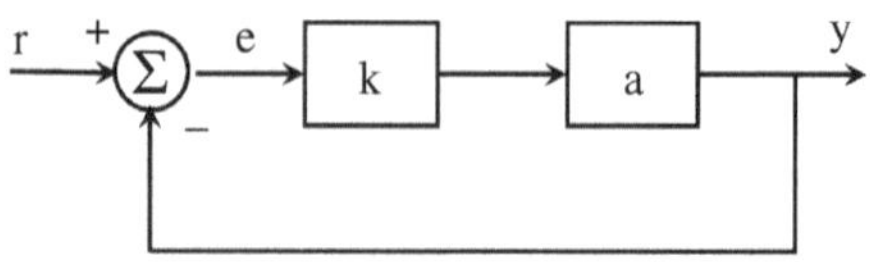

11.3 Derive Equation (11.3) as the step response of the damped oscillator system.

11.4 Verify Equation (11.14) showing that the system in Equation (11.13) is controllable.

11.5 Consider the problem of balancing the inverted pendulum on a cart. For suitable parameter values, a simple model for a linearized cart-pole system is

$$\ddot{x} = \theta + u$$
$$\ddot{\theta} = -2\theta - u,$$

where θ is the pendulum angle about the vertical, x is the cart position, and u is the acceleration of the cart, which is taken as the control input.

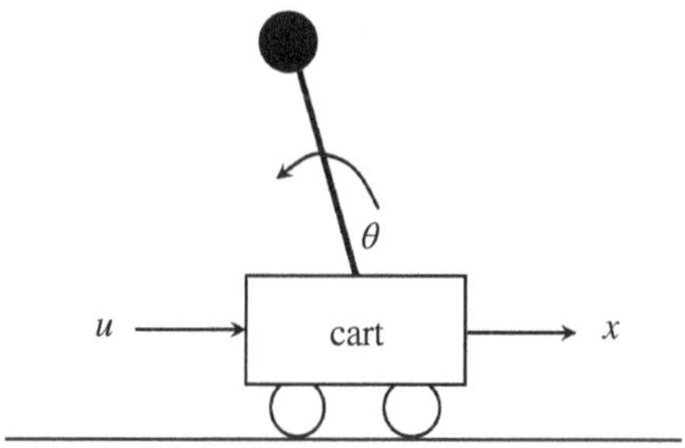

Design a control law to balance the inverted pendulum. Plot the pendulum angle, cart position, and control input starting from initial conditions $x(0) = 0$, $\theta(0) = 5$ degrees.

11.6 Suppose that only the cart position is measured. Show that the system is observable with output $y = x_1$ and design an observer/state-feedback controller to stabilize the system. Plot the pendulum angle, cart position, and control input starting from initial conditions $x(0) = 0$, $\theta(0) = 5$ degrees.

11.7 Consider a first-order system with transfer function $\frac{1}{s+1}$. Write a Matlab script using the functions `tf`, `step`, and `feedback` to simulate the step response for the system using a PI controller $u = k_p + \frac{k_i}{s}$. Choose the PI gains so that the closed-loop system reaches steady state in less than 2 seconds with minimal overshoot. Plot the step response from 0 to 2 seconds.

11.8 Consider the system in Example 11.9, equations (11.11)–(11.12). Take as output the angle θ_1. Using Simulink, design and simulate the step response using a PD controller for this system. Justify your choice of PD gains based on the system response.

11.9 Consider again the system in Example 8.6. Show that the system is observable if the output is $y = \theta_1$. Using Simulink, design an observer/state feedback controller and simulate the step response. Justify your choice of gains.

11.10 Consider the Simulink model below.

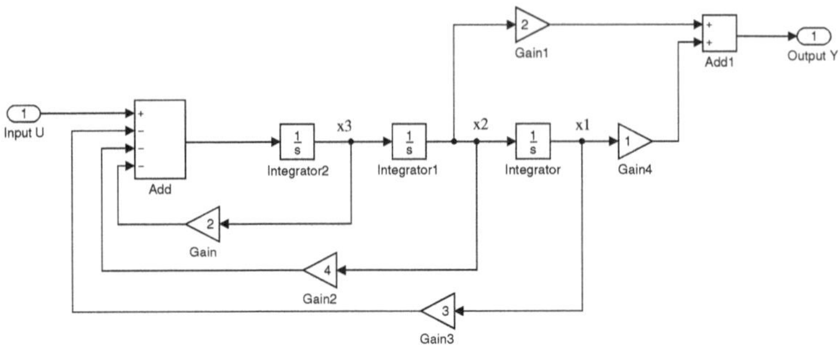

Write the state equations for this model. In other words, define the matrices A, B, and C so that

$$\dot{x} = Ax + Bu$$

$$y = Cx$$

11.11 Verify Equation (11.7) by direct calculation.

11.12 The model below is called the *quarter car model* and represents a model of an automobile suspension system.

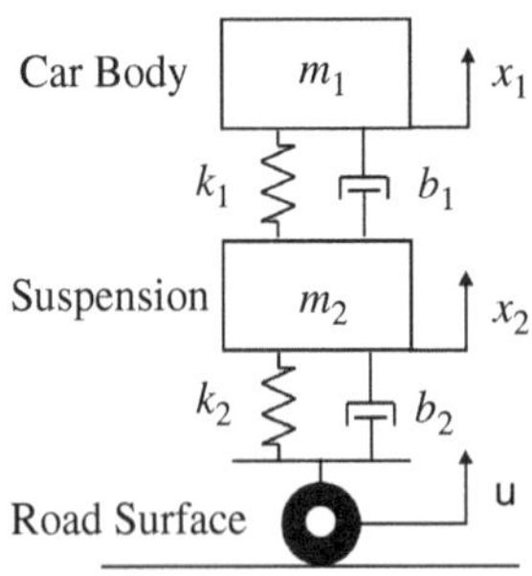

Derive the model for this system as a set of coupled second-order differential equations. Take the position of the road surface as input and the position of the car body as output. Compute the transfer function $Y(s)/U(s)$.

11.13 Create a Simulink diagram of the quarter car model above. Use the following parameters: $m_1 = 500$, $m_2 = 20$, $k_1 = 2500$, $b_1 = 1000$, $k_2 = 20,000$, $b_2 = 2000$. Plot the open-loop step response.

11.14 Show that the system below is controllable and observable if and only if $a \neq b$.

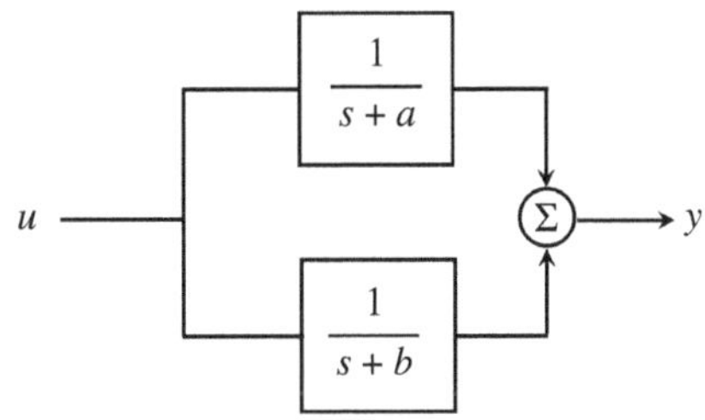

11.15 Consider the reaction-wheel pendulum from Example 11.17. Write the system in the form

$$\dot{x} = f(x) + g(x)u.$$

Linearize the system about the origin and design a linear state feedback controller to stabilize the system.

11.16 Consider a one-link robot arm as shown with input u equal to the torque at the joint, output $y = \theta$ the link angle, and $g = 9.8$ the acceleration due to gravity.

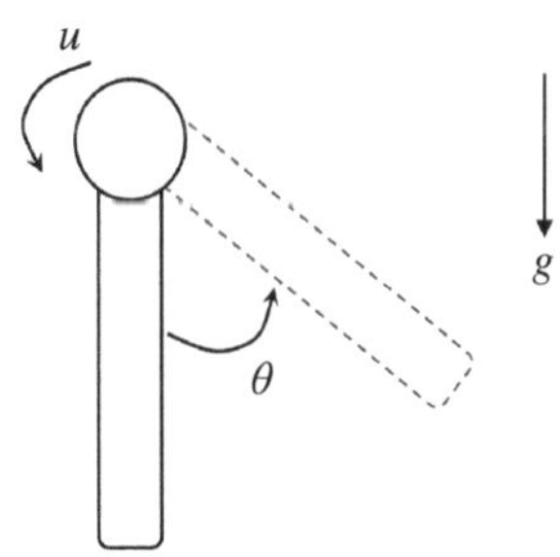

The equation governing this system is

$$\ddot{\theta} + g\sin(\theta) = u.$$

Using Simulink, model this system and design a PD control to move the arm to 45 degrees in 1 second. What is the steady-state error using your controller?
Now add an integral control to make the steady-state error zero.

12

Partial Differential Equation Models

12.1 Introduction

In this chapter we discuss systems modeled by partial differential equations (PDEs), also called distributed-parameter systems. Whereas ordinary differential equations (ODEs) describe the time evolution of a single vector function $x(t)$, PDEs involve functions of multiple variables $u(x_1, \dots, x_m)$, that may include both time and spatial variables. PDEs arise in applications involving sound, heat transfer, vibrating strings and plates, fluid flow, electromagnetics, transmission lines, gravity and other physical phenomena, including quantum mechanics.

12.1.1 Existence and Uniqueness of Solutions

The questions of existence and uniqueness of solutions of PDEs are more difficult than for ODEs. A general mathematical discussion of existence and uniqueness of solutions of PDEs is beyond the scope of this text. We will instead describe a few of the most important classical linear PDE systems, their applications, and their solutions.

In particular, we will discuss the wave equation, the heat equation, and Laplace's equation as important examples of linear second-order PDEs of the form

$$A\frac{\partial^2 u}{\partial x^2} + B\frac{\partial^2 u}{\partial x \partial y} + C\frac{\partial^2 u}{\partial y^2} = g\left(x, y, \frac{\partial u}{\partial x}, \frac{\partial u}{\partial y}\right), \tag{12.1}$$

where the unknown function $u(x, y)$ is a function of two variables x and y and A, B, C are constants. These equations are amenable to solution using several methods.

- Separation of variables. In this method a linear PDE can be reduced to two systems of ODEs, which are easier to solve.
- Method of characteristics. In this method, one tries to find special curves, called characteristics, on which the PDE can be reduced to an ODE.

Introduction to Modeling and Simulation: A Systems Approach, First Edition. Mark W. Spong.

Companion Website: www.wiley.com/go/spong/engineeringmodeling

- Integral transforms. Integral transforms, such as Laplace transforms or Fourier transforms, are applicable to linear PDEs.
- Numerical methods. A common numerical method for solutions of PDEs is the finite-element method (FEM). Software packages are readily available for FEM analysis of systems.

We first note that PDEs have a much richer set of possible solutions than ODEs. For example, consider the equation

$$\frac{\partial^2 u}{\partial x^2} + \frac{\partial^2 u}{\partial y^2} = 0. \tag{12.2}$$

This is Laplace's equation, which we will discuss in more detail later in the chapter. It is easy to check by direct calculation (Problem 12.4) that each of the functions below satisfies Equation (12.2):

$$u(x,y) = 1, \quad x + y, \quad x^3 - 3xy^2, \quad x^2 - y^2, \quad \ln(x^2 + y^2), \quad e^x \cos(y). \tag{12.3}$$

Thus, meaningful problems, for which we can obtain unique solutions, must include additional conditions, such as initial conditions and boundary conditions. Initial conditions, as in the case of ODEs, arise when time t is one of the variables of interest and specify the values of u and its time derivatives at $t = 0$. Boundary conditions arise when a given region R, defined by the spatial variables, is given and the values of a solution u are specified on the boundary of the region R.

12.1.2 Classification of Linear Second-Order PDEs

The solutions of PDEs of the form (12.1) depend fundamentally on the sign of the *discriminant* $B^2 - 4AC$. Equation (12.1) is called

$$\begin{aligned} \text{hyperbolic if } & B^2 - 4AC > 0, \\ \text{parabolic if } & B^2 - 4AC = 0, \\ \text{elliptic if } & B^2 - 4AC < 0. \end{aligned}$$

The most important examples of hyperbolic, parabolic, and elliptic PDEs are, respectively,

$$\begin{aligned} \text{the wave equation, } & a^2 \frac{\partial^2 u}{\partial x^2} - \frac{\partial^2 u}{\partial t^2} = 0 \text{ (hyperbolic)}, \\ \text{the heat equation, } & \frac{\partial^2 u}{\partial x^2} - a^2 \frac{\partial u}{\partial t} = 0 \text{ (parabolic)}, \\ \text{Laplace's equation, } & \frac{\partial^2 u}{\partial x^2} + \frac{\partial^2 u}{\partial y^2} = 0 \text{ (elliptic)}. \end{aligned}$$

12.2 The Wave Equation

In this section we consider the one-dimensional wave equation

$$\frac{\partial^2 u}{\partial t^2} = a^2 \frac{\partial^2 u}{\partial x^2},$$

which can describe, for example, the motion of a vibrating string, the vibrations in a linear bar, or pressure waves (sound) in a tube.

Derivation of the Wave Equation

A derivation of the one-dimensional wave equation can be found by imagining a bar or string as an array of equally spaced masses, m, interconnected by linear springs with stiffness k as shown in Figure 12.1. Let h be the spacing between the masses m and let $u(x, t)$ be the displacement of the mass at position x at time t. By Newton's second law, the force acting on the mass located at position $x + h$ is

$$F = ma(t) = m\frac{\partial^2}{\partial t^2}u(x + h, t).$$

By Hooke's law, the spring force on the same mass is the sum of the spring forces at the two ends of the mass,

$$F = k([u(x + 2h, t) - u(x + h)] + [u(x, t) - u(x + h, t)]).$$

Equating the inertial and spring forces results in

$$\frac{\partial^2}{\partial t^2}u(x + h, t) - \frac{k}{m}(u(x + 2h, t) - 2u(x + h, t) + u(x, t)).$$

Now, let N be the total number of such masses and let $L = Nh$ and $M = Nm$ be the total length and total mass, respectively, of the bar. Then we can write

$$\frac{\partial^2}{\partial t^2}u(x + h, t) = \frac{KL^2}{M}\frac{u(x + 2h, t) - 2u(x + h, t) + u(x, t)}{h^2},$$

where $K = k/N$. Taking the limit as $N \to \infty$ and $h \to 0$, we get

$$\frac{\partial^2 u(x, t)}{\partial t^2} = \frac{KL^2}{M}\frac{\partial^2 u(x, t)}{\partial x^2} = a^2\frac{\partial^2 u(x, t)}{\partial x^2}.$$

The constant $a = \sqrt{KL^2/M}$ is the *propagation speed*.

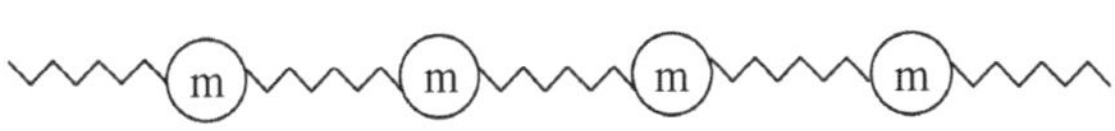

Figure 12.1 Coupled mass spring system.

12.2.1 The D'Alembert Solution

In this section we present a solution of the one-dimensional wave equation

$$\frac{\partial^2 u}{\partial t^2} = a^2 \frac{\partial^2 u}{\partial x^2}, \tag{12.4}$$

called the *D'Alembert solution*. In subsequent sections we will present an alternative solution based on the method of separation of variables, which leads to an infinite series solution.

Let f be any C^2 function. (Recall that C^2 means that the function has continuous second derivatives.) Then $f(x - at)$ is a solution of (12.4) since, by direct calculation

$$\frac{\partial^2 y}{\partial t^2} = a^2 f''(x - at), \quad \frac{\partial^2 y}{\partial x^2} = f''(x - at).$$

Likewise, if g is a C^2 function, then $g(x + at)$ is also a solution. In fact, it can be shown that any solution can be expressed as a sum

$$u(x, t) = f(x - at) + g(x + at).$$

To show this, we note that, with the change of variables

$$v = x - at, \quad w = x + at, \tag{12.5}$$

a direct calculation (Problem 12.8) shows that the wave equation becomes

$$\frac{\partial^2 u}{\partial v \partial w} = 0. \tag{12.6}$$

Thus, any function $u(v, w)$ satisfying the above must be of the form $f(v) + g(w)$, which is to say

$$u(x, t) = f(x + at) + g(x - at).$$

The interpretation of D'Alembert's solution is that the solution is the superposition of a right-traveling wave and a left-traveling wave.

12.2.2 Initial-Value Problem

Suppose we are now given initial conditions

$$u(x, 0) = f(x), \quad \frac{\partial u}{\partial t}(x, 0) = g(x),$$

specifying the initial shape and transverse velocity of the wave. Then the solution satisfying these conditions is

$$u(x, t) = \frac{1}{2}(f(x + at) + f(x - at)) + \frac{1}{2a} \int_{x-at}^{x+at} g(s)ds, \tag{12.7}$$

which is easily shown by direct calculation (Problem 12.9).

Example 12.1 Suppose that the initial shape of a vibrating string is a parabola on $[0,1]$,

$$f(x) = x(1-x), \quad 0 \leq x \leq 1,$$

and suppose $g(x) = 0$. Then a plot of D'Alembert's solution of the wave equation for several points in time with these initial conditions is shown in Figure 12.2.

D'Alembert Solution of a Vibrating String

```
clear
x = linspace(-1,1);
f0 = [@(x)x.*(1-x).*(0<=x & x<1)+x.*(1+x).*(-1<=x & x<0)];
f = @(x) f0(x - floor((x+1)/2)*2);
axis([0 1 -0.5 0.5]);hold on
title("D'Alembert Solution of a Vibrating String");
xlabel('x'); ylabel('y');
for t=-10:0.3:10
    solution = (1/2)*(f(x+t) + f(x-t));
    plot(x,solution,'k')
end
```

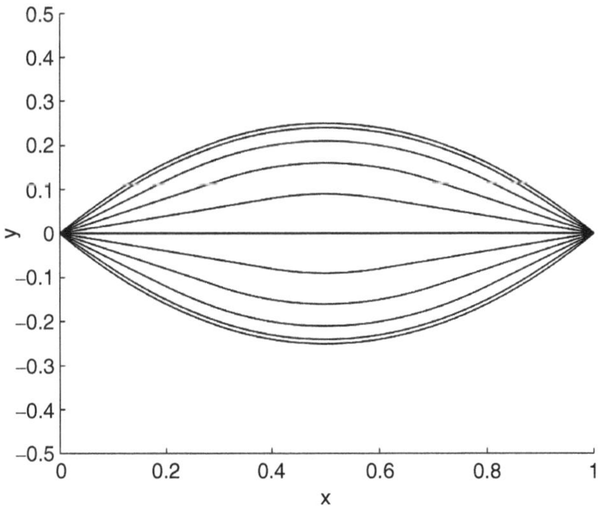

Figure 12.2 Snapshots of the vibrating spring with initial shape $f(x) = x(1-x)$. The solution is called a *standing wave*.

12.2.3 Separation of Variables

In this subsection we present a solution of the one-dimensional wave equation using the method of *separation of variables*. This leads to a solution expressed as an infinite series.

Background on Fourier Series

We first need to give some background on Fourier series that will be used to compute series solutions of the wave equation in this section and also of the heat equation and Laplace's equation in later sections. Suppose that a given function $f(x)$ is periodic with period $2L$. A Fourier series expresses $f(x)$ using sine and cosine functions as

$$f(x) = \frac{a_0}{2} + \sum_{n=1}^{\infty} a_n \cos\left(\frac{n\pi x}{L}\right) + \sum_{n=1}^{\infty} b_n \sin\left(\frac{n\pi x}{L}\right) \tag{12.8}$$

for suitable coefficients a_n and b_n. The main question is how to obtain the coefficients a_n, b_n.

Remark 12.1 Jean-Baptiste Joseph Fourier (1768–1830) introduced the series expansion that bears his name in his study of the heat equation. It is, of course, now recognized as a general method broadly applicable. A familiar special case is a musical note consisting of a superposition of a fundamental frequency and higher harmonics, each of which is a pure sinusoidal sound. The relative frequencies and amplitudes of the harmonics are what distinguish, for example, a middle C note on a trumpet from a middle C note on a violin.

The key to deriving the Fourier series expansion of a function $f(x)$ is to notice that the sine and cosine functions satisfy an orthogonality condition, namely,

$$\int_{-L}^{L} \sin\left(\frac{m\pi x}{L}\right) \sin\left(\frac{k\pi x}{L}\right) dx = 0 \quad \text{if } k \neq m,$$
$$\int_{-L}^{L} \cos\left(\frac{m\pi x}{L}\right) \cos\left(\frac{k\pi x}{L}\right) dx = 0 \quad \text{if } k \neq m,$$
$$\int_{-L}^{L} \sin\left(\frac{m\pi x}{L}\right) \cos\left(\frac{k\pi x}{L}\right) dx = 0 \quad \text{if } k \neq m.$$

The orthogonality conditions are not difficult to prove using various trigonometric identities. For example, since

$$\int_{-L}^{L} \cos\left(\frac{m\pi x}{L}\right) dx = \frac{L}{m\pi} \sin\left(\frac{m\pi x}{L}\right)\Bigg|_{-L}^{L} = 0,$$

then, using the trigonometric identity

$$2\sin(ax)\sin(bx) = \cos((a-b)x) - \cos((a+b)x),$$

we have that

$$\int_{-L}^{L} \sin\left(\frac{m\pi x}{L}\right)\sin\left(\frac{k\pi x}{L}\right)dx$$
$$= \int_{-L}^{L}\left(\frac{1}{2}\cos\left(\frac{(m-k)\pi x}{L}\right) - \frac{1}{2}\cos\left(\frac{(m+k)\pi x}{L}\right)\right)dx = 0.$$

The other orthogonality conditions follow similarly. The exception is when $m = k$. In this case, we have

$$\int_{-L}^{L} \sin\left(\frac{m\pi x}{L}\right)\sin\left(\frac{k\pi x}{L}\right)dx = \int_{-L}^{L}\sin^2\left(\frac{k\pi x}{L}\right)dx$$
$$= \int_{-L}^{L}\frac{1}{2}dx - \int_{-L}^{L}\cos\left(\frac{2k\pi x}{L}\right)dx = L.$$

Likewise,

$$\int_{L}^{L} \cos\left(\frac{m\pi x}{L}\right)\cos\left(\frac{k\pi x}{L}\right)dx = L$$

for $m = k$ and $k > 0$. For $k = 0$, we have

$$\int_{-L}^{L}\cos(0x)\cos(0x)dx = \int_{-L}^{L}dx = 2L.$$

With the preceding results as background, we can find the coefficients a_n and b_n in the Fourier series expansion in Equation (12.8) as follows. For $n = 0$, integrating both sides of Equation (12.8) from $-L$ to L gives

$$a_0 = \frac{1}{L}\int_{-L}^{L} f(x)dx,$$

since the integrals of the sine and cosine functions integrate to zero. The coefficient a_0 is the average value of $f(x)$ on the interval $[-L, L]$. For $n > 0$ we multiply both sides of Equation (12.8) by $\cos\left(\frac{n\pi x}{L}\right)$ and integrate from $-L$ to L. Using the orthogonality conditions we obtain

$$a_n = \frac{1}{L}\int_{-L}^{L} f(x)\cos\left(\frac{n\pi x}{L}\right)dx.$$

Likewise, the coefficients b_n are found by multiplying both sides of Equation (12.2) by $\sin\left(\frac{n\pi x}{L}\right)$ and integrating from $-L$ to L, resulting in

$$b_n = \frac{1}{L}\int_{-L}^{L} f(x)\sin\left(\frac{n\pi x}{L}\right)dx.$$

A result known as *Dirichlet's theorem* [18] gives one characterization of the class of functions that can be expressed in terms of Fourier series.

Theorem 12.1 ***(Dirichlet's Theorem)*** *Suppose $f(x)$ is periodic with period $2L$ and is bounded on $[-L, L]$. If $f(x)$ has at most a finite number of maxima and minima and a finite number of discontinuities on $[-L, L]$, then the finite sum*

$$\frac{a_0}{2} + \sum_{n=1}^{N} a_n \cos\left(\frac{n\pi x}{L}\right) + \sum_{n=1}^{N} b_n \sin\left(\frac{n\pi x}{L}\right)$$

converges to $f(x)$ as $N \to \infty$ at points where f is continuous and to the average value $\frac{1}{2}[f(t^+) + f(t^-)]$ at points where f is discontinuous.

The points t^+ and t^- respectively denote the right- and left- hand limits of $f(x)$ at points of discontinuity.

Even and Odd Functions

A function $f(x)$ is odd if $f(-x) = -f(x)$ and even if $f(-x) = f(x)$. Note that $\sin(x)$ is an odd function and $\cos(x)$ is an even function. If $f(x)$ is an odd function, then we can express $f(x)$ as a Fourier sine series

$$f(x) = \sum_{n=1}^{\infty} b_n \sin\left(\frac{n\pi x}{L}\right),$$

and if $f(x)$ is an even function, then we can express $f(x)$ as a Fourier cosine series

$$f(x) = \frac{a_0}{2} + \sum_{n=1}^{\infty} a_n \cos\left(\frac{n\pi x}{L}\right).$$

This is a useful observation since any function $f(x)$ can always be written as the sum of an even and an odd function as

$$f(x) = \frac{1}{2}(f(x) + f(-x)) + \frac{1}{2}(f(x) - f(-x)).$$

If $f(x)$ is an odd function, then the Fourier coefficients become

$$a_0 = a_n = 0$$
$$b_n = \frac{1}{L}\int_{-L}^{L} f(x) \sin\left(\frac{n\pi x}{L}\right) dx = \frac{2}{L}\int_{0}^{L} f(x) \sin\left(\frac{n\pi x}{L}\right) dx.$$

Likewise, if $f(x)$ is an even function, the Fourier coefficients become

$$a_0 = \frac{1}{L}\int_{-L}^{L} f(x)dx = \frac{2}{L}\int_{0}^{L} f(x)dx$$
$$a_n = \frac{1}{L}\int_{-L}^{L} f(x) \cos\left(\frac{n\pi x}{L}\right) dx, = \frac{2}{L}\int_{0}^{L} f(x) \cos\left(\frac{n\pi x}{L}\right) dx$$
$$b_n = 0.$$

Example 12.2 ***(Square Wave)***
Let $f(x)$ be the square wave function

$$f(x) = \begin{cases} 1, & \text{if } 0 < x < \pi, \\ -1, & \text{if } -\pi < x < 0, \end{cases}$$

which is periodic with period $L = 2\pi$ (Figure 12.3).

Since $f(x)$ is an odd function, we express it as a Fourier sine series

$$f(x) = \sum_{n=1}^{\infty} b_n \sin(nx).$$

The coefficients b_n are given by

$$b_n = \frac{2}{\pi}\int_0^{\pi} f(x)\sin(nx)dx$$

$$= \frac{2}{\pi}\left[\frac{-\cos(nx)}{n}\right]_0^{\pi} = \frac{2}{\pi}\left[\frac{2}{1}, \frac{0}{2}, \frac{2}{3}, \frac{0}{4}, \frac{2}{5}, \dots\right].$$

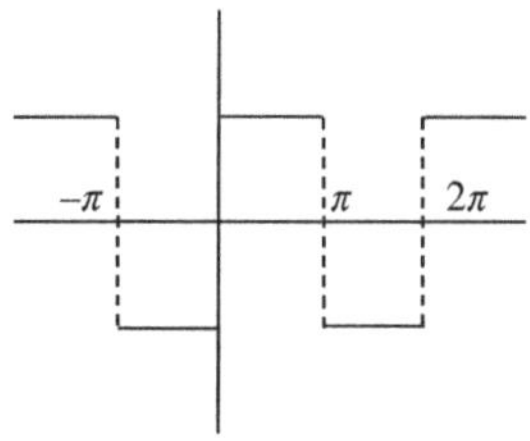

Figure 12.3 The square wave function is a piecewise-continuous, odd function, which takes values ± 1, and which is discontinuous at multiples of π.

Therefore, the Fourier sine series expansion of the square wave function (see Figure 12.4) is

$$f(x) = \frac{4}{\pi}\left(\sin(x) + \frac{1}{3}\sin(3x) + \frac{1}{5}\sin(5x) + \dots\right).$$

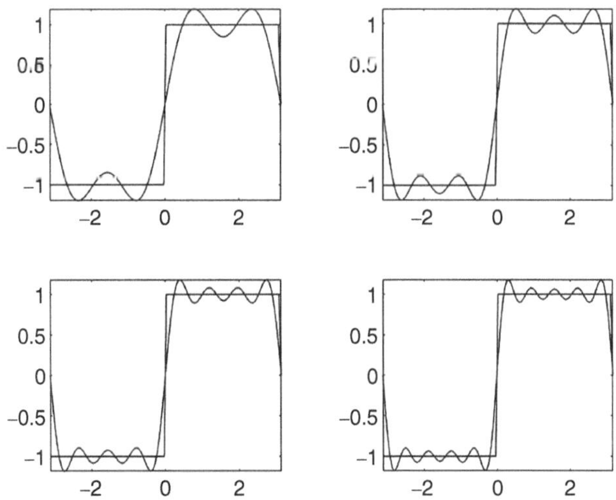

Figure 12.4 Fourier series approximation of the square wave function with $n = 3, 5, 7, 9,$ terms in the series, respectively.

Example 12.3 ***(Delta Function)***
As a second example, consider the delta function $f(x) = \delta(x)$ defined as

$$\delta(x) = \begin{cases} 0, & \text{if } x \neq 0 \\ \infty, & \text{if } x = 0, \end{cases}$$

together with the condition that

$$\int_{-\infty}^{\infty} \delta(x)dx = 1.$$

The delta function does not satisfy the conditions of Dirichlet's theorem, and the Fourier series expansion of the delta function is not a convergent series. We can nevertheless approximate the delta function by a finite series consisting of the first few terms of the Fourier series expansion (Figure 12.5). Since $\delta(x)$ is an even function we choose a Fourier cosine series defined on $[-\pi, \pi]$:

$$\delta(x) = a_0 + \sum_{n=1}^{\infty} a_n \cos(nx).$$

So, in this case we have

$$a_0 = \frac{1}{2\pi} \int_{-\pi}^{\pi} \delta(x)dx = \frac{1}{2\pi},$$

and, for $n > 0$,

$$a_n = \frac{1}{\pi} \int_{-\pi}^{\pi} \delta(x) \cos(nx)dx = \frac{1}{\pi}.$$

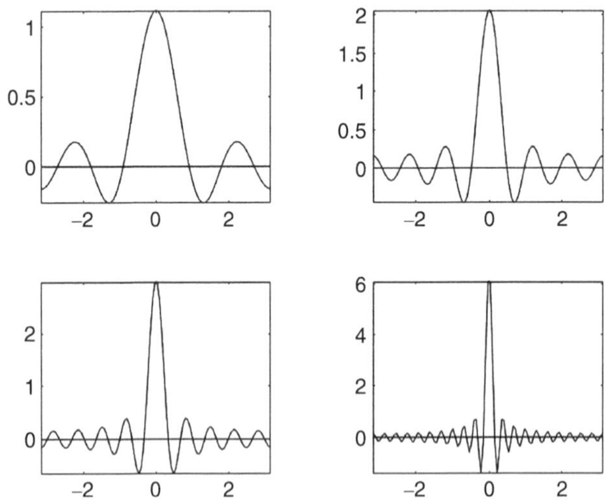

Figure 12.5 Fourier series approximation of the delta function for $n = 3, 6, 9, 20$, respectively.

Therefore, we have

$$\delta(x) = \frac{1}{2\pi} + \frac{1}{\pi}(\cos(x) + \cos(2x) + \cos(3x) + \cos(4x) + \ldots).$$

Let us return now to the one-dimensional wave equation

$$\frac{\partial^2 u}{\partial t^2} = a^2 \frac{\partial^2 u}{\partial x^2}$$

with initial conditions

$$u(x, 0) = f(x), \quad \frac{\partial u}{\partial t}(x, 0) = g(x),$$

where $f(x)$ and $g(x)$ are given functions of x. In the method of separation of variables we search for a solution expressed as a product of a function of x and a function of t as

$$u(x, t) = X(x)T(t).$$

In this case, the partial derivatives with respect to x and t can be expressed as ordinary derivatives

$$\frac{\partial^2 u}{\partial x^2} = X''(x)T(t), \quad \frac{\partial^2 u}{\partial t^2} = X(x)\ddot{T}(t).$$

Therefore, suppressing the arguments of X and T, the wave equation becomes

$$X\ddot{T} = a^2 X'' T,$$

which implies that

$$\frac{\ddot{T}}{T} = a^2 \frac{X''}{X}. \tag{12.9}$$

Now, since the left-hand side of Equation (12.9) is independent of x and the right-hand side is independent of t, the only way for Equation (12.9) to hold for arbitrary x and t is if both sides are identically equal to a constant. Thus, we can write

$$\frac{\ddot{T}}{T} = a^2 \frac{X''}{X} = \mu = \text{constant}.$$

This leads to two separate ODEs,

$$\ddot{T} = \mu T, \quad X'' = \frac{\mu}{a^2} X.$$

A value of $\mu \geq 0$ does not yield meaningful solutions, so we suppose $\mu = -\lambda^2$ for $\lambda > 0$, and write

$$\ddot{T} = -\lambda^2 T, \quad X'' = -\frac{\lambda^2}{a^2} X.$$

These equations both have the form of the harmonic oscillator equation and so the general solutions are, respectively,

$$T(t) = A\cos(\lambda t) + B\sin(\lambda t)$$
$$X(x) = C\cos\left(\frac{\lambda}{a}x\right) + D\sin\left(\frac{\lambda}{a}x\right).$$

Thus, the solution can be written as the product

$$u(x,t) = \left(C\cos\left(\frac{\lambda}{a}x\right) + D\sin\left(\frac{\lambda}{a}x\right)\right)(A\cos(\lambda t) + B\sin(\lambda t)). \tag{12.10}$$

The four parameters in Equation (12.10) are determined by the initial conditions and boundary conditions. We will illustrate this in the example below.

Example 12.4 Let us consider a vibrating string with both ends clamped on the interval $[0, 1]$. This leads to the boundary conditions

$$u(0,t) = u(1,t) = 0, \quad \text{for all } t \geq 0. \tag{12.11}$$

Suppose that we take the same initial conditions as in the D'Alembert solution example above, namely,

$$u(x,0) = f(x) = x(1-x)$$
$$u_t(x,0) = g(x) = 0.$$

Using the given boundary conditions in Equation (12.11) results in

$$C = 0, \quad \text{and} \quad \sin\left(\frac{\lambda}{a}\right) = 0,$$

which implies that $\lambda = \lambda_n = n\pi a$, $n = 1, 2, \ldots$. Therefore, any function of the form

$$u(x,t) = \sin(n\pi x)(A_n\cos(n\pi at) + B_n\sin(n\pi at))$$

is a solution. By superposition, the sum of such solutions is also a solution, so we can write

$$u(x,t) = \sum_{n=1}^{\infty}\sin(n\pi x)(A_n\cos(n\pi at) + B_n\sin(n\pi at)).$$

Now we can determine the remaining constants A_n and B_n by applying the given initial conditions. The condition $u(x,0) = f(x)$ gives

$$f(x) = \sum_{n=1}^{\infty}A_n\sin(n\pi x).$$

Thus, $f(x)$ is expressed as a Fourier sine series and so the coefficients A_n are given by

$$A_n = 2\int_0^1 f(x)\sin(n\pi x)dx.$$

The condition $\frac{\partial u}{\partial t}(x,0) = 0$ leads to

$$g(x) = \sum_{n=1}^{\infty}n\pi aB_n\sin(n\pi x),$$

and so

$$B_n = \frac{2}{n\pi a}\int_0^1 g(x)\sin(n\pi x)dx = 0.$$

With $f(x) = x(1-x)$ and $a = 1$ it is a straightforward exercise in calculus (Problem 12.14) to compute the integral

$$A_n = 2\int_0^1 x(1-x)\sin(n\pi x)dx = \frac{8}{n^3\pi^3}\sin^2(n\pi/2). \tag{12.12}$$

Notice that

$$\sin^2(n\pi/2) = \begin{cases} 0, & \text{if } n \text{ is even} \\ 1, & \text{if } n \text{ is odd,} \end{cases}$$

and therefore we can write $u(x, t)$ as

$$u(x,t) = \sum_{k=1}^{\infty} \frac{8}{(2k-1)^3\pi^3}\sin((2k-1)\pi x)\cos((2k-1)\pi t).$$

Computing the first few coefficients gives

$$\begin{aligned} u(x,t) = {} & 0.2580\sin(\pi x)\cos(\pi t) + 0.0096\sin(3\pi x)\cos(3\pi t) \\ & + 0.0021\sin(5\pi x)\cos(5\pi t) + \dots \end{aligned}$$

Symbolic Solution of the Wave Equation

The Matlab script below computes the series solution for the above example using symbolic integration and plots an animation of the response (from which Figure 12.6 is taken). The number of terms in the series approximation is equal to num.

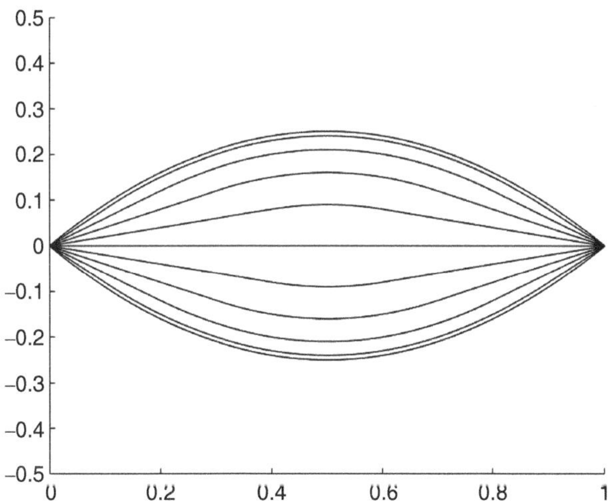

Figure 12.6 Series solution of the wave equation. Note that the response is the same as the previous response computed via the D'Alembert solution.

Solution of a Vibrating String via Separation of Variables

```
clear
syms t x n
integrand = 2*x*(1-x)*sin(n*pi*x);
h = int(integrand,x,0,1);  % symbolic integration
a1 = matlabFunction(h);
num=20;An=zeros(num,1);
for i=1:num
    k=2*i-1;
    An(i)=a1(k);                % compute An for n=odd
end
term = 0;
for i= 1:length(An)
    k=2*i-1;
    term = term+An(i)*sin(k*pi*x)*cos(k*pi*t);
end
    u=matlabFunction(term);
    z=linspace(0,1);            % begin the animation
axis([0 1 -1/2 1/2]);hold on
for t=0:0.1:10
    h = plot(z,u(t,z),'k')
    pause(0.1);
    delete(h);
end
```

12.3 The Heat Equation

We next consider the one-dimensional heat equation

$$\frac{\partial u}{\partial t} = \alpha \frac{\partial^2 u}{\partial x^2}.$$

The heat equation describes how the temperature in a solid medium evolves over time. However, the same equation arises in other applications as well. In probability theory, the heat equation arises in studies of random walks and Brownian motion. In financial systems, it is related to the dynamics of financial systems, the so-called Black–Scholes equation [11].

Derivation of the Heat Equation

Let us consider a thin, insulated rod so that heat flows only along the length of the rod (Figure 12.7). Consider a small section of length $2h$ and let

$$\frac{u(x+h,t)-u(x-h,t)}{2}$$

be the average temperature over this section at time t.

Since heat always flows from areas of high temperature to areas of low temperature, the temperature at $u(x,t)$ will increase or decrease depending on the sign of

$$\frac{2}{h^2}\left(\frac{u(x+h,t)-u(x-h,t)}{2}-u(x,t)\right)$$

and rate of change of temperature will be proportional to the magnitude of this difference. Therefore, we can approximate the rate of change of temperature as

$$\frac{\partial u(x,t)}{\partial t}=\alpha\frac{2}{h^2}\left(\frac{u(x+h,t)-u(x-h,t)}{2}-u(x,t)\right).$$

Taking the limit as $h\to 0$ gives the one-dimensional heat equation

$$\frac{\partial}{\partial t}u(x,t)=\alpha\frac{\partial^2}{\partial x^2}u(x,t).$$

The constant α is called the *thermal conductivity*.

The method of separation of variables can be used to solve the heat equation in the same way as we showed for the wave equation. In this case, with $u(x,t)=X(x)T(t)$ we get

$$\frac{\dot{T}}{\alpha T}=\frac{X''}{X}=-\lambda^2,$$

which leads to the pair of ODEs

$$\dot{T}=-\lambda^2\alpha T,\quad X''=-\lambda^2 X.$$

Therefore, since the equation in T is a first-order equation, we have

$$T(t)=Ae^{-\lambda^2\alpha t}\quad\text{and}\quad X(x)=B\sin(\lambda x)+C\cos(\lambda x),$$

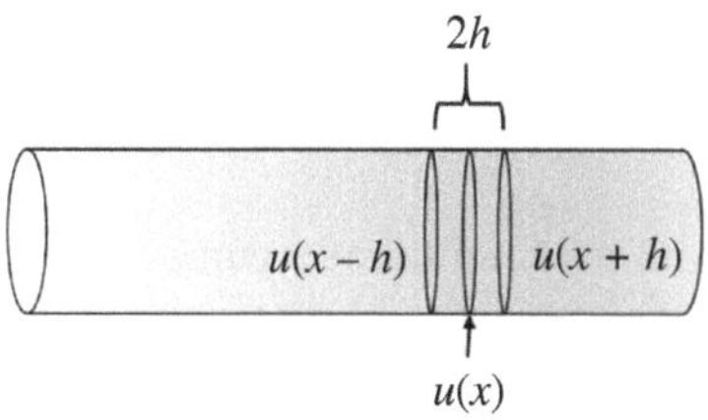

Figure 12.7 Insulate rod with heat flowing in one dimension along the rod.

and, as before, by superposition we look for solutions of the form

$$u(x,t) = \sum_{n=1}^{\infty} e^{-\lambda_n^2 a t}(B_n \sin(\lambda_n x) + C_n \cos(\lambda_n x)),$$

where we have combined the constant A with B_n and C_n.

As an example, suppose that we take the initial and boundary conditions as

$$u(x,0) = f(x), \quad \text{for } x \in [0, L]$$

and

$$u(0,t) = 0 = u(L,t), \quad \text{for } t \geq 0.$$

The first condition specifies the initial temperature distribution along the bar of length L. The second condition says that the temperature at each end of the bar is kept at zero degrees Celsius for all time. Under these conditions, it is now straightforward to determine that

$$C = 0 \quad \text{and} \quad \lambda = \frac{n\pi}{L}.$$

Therefore, the solution is

$$u(x,t) = \sum_{n=1}^{\infty} B_n \sin\left(\frac{n\pi}{L}x\right) e^{-\frac{n^2\pi^2 a}{L^2}t},$$

where

$$B_n = \frac{2}{L}\int_0^L f(x) \sin\left(\frac{n\pi}{L}x\right) dx.$$

Example 12.5 ***(Adapted from Duffy [20])***
Suppose we take $L = \pi$ and $f(x) = x(\pi - x)$. Then a straightforward calculation shows

$$\begin{aligned} B_n &= \frac{2}{\pi}\int_0^{\pi} x(\pi - x)\sin(nx)dx \\ &= 2\int_0^{\pi} x\sin(nx)dx - \frac{2}{\pi}\int_0^{\pi} x^2 \sin(nx)dx \\ &= 4\frac{1-(-1)^n}{n^3\pi}, \end{aligned}$$

and so the solution becomes

$$u(x,t) = \frac{8}{\pi}\sum_{k=1}^{\infty} \frac{1}{(2k-1)^3} \sin((2k-1)x) \exp -(2k-1)^2 a^2 t.$$

Figure 12.8 shows the temperature profile along the bar over time with $a = 1$.

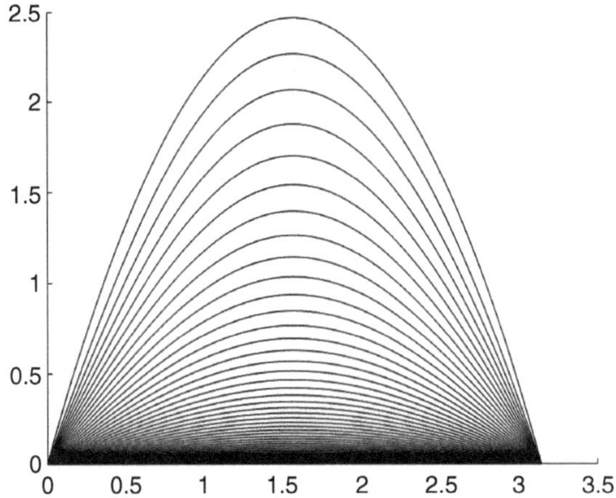

Figure 12.8 The temperature $u(x, t)$ is a thin bar with both ends held at zero temperature and initial temperature distribution given by $x(\pi - x)$. The temperature converges exponentially to zero over time.

12.4 Laplace's Equation

The third equation we consider is Laplace's equation

$$\frac{\partial^2 u}{\partial x^2} + \frac{\partial^2 u}{\partial y^2} = 0.$$

Laplace's equation arises in numerous applications. First, we can see that it represents the steady-state heat equation in two spatial dimensions, i.e., with

$$\frac{\partial u}{\partial t} = 0.$$

Laplace's equation also arises in problems in electrostatics, gravitation, and fluid dynamics, as well as in image processing. Solutions of Laplace's equation are known as *harmonic functions*. The general theory of solutions to Laplace's equation is known as potential theory, and is a vast subject in its own right [42].

Laplace's Equation on a Rectangle

Let us consider the following problem. Find a function $u(x, y)$ satisfying Laplace's equation on a rectangle with sides of length L and K with the boundary conditions

$$u(x, 0) = f_1(x), \quad u(x, K) = f_2(x), \quad 0 \le x \le L$$
$$u(0, y) = g_1(y), \quad u(L, y) = g_2(y), \quad 0 \le y \le K,$$

as illustrated in Figure 12.9.

We can solve this problem using superposition by solving four simpler problems, each of which takes three of the boundary conditions equal to zero, for example,

$$u(x, 0) = 0, \quad u(x, K) = f(x), \quad 0 \le x \le L,$$
$$u(0, y) = 0, \quad u(L, y) = 0, \quad 0 \le y \le K.$$

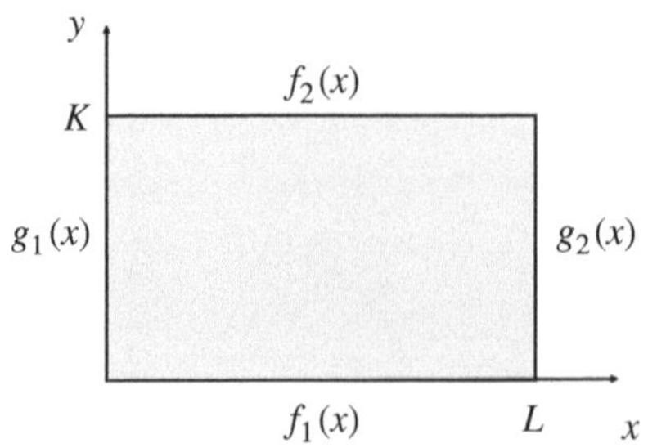

Figure 12.9 Rectangular region with values of $u(x, y)$ specified on the boundary.

The general solution will then be the sum of solutions of the four simpler problems. As before, we use the method of separation of variables and write $u(x, y) = X(x)Y(y)$ and substitute this expression into Laplace's equation to obtain

$$\frac{\partial^2 u}{\partial x^2} + \frac{\partial^2 u}{\partial y^2} = X''Y + XY'' = 0,$$

which leads to the expression

$$\frac{X''}{X} = -\frac{Y''}{Y} = -\lambda^2$$

and therefore the two equations

$$X'' + \lambda^2 X = 0, \quad Y'' - \lambda^2 Y = 0.$$

These two equations have solutions

$$X(x) = A\sin(\lambda x) + B\cos(\lambda x)$$
$$Y(y) = C\sinh(\lambda y) + D\cosh(\lambda y).$$

Using the given boundary conditions immediately gives us $B = 0$ and $D = 0$. The condition $u(L, y) = 0$ implies that $A\sin(\lambda L) = 0$ and therefore, $\lambda = \frac{n\pi}{L}$ for $n = 1, 2, \ldots$. Thus, we look for solutions of the form

$$u_n(x, y) = A_n \sin\left(\frac{n\pi}{L}x\right)\sinh\left(\frac{n\pi}{L}y\right),$$

where we combine the constants A_n and C_n into a single constant. Applying the remaining boundary condition yields

$$u_n(x, K) = f(x) = A_n \sin\left(\frac{n\pi}{L}x\right)\sinh\left(\frac{n\pi}{L}K\right).$$

Therefore, we have

$$A_n \sinh\left(\frac{n\pi}{L}K\right) = \frac{2}{L}\int_0^L f(x)\sin\left(\frac{n\pi}{L}x\right)dx,$$

which implies, finally, that

$$A_n = \frac{2}{L\sinh\left(\frac{n\pi}{L}K\right)}\int_0^L f(x)\sin\left(\frac{n\pi}{L}x\right)dx.$$

We leave it as an exercise (Problem 12.15) to compute the other three solutions to complete the problem.

Laplace's Equation on a Disk

As a second example, we consider Laplace's equation on the unit disk, where we specify boundary conditions $u_0(x, y)$ on the unit circle $x^2 + y^2 = 1$. For this example, given the circular symmetry, it is useful to use polar coordinates $x = r\cos(\theta)$, $y = r\sin(\theta)$. Laplace's equation in polar coordinates is given by [20]

$$\frac{\partial^2 u}{\partial r^2} + \frac{1}{r}\frac{\partial u}{\partial r} + \frac{1}{r^2}\frac{\partial^2 u}{\partial \theta^2}.$$

We therefore look for a solution of the form

$$u(x, y) = u(r, \theta) = a_0 + \sum_{n=1}^{\infty} a_n r^n \cos(n\theta) + \sum_{n=1}^{\infty} b_n r^n \sin(n\theta)$$

with boundary condition $u(1, \theta) = g(\theta)$ for some function $g(\theta)$, in which case

$$g(\theta) = a_0 + \sum_{n=1}^{\infty} a_n \cos(n\theta) + \sum_{n=1}^{\infty} b_n \sin(n\theta).$$

The above equation is then the usual Fourier series expansion for $g(\theta)$. As a specific example, let us suppose that $g(\theta) = \delta(\theta)$, the δ function. Since $\delta(\theta)$ is an even function (i.e., $\delta(-\theta) = \delta(\theta)$), it follows that the coefficients b_n are all equal to zero so we have

$$\delta(\theta) = a_0 + \sum_{n=1}^{\infty} a_n \cos(n\theta),$$

from which it follows, from the usual formulae for Fourier series, that $a_0 = \frac{1}{2\pi}$ and $a_n = \frac{1}{\pi}$ for $n > 1$. With some additional calculation it can be shown that the complete solution is found by summing the infinite series for $u(r, \theta)$ as

$$u(r, \theta) = \frac{1}{2\pi}\frac{1 - r^2}{1 + r^2 - 2r\cos(\theta)}. \tag{12.13}$$

12.5 Numerical Solution of PDEs

The numerical solution of ordinary differential equations in Chapter 7 started by discretizing the time variable. For numerical solution of partial differential equations one must discretize multiple variables. As an example, let us consider the one-dimensional heat equation

$$\frac{\partial u}{\partial t} = \frac{\partial^2 u}{\partial x^2}, \quad 0 \leq x \leq 1,\ t \geq 0$$

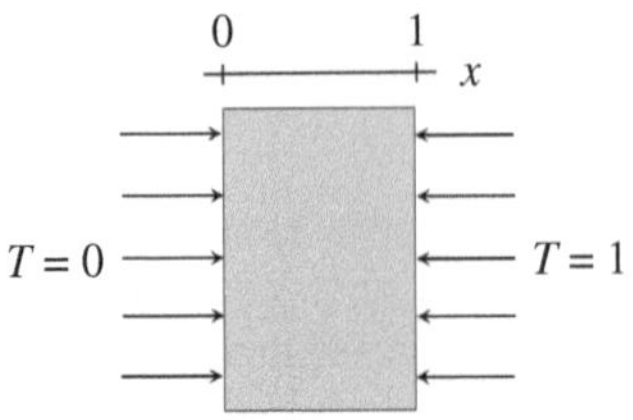

Figure 12.10 Slab of width 1. The left-hand side is held at temperature $T = 0$ and the right-hand side is held at temperature $T = 1$.

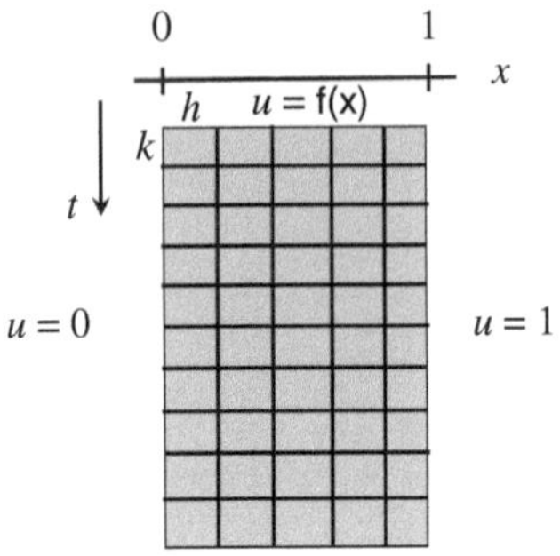

Figure 12.11 Grid of mesh points discretizing time in increments of k and space in increments of h.

with initial condition

$$u(x, 0) = f(x)$$

and boundary conditions

$$u(0, t) = 0, \quad u(1, t) = 1.$$

The variable $u(x, t)$ represents the temperature distribution of a slab of width 1 in which the left- and right-hand ends are kept at temperature $T = 0$ and $T = 1$, respectively (Figure 12.10).

To solve the equation numerically we discretize both time and space. A finite-difference approximation of the equation is

$$\frac{1}{k}(u_{i,j+1} - u_{ij}) = \frac{1}{h^2}(u_{i+1,j} - 2u_{ij} + u_{i-1,j}).$$

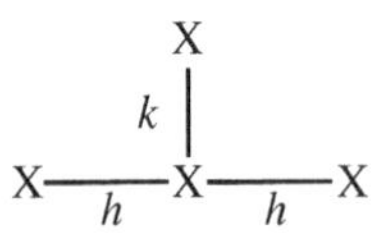

Figure 12.12 Points used in the finite-difference approximation.

Using a grid of mesh points as shown in Figure 12.11, the finite-difference approximation involves four points in the grid (Figure 12.12).

Solving for $u_{i,j+1}$ gives

$$u_{i,j+1} = (1 - 2r)u_{ij} + r(u_{i+1,j} + u_{i-1,j}), \tag{12.14}$$

where $r = \frac{k}{h^2}$. It can be shown that the iterative procedure defined by the above will converge if $r \leq 1/2$.

The Matlab pdepe function

The Matlab function pdepe solves parabolic partial differential equations such as the above heat equation. The general form of the function is

```
sol = pdepe(m,PDEFUN,ICFUN,BCFUN,XMESH,TSPAN).
```

The function PDEFUN specifies the PDE to be solved, which should take the form

$$\mathtt{c(x,t,u,dudx)dudt} = \mathtt{x}^{-\mathtt{m}}\mathtt{D(x}^{\mathtt{m}}\mathtt{f(x,t,u,dudx))/dx} + \mathtt{s(x,t,u,dudx)}.$$

For the above heat equation example we have, therefore,

$$\mathtt{c} = 1, \quad \mathtt{m} = 0, \quad \mathtt{f} = \mathtt{dudx}, \quad \mathtt{s} = 0.$$

The function ICFUN specifies the initial conditions, BCFUN specifies the boundary conditions, XMESH specifies the discretization in the x-variable, and TSPAN is the simulation time.

The boundary conditions are specified in the form

$$\mathtt{p(x,t,u)} + \mathtt{q(x,t)f(x,t,u,dudx)} = 0.$$

The boundary condition function should be of the form

$$\mathtt{[pl,ql,pr,qr]} = \mathtt{BCFUN(xl,ul,xr,ur,t)}.$$

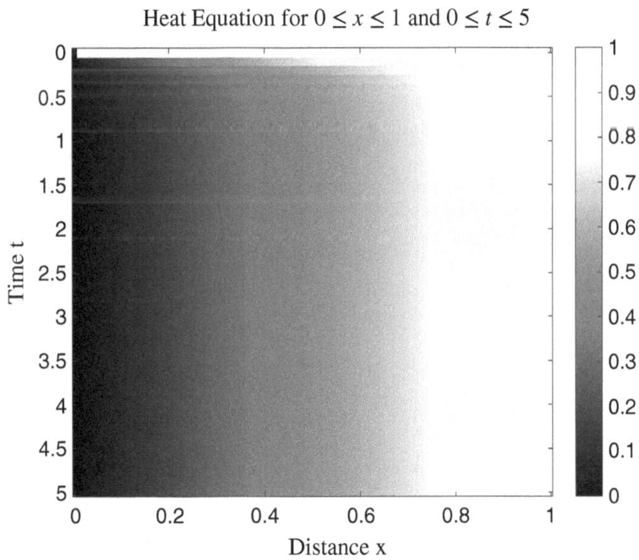

Figure 12.13 Temperature distribution generated by the Matlab simulation using the function pdepe.

In our example above,

$$\texttt{pl} = \texttt{ul}, \quad \texttt{ql} = 0, \; \texttt{pr} = \texttt{ur} - 1, \quad \texttt{qr} = 0.$$

The Matlab script below implements this example and plots the temperature distribution over time along the bar (see Figure 12.13).

Heat Equation Example

```
function heateq
L=1;           % length of the bar
tend=5;        % simulation time
m = 0;
x = linspace(0,L,100);
t = linspace(0,tend,50);
sol = pdepe(m,@pdex1pde,@pdex1ic,@pdex1bc,x,t);
% The next few commands generate the temperature plot
colormap hot
imagesc(x,t,sol)
colorbar
xlabel('Distance x','interpreter','latex')
ylabel('Time t','interpreter','latex')
title('Heat Equation for $0 \le x \le 1$ and
$0 \le t \le 5$','interpreter','latex')
% ---------------
function [c,f,s] = pdex1pde(x,t,u,DuDx)
c = 1;
f = DuDx;
s = 0;
% ---------------
function u0 = pdex1ic(x)
u0 = 1;
% ---------------
function [pl,ql,pr,qr] = pdex1bc(xl,ul,xr,ur,t)
pl = ul;
ql = 0;
pr = ur-1;
qr = 0;
```

Problems

12.1 Compute the Fourier cosine series expansion for the function shown below.

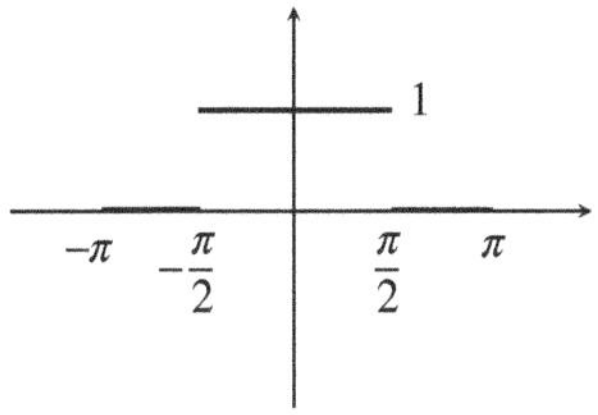

Assume that the function has period 2π. Use Matlab to plot the first three partial sums.

12.2 Computer the Fourier series for the function

$$f(t) = t\cos\left(\frac{\pi t}{L}\right), -L < t < L.$$

Use Matlab to plot several of the partial sums.

12.3 Show that if the function $f(x)$ has Fourier coefficients a_n and b_n and the function $g(x)$ has Fourier coefficients a_n^* and b_n^*, then the function $h(x) = \alpha f(x) + \beta g(x)$, where α and β are constant, has Fourier coefficients $\alpha_n + \alpha_n^*$ and $\beta_n + \beta_n^*$.

12.4 Verify by direct substitution that the following functions satisfy Laplace's equation:

$$u(x,y) = 1, \quad xy, \quad x+y, \quad x^3 - 3xy^2, \quad x^2 - y^2, \quad \ln(x^2 + y^2), \quad e^x\cos(y).$$

12.5 Verify that the following functions are solutions of the wave equation:

$$u = x^2 + 4t^2,$$
$$u = \cos(4t)\sin(x).$$

12.6 Verify that the following functions are solutions of the heat equation:

$$u = e^{-t}\sin(3x),$$
$$u = e^{-2t}\cos(x).$$

12.7 Show by explicitly evaluating the discriminant that the wave equation is hyperbolic, the heat equation is parabolic, and Laplace's equation is elliptic.

12.8 Show that the change of variables in the wave equation (12.4) given by (12.5) leads to Equation (12.6).

12.9 Verify Equation (12.7) by direct calculation.

12.10 Consider the wave equation describing the deflection $u(x, t)$ of a vibrating string of $L = 1$ with ends fixed. Suppose the initial deflection is given by $u(x, 0) = f(x)$ and initial velocity $u_t(x, 0) = 0$. Write a Matlab program to animate D'Alembert's solution of the equation for time $t = 0$ to $t = 10$ seconds. Test your program for the function $f(x) = \sin(2\pi x)$.

12.11 Complete Example 7.1 by deriving solutions for the torsional vibrations of a one-dimensional shaft in the case of fixed–free and free–free ends.

12.12 Write a Matlab script to solve the problem in Example 12.5 and generate Figure 12.8.

12.13 Solve the heat equation for $0 < x < \pi$ and $t > 0$ subject to the boundary conditions $u(0, t) = u(\pi, t) = 0$ and the following initial conditions:

(1) $u(x, 0) = A,\ A = \text{constant},$

(2) $u(x, 0) = \sin^2(x),$

(3) $u(x, 0) = x.$

12.14 Verify the integral in Equation (12.12).

12.15 Complete the solution of Laplace's equation on the rectangle with sides of length K and L by solving the three remaining subproblems where the boundary conditions are zero on three of the four sides.

12.16 Derive Equation (12.13).

12.17 Write a Matlab program to solve the one-dimensional heat equation using the discretization method in Equation (12.14).

13

Complex Networks

13.1 Introduction

This chapter deals with the analysis and applications of *complex networks*. Complex networks are ubiquitous in natural and engineered systems. Roughly speaking, a *network* is a collection of entities or *agents* that interact, i.e., that are *interconnected*, as in Figure 13.1.

Mathematically, we will represent networks as graphs with agents represented as nodes or vertices of the graph. Edges or links connecting the vertices represent the interactions among the agents. The agents may represent, for example, people in a social network, workstations in a computer network, power plants in an electrical grid, or proteins in a cell, to name but a few. The interconnections may represent friendships and acquaintances, fiber-optic cable, transmission lines, or protein–protein interactions in such examples. We discuss general questions applicable to a wide range of complex networks, such as:

- Connectedness – with which agent can a given agent communicate?
- Centrality – which agents are the most important?
- Distance – how closely are agents related?
- Clustering – how do agents form groups and cliques?
- Degree distribution – how many neighbors does each agent have?
- Synchronization – how does global behavior emerge from local interactions?

A key feature of complexity is *emergent behavior*, which is the idea that the whole is more than the sum of the parts. For example, in biology, how does life emerge from the interactions of atoms and molecules, none of which is individually alive? How does intelligence emerge from brain neurons, each of which is a cell no more individually intelligent than a muscle cell or liver cell? How do schools of fish and flocks of birds seem to move as a single entity when each individual animal has only limited sensing information and cannot coordinate its own motion with

Introduction to Modeling and Simulation: A Systems Approach, First Edition. Mark W. Spong.

Companion Website: www.wiley.com/go/spong/engineeringmodeling

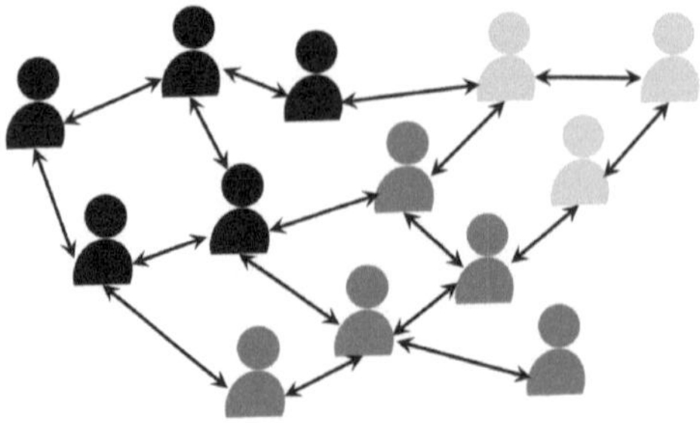

Figure 13.1 A Network of interconnected agents. Depending on the application, the agents represent entities such as individuals in social networks, computers in a wireless network, neurons in the brain, banks in a financial network, drones in a formation, and so on. The arrows connecting agents represent exchange of information between agents.

all others? The answer lies not only in the individuals (molecule, neuron, animal) but also in the way they are interconnected, in other words, in the structure of the network to which they belong.

13.1.1 Examples of Complex Networks

In this section we briefly describe several examples of complex networks. It is a striking observation that many seemingly disparate networks, such as a cell metabolic network and the World Wide Web, share many common attributes in their structure that can be abstracted and analyzed by the same mathematical tools.

The Internet

Most computers are now connected to the internet, which represents the collection of computers and routers exchanging information over wired and wireless networks, typically using the *Internet Protocol*. The internet is a rapidly growing network and the manner in which it grows fundamentally determines its large-scale structure.

Telecommunications Networks

Computer networks and the internet are both examples of telecommunications networks, which also includes phone and data networks, increasingly involving the internet protocol for voice and data transmission.

The World Wide Web

The World Wide Web (WWW) is the network of (HTML) webpages. It is a *logical network*, containing hundreds of billions of pages, and is distinct from the internet, which is a *physical network*. Each webpage in the WWW is a node and an edge is a hyperlink to another page. Such edges are called directed since there is a direction associated to the edge, i.e., page A points to page B but the reverse is not necessarily true. The WWW is therefore an example of a directed graph or digraph. Like the internet, the WWW is also highly dynamic in that it is constantly and rapidly changing.

The Electric Power Grid

The electric power grid is a large-scale network consisting of power generating plants, transmission lines, substations, and end-users and is an example of a complex network where dynamic effects are important to consider. The problem of *cascading failure* in power networks is caused when failure of one node causes successive overload of other nodes, leading to multiple failures. Cascading failure can also occur in many other systems, including financial networks and computer networks.

Social Networks

Social networks represent the social interactions among individuals and are increasingly important in business applications, in epidemiology, as well as in politics. Among social networks that have been extensively studied are the network of Hollywood actors, coauthorship networks, and scientific citation networks. Social networks, such as Facebook, Twitter, Instagram, and others are having an enormous impact on all aspects of society.

Financial Networks

Financial networks are also a type of social network involving financial transactions and flow of money between banks and can be studied as complex networks. In this case, the agents are the financial institutions and the links are financial transactions between financial institutions.

Biological Networks

Biological networks are another important class of complex networks. Examples of biological networks include gene regulation networks, protein networks, and cell metabolic networks.

Ecological networks, such as food webs, can also be classified under biological networks. Food webs are directed networks of interaction among species, essentially "who eats whom." The dynamics of such networks can be complex. For example, populations of species may remain stable, or may oscillate as we saw with the predator–prey model previously. Also, the introduction of a new species into an ecosystem can disrupt an otherwise stable system.

Transportation Networks

Transportation networks include highway networks, airline networks, subway networks, railroad networks, canal and river networks, shipping networks, and pipeline networks. There are important structural differences among such networks in terms of degree distribution and other network metrics that we will study. For example, in a highway network connecting cities, most cities have a relatively small number of highways entering and leaving the city. By contrast, the airline network has a few airports that are hubs, meaning they have huge numbers of connections, while most other airports have a very small number of connections. Thus highway networks and airline networks are fundamentally different in how they operate, how resilient they are to failures in any portion of network, and so on.

13.2 Graph Theory: Basic Concepts

In this section we give basic definitions and properties of graphs and networks [74].

Definition 13.1 ***(Graph)*** By a graph G, we mean a finite set $V = \{v_1, v_2, \dots, v_n\}$, called the *vertex set*, together with a set $E = \{e_1, e_2, \dots, e_m\}$, called the *edge set*. An edge e_k in the edge set is associated with an ordered pair $\langle v_i, v_j \rangle$ in the vertex set, meaning that edge e_k connects vertices v_i and v_j. If both $\langle v_i, v_j \rangle$ and $\langle v_j, v_i \rangle$ are edges for every i and j, the graph is *undirected*. Otherwise, it is *directed*.

- The number of vertices, n, is called the *order of* G.
- The number of edges, m, is called the *size of* G.

Example 13.1 ***(An Undirected Graph)***
The graph G in Figure 13.2 is an undirected graph with vertex set $\{v_1, v_2, v_3, v_4\}$, edge set $\{e_1, e_2, e_3, e_4, e_5\}$, and therefore the order is 4 and the size is 5.

Definition 13.2 ***(Adjacency)*** Two vertices, v_i and v_j, are said to be *adjacent* if the edge $\langle v_i, v_j \rangle \in E(G)$. Otherwise v_i, v_j are *nonadjacent*. If $e = \langle v_i, v_j \rangle$ is an edge in $E(G)$ then v_i and v_j are *incident to* the edge e. If $\langle u, v \rangle$ and $\langle u, w \rangle$ are distinct edges of G, then $\langle u, v \rangle$ and $\langle u, w \rangle$ are called *adjacent edges*.

For example, in Figure 13.2, v_1 and v_2 are adjacent vertices, while v_1 and v_4 are nonadjacent. Also, e_1 and e_3 are adjacent edges, while the edges e_1 and e_4 are non-adjacent.

Definition 13.3 ***(Simple Graph and Multigraph)*** A graph is a *simple graph* if it has no loops, i.e., no edges connecting a vertex to itself. A *multigraph* is a graph that may have multiple edges connecting nodes.

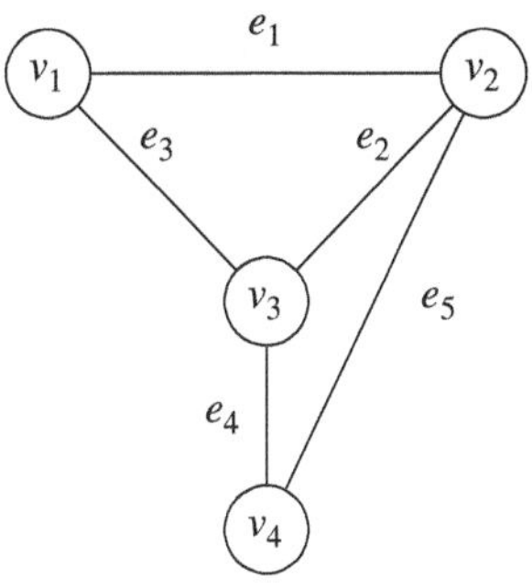

Figure 13.2 An undirected graph G of order 4 and size 5. We say that G is a $(4, 5)$ graph in order to specify its order and size.

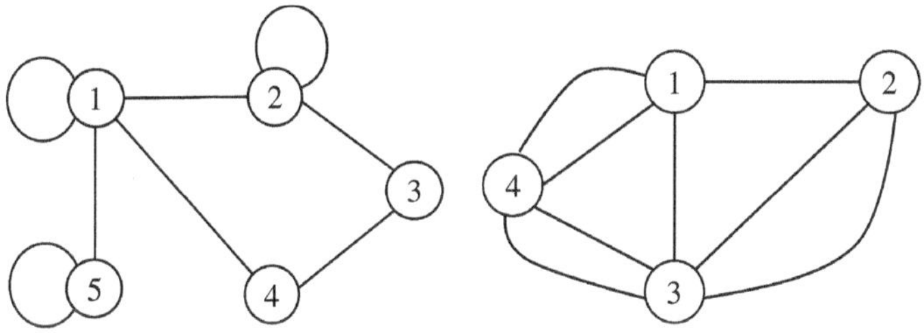

Figure 13.3 A nonsimple graph (left) containing loops and a multigraph (right) with multiple edges between nodes.

Figure 13.2 is therefore a simple graph. Figure 13.3 shows a nonsimple graph and a multigraph.

Weighted Graphs and Networks

In many applications it is useful to assign values or weights to the edges of a graph. The weights represent quantities like distance, cost, probability, or other values associated with the edges.

Definition 13.4 ***(Weighted Graph)*** A *weighted graph* is a graph together with a function $f : E \to \mathbb{R}$. The function f assigns a real number (a weight) to each edge of the graph.

Figure 13.4 shows a weighted, directed graph.

Remark 13.1 A weighted graph is also called a *network*. Since, given any graph, we can trivially assign a weight of 1 to each edge, we use the terms "graph" and "network" interchangeably. Note that a transition diagram for a Markov chain in

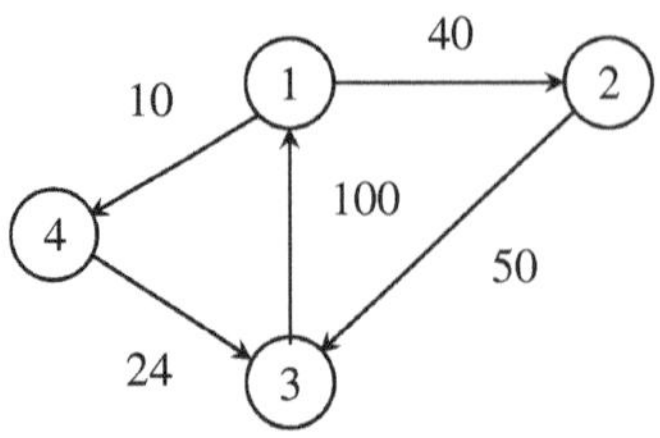

Figure 13.4 Example of a weighted, directed graph. The edge weights may represent cost, distance, or other values, depending on the application.

Chapter 10 is an example of a weighted, directed network. The graph of a Markov chain is generally not a simple graph due to the presence of loops.

Definition 13.5 ***(Vertex Degree)*** The number of edges incident to a given vertex $v \in G$ is the *degree* of the vertex. For a directed graph the *in-degree* is the number of edges entering v. The *out-degree* is the number of edges leaving v. The degree of v in a directed graph is then the sum of the in-degree and out-degree.

Definition 13.6 ***(Regular Graph)*** Let G be a simple, undirected graph. If every vertex of G has the same degree r, we say that the graph is *r-regular*.

Figure 13.5 shows a 3-regular graph.

Definition 13.7 ***(The Complete Graph K_p)*** A simple undirected graph is *complete* if every two vertices are adjacent. This means that the degree of each vertex is the maximum possible. For a graph with p vertices, the degree of each vertex of a complete graph is therefore $p - 1$. The complete graph with p vertices is typically denoted by K_p.

The first five complete graphs $K_1, \dots, K_5$ are shown in Figure 13.6.

Proposition 13.1 A complete graph of order p is $(p - 1)$-regular. The complete graph K_p has $\frac{p(p-1)}{2}$ edges.

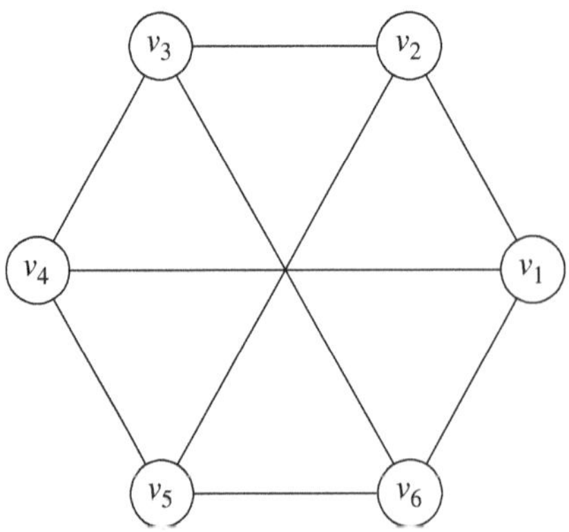

Figure 13.5 A 3-regular graph. Each vertex in the graph has degree 3.

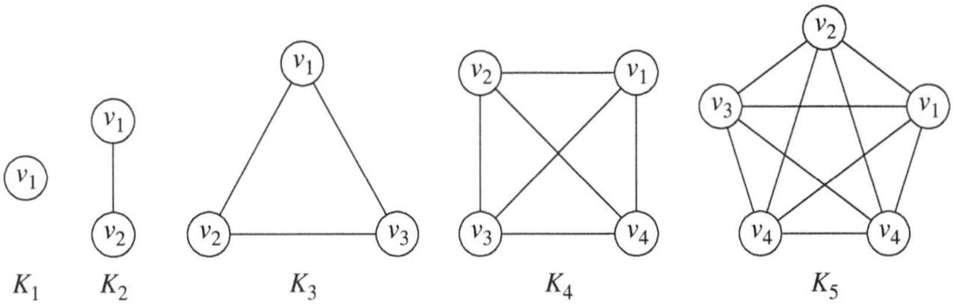

Figure 13.6 The first five complete graphs $K_1, \ldots, K_5$.

Proof: The proof is left as an exercise (Problem 13.17).

13.2.1 Graph Isomorphism

Intuitively, two graphs are the same if it is possible to relabel and redraw one to appear identical to the other. This is made precise by the notion of *isomorphism*. Recall our definition of the standard form for the transition matrix of a Markov chain.

Definition 13.8 ***(Graph Isomorphism)*** Let G_1 and G_2 be graphs. An *isomorphism* from G_1 to G_2 is a one-to-one mapping $\phi : V(G_1) \to V(G_2)$ onto $V(G_2)$ such that two vertices u_1, v_1 are adjacent in G_1 if and only if vertices $\phi(u_1)$, $\phi(v_1)$ are adjacent in G_2. We say that G_1 and G_2 are isomorphic graphs if there exists an isomorphism between them. If $\phi : V(G_1) \to V(G_2)$ is an isomorphism, then the inverse mapping ϕ^{-1} exists and is an isomorphism from G_2 to G_1. ϕ^{-1} is defined as follows. $\phi^{-1}(v_2) = v_1 \in G_1$ such that $\phi(v_1) = v_2$.

The isomorphism relation is an *equivalence relation* on the set of graphs. This means that the isomorphism relation partitions the set of all graphs into *equivalence classes.*

- It is easy to see that if G_1 and G_2 are isomorphic graphs then they have the same order and same size. Also, the degrees of the vertices are the same.
- However, if two graphs have the same order and size and the vertices have the same degrees, they are not necessarily isomorphic.

Example 13.2 The graphs shown in Figure 13.7 each have order 6 and size 9. Moreover, the degree of each vertex is the same in each graph. It is left as an exercise (Problem 13.9) to show that the two graphs are not isomorphic.

13.2.2 Connectivity

Connectivity is one of the most important properties of a network. Intuitively, a graph is connected if it is possible to travel between any two vertices along edges of the graph. More precisely, we can state the following definition.

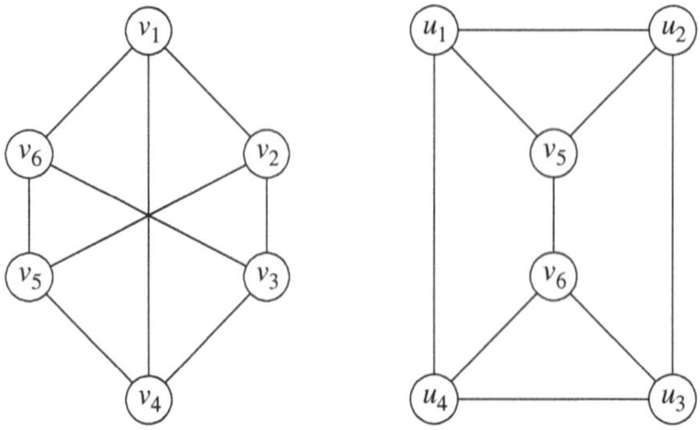

Figure 13.7 The two graphs above are not isomorphic despite the fact that they have the same order and size and the their respective vertices have the same degrees. In order to be isomorphic, two graphs must have the same adjacency structure.

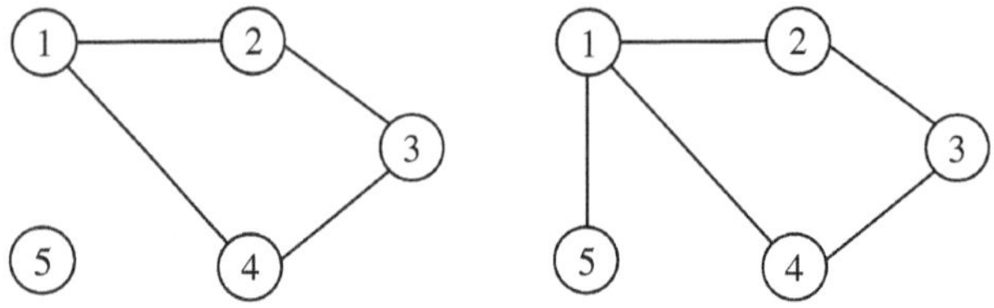

Figure 13.8 The graph on the right is connected. It is possible to reach every node from every other node by following edges of the graph. The graph on the left is not connected. Node 5 cannot be reached from the other nodes nor can any of the other nodes be reached from node 5.

Definition 13.9 ***(Connectedness)*** We say that two vertices are connected if there is a sequence of distinct vertices and edges, called a *path*, connecting them. We say that a graph is connected if every two vertices are connected; otherwise it is disconnected.

Example 13.3 In Figure 13.8, the graph on the right is connected. It is possible to start at any vertex of the graph and reach any other vertex by following edges of the graph. The graph on the left is not connected since node 5 is not connected by an edge to any other node in the graph.

We also need to introduce the notions of subgraph and graph components as follows.

Definition 13.10 ***(Subgraph)*** Let G be a graph. A graph H is a *subgraph* of G if $V(H) \subset V(G)$ and $E(H) \subset E(G)$. If a graph F is isomorphic to a subgraph H of G, then F is also called a subgraph of G.

Example 13.4 In Figure 13.9 the graph on the right is a subgraph of the graph on the left.

Figure 13.9 Example of a subgraph. The graph on the right is a subgraph of the graph on the left.

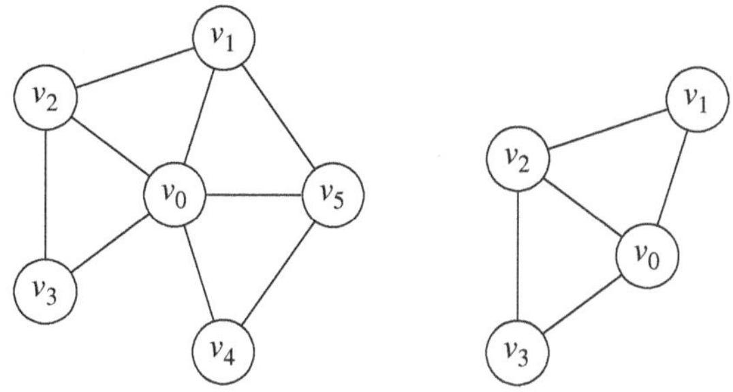

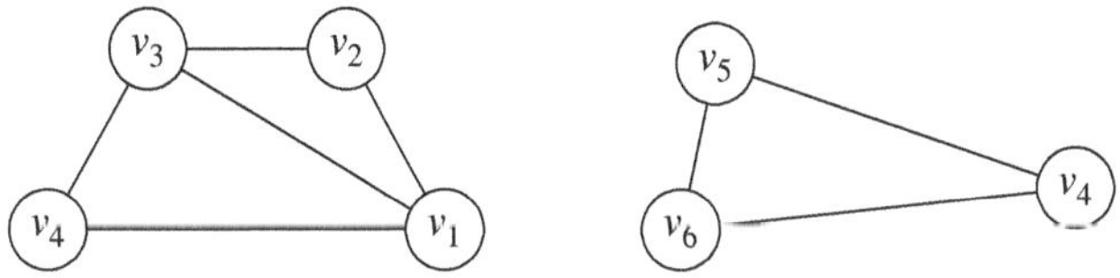

Figure 13.10 An example of a graph containing two connected components

Definition 13.11 ***(Component)*** A connected subgraph H of a graph G is a *component* of G if H is not contained in any connected subgraph of G having more vertices or edges than H. Thus, a component H of G is a maximally connected subgraph of G.

Figure 13.10 shows a graph with two connected components. A connected graph, therefore, has only one component. For a directed graph the notion of connectivity is different from that of an undirected graph. In a digraph, two vertices v_i and v_j may be connected by edges but there may not necessarily be a directed path from v_i to v_j. We would like to distinguish such a situation from one where there is no edge at all connecting v_i and v_j. We do this by introducing the notions of strong and weak connectivity.

Definition 13.12 ***(Strong and Weak Connectivity)*** A directed graph is *strongly connected* if there is a directed path between any two vertices. The *strong components* are the maximal strongly connected subgraphs. A directed graph is *weakly connected* if there is an undirected path connecting any two vertices).

In other words, strongly connected means that one can find a path between vertices respecting the direction of the arrows. Weakly connected means that one can find a path, possibly ignoring the sense of the arrows. Figure 13.11 illustrates the differences between weak connectivity and strong connectivity for directed graphs.

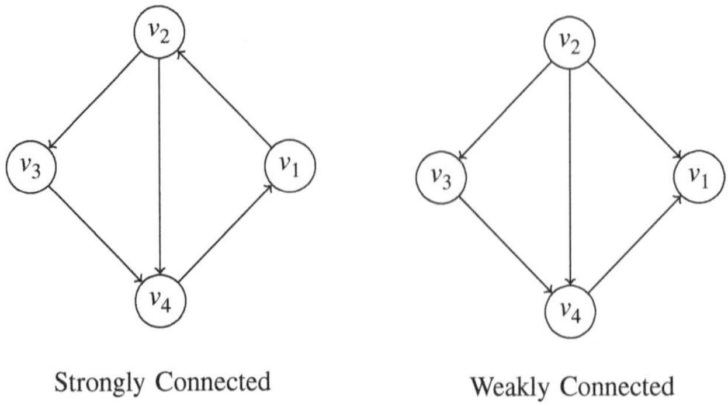

Figure 13.11 The graph on the left is strongly connected. The graph on the right is weakly connected. There is no path from v_1 to another vertex that respects the direction of the arrows.

Cut Vertices and Bridges

It is often important to know if a connected graph remains connected if a given vertex or edge is removed. For example, if a certain router fails in a computer network, is it still possible to reroute packets from any computer to any other? If not, then that particular router has a special importance in the network. This notion is captured by the notion of *bridges* and *cut vertices.*

Definition 13.13 ***(Bridges and Cut Vertices)***

- If e is an edge of a graph G then $G - e$ is the subgraph of G having the same vertex set and all edges of G except e.
- If v is a vertex of G then $G - v$ is the subgraph of G whose vertex set consists of all vertices of G except v and all edges of G except those edges that are incident with v.
- An edge e is a *bridge* if $G - e$ is disconnected.
- A vertex v is a *cut vertex* if $G - v$ is disconnected.

Theorem 13.1 *Let G be a connected graph. An edge e of G is a bridge if and only if e does not lie on any cycles (i.e., closed paths) of G.*

Example 13.5 Consider the graph in Figure 13.12. We see that only edge $\langle v_4, v_5\rangle$ does not lie on any cycle and therefore is a bridge. Any edge other than $\langle v_4, v_5\rangle$ can be removed without affecting connectivity of the graph. Also note that vertices v_4 and v_5 are cut vertices.

Figure 13.12 A graph with cut vertices and a bridge.

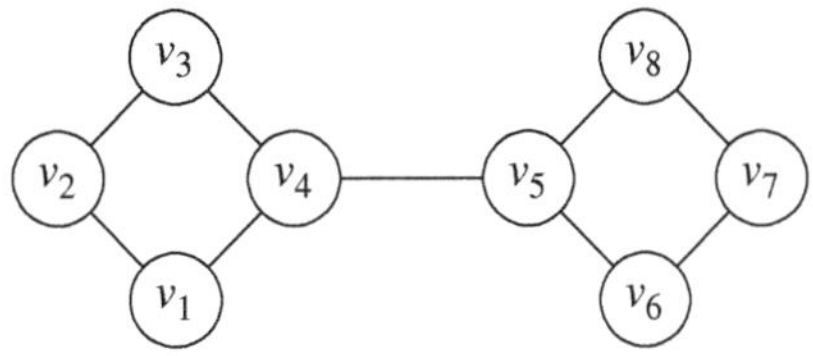

Figure 13.13 A tree contains no cycles (i.e., no closed paths). This also means that every edge is a bridge.

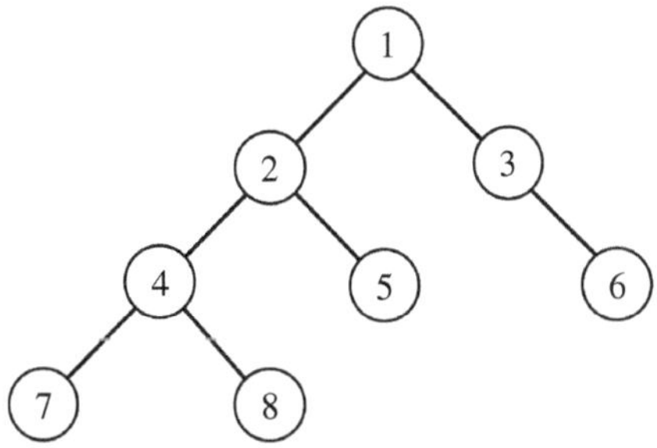

13.2.3 Trees

In this subsection we will discuss *trees*, which form an important subclass of graphs. Trees find applications in efficient data storage and search methods in computer science, in transportation networks, in project management (project management and review technique), and a host of other applications.

Definition 13.14 ***(Tree)*** A *tree* is a connected simple graph with no cycles (Figure 13.13). Without the requirement of being connected, the graph is called a *forest*.

Definition 13.15 ***(Spanning Tree)*** Let G be a connected graph. An acyclic connected subgraph S of G containing all of the vertices of G is called a *spanning tree* (Figure 13.14).

Figure 13.14 Graph G and spanning tree T (bold edges).

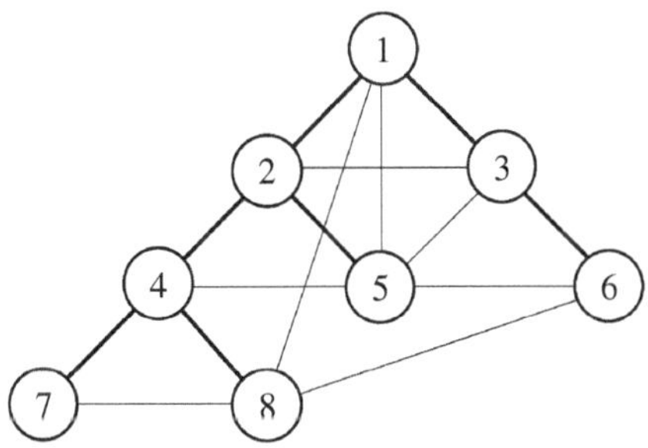

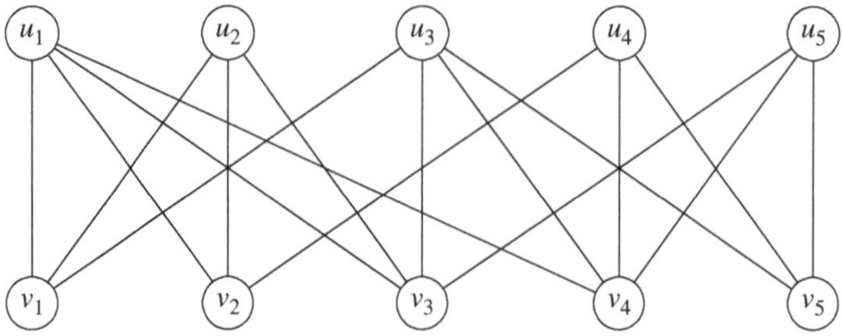

Figure 13.15 Example of a bipartite graph. The sets partitioning the graph are $V_1 = \{u_1, \ldots, u_5\}$ and $V_2 = \{v_1, \ldots, v_5\}$. There are no edges connecting nodes within each set, only between the two sets.

13.2.4 Bipartite Graphs

Definition 13.16 ***(Bipartite Graph)*** A *bipartite graph* is a graph G whose vertices can be partitioned into two sets V_1 and V_2 so that every edge of G joins a vertex of V_1 to a vertex of V_2 and no vertex is connected to another vertex in its own set.

Example 13.6 Figure 13.15 is an example of a bipartite graph.

Theorem 13.2 *A graph is bipartite if and only if it contains no odd cycles.*

Corollary 13.1 *Since trees have no cycles we can state immediately that every tree is bipartite.*

Example 13.7 Figure 13.16 shows a tree and the corresponding bipartite representation.

Bipartite graphs are useful for modeling so-called *matching problems*. An example of bipartite graph matching problem is a job matching problem. Suppose we have a set N of people and a set M of jobs, with not all people suitable for all jobs. We can model this as a bipartite graph (N, M, E). If a person n_i is suitable for a certain job m_j, there is an edge between n_i and m_j in the graph.

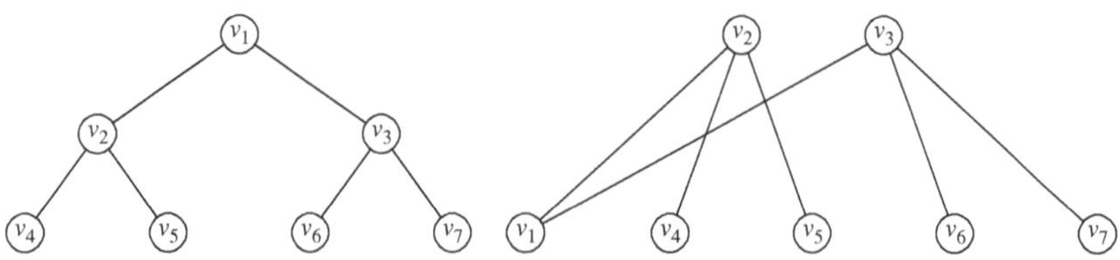

Figure 13.16 A tree (left) redrawn as a bipartite graph (right).

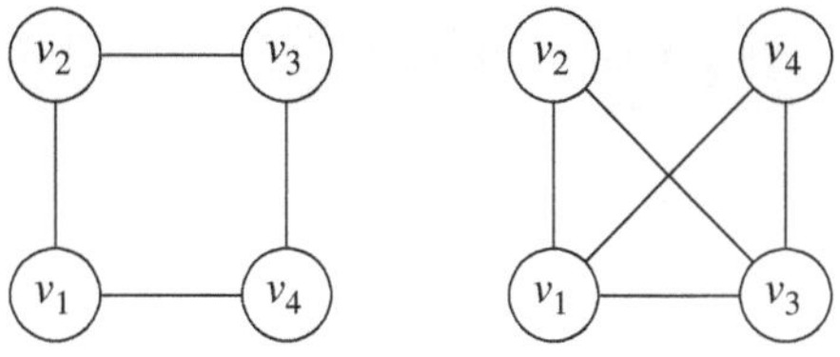

Figure 13.17 The above two graphs are isomorphic and so both are planar. The graph on the right can be redrawn so that no edges cross.

Bipartite graphs are also extensively used in modern coding theory, especially to decode codewords received from the channel. In computer science, a Petri net is a mathematical modeling tool used in analysis and simulations of concurrent systems. A system is modeled as a bipartite directed graph with two sets of nodes: a set of place nodes that contain resources, and a set of event nodes which generate and/or consume resources. Petri nets utilize the properties of bipartite directed graphs and other properties to allow mathematical proofs of the behavior of systems while also allowing easy implementation of simulations of the system.

13.2.5 Planar Graphs

In this section we discuss planar graphs. A *planar graph* is a graph that can be drawn such that no two edges intersect. Planar graphs arise in transportation networks, among other applications. If a graph corresponding to a transportation network is planar then there is no need for tunnels and bridges. For example, in a subway system if distinct lines cross it requires tunnels to be constructed farther below existing tunnels, which should be avoided to save cost. In the design of electrical circuits, planar networks mean that wires do not have to cross. Complex computer chips are often designed in layers so that each layer is represented as a planar graph.

Example 13.8 Note that one must be careful when determining whether or not a graph is planar. A seemingly nonplanar graph can sometimes be redrawn so that no two edges intersect and is therefore planar (see Figure 13.17).

An interesting problem related to planar graphs is the problem of connecting utilities (electricity, gas, water) to each of three houses in such a way that no cable or pipe crosses another. Figure 13.18 shows the bipartite graph $K_{3,3}$. It turns out that $K_{3,3}$ is not planar.

Euler's Formula

A planar graph encloses a number of regions, $r_1, \dots, r_n$, where we count the region exterior to the graph as region r_1. For example, the graph in Figure 13.19 is a planar

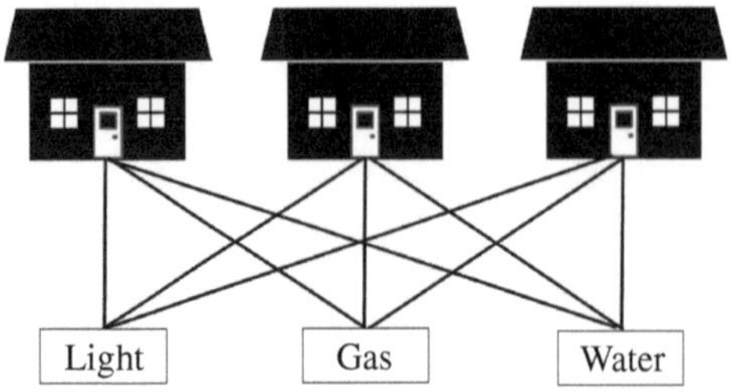

Figure 13.18 Utility connections to three houses. The bipartite graph representing the connections of the three utilities to the three houses is called $K_{3,3}$. It is not possible to redraw $K_{3,3}$ so that no edges cross.

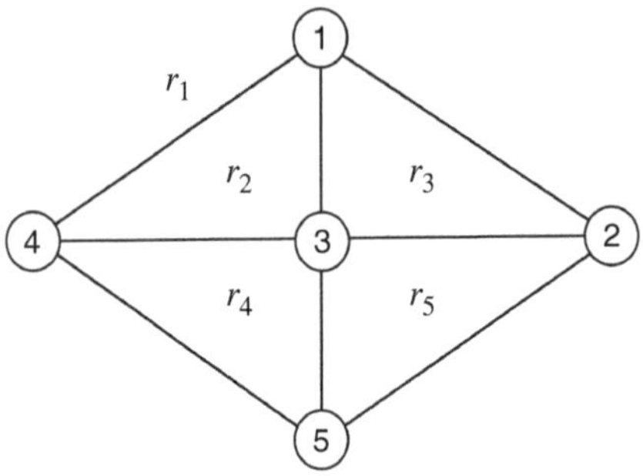

Figure 13.19 Example of a planar graph with $r = 5$ regions, $v = 5$ vertices, and $e = 8$ edges. We see that $v + r - e = 2$.

graph with five regions. Given such a planar graph with v vertices, e edges and r regions, the relationship among these quantities is given by Euler's formula.

Theorem 13.3 ***(Euler's Formula)*** *Let G be a simple connected planar graph of order v and size e. Let r be the number of regions created by a planar representation of G. Then* $v - e + r = 2$.

Proof: The proof is by induction on the number of regions r. See Appendix D.4.

Euler's formula can be used for, among other things, determining whether a given graph is planar.

Corollary 13.2 *For any connected simple planar graph G with* $v \geq 3$ *vertices and e edges, we have that* $e \leq 3v - 6$.

Proof: See Appendix D.5.

Note that the above result means that, if we have a simple graph in which $e > 3v - 6$, then it cannot be planar.

Example 13.9 The complete graph K_5 on five vertices is nonplanar.

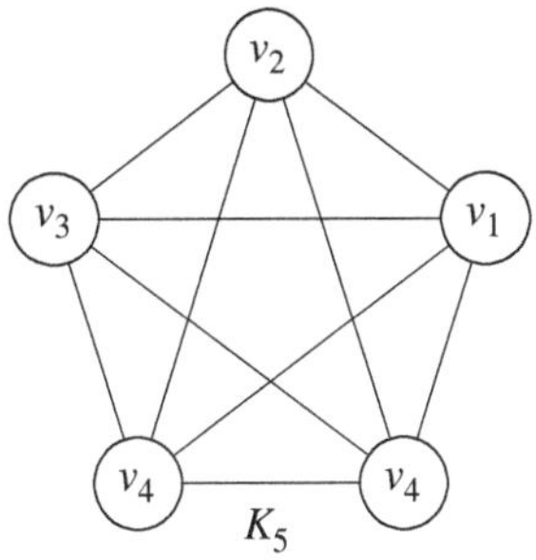

K_5

In the graph K_5, $v = 5$ and $e = 10$ so $3v - 6 = 9 < e$ and so the graph cannot be planar.

Corollary 13.3 *Any connected planar graph with v vertices and no triangles contains at most* $2v - 4$ *edges.*

As a consequence the graph $K_{3,3}$ is nonplanar. Since $K_{3,3}$ is bipartite, it contains no triangles. Examining $K_{3,3}$, we have $v = 6$, $e = 9$ and so $2v - 4 = 8 < e$ and so $K_{3,3}$ cannot be planar.

The graphs K_5 and $K_{3,3}$ have a special place in network theory. Roughly speaking, all nonplanar graphs must contain a subgraph that "looks like" either $K_{3,3}$ or K_5. This statement is made precise by two results, known as Kuratowski's theorem and Wagner's theorem. A complete understanding of these results would require a digression into graph subdivisions and graph minors, which are outside the scope of this text.

13.2.6 Graphs and Matrices

Here we introduce matrix representations of graphs, including the adjacency matrix, Laplacian matrix, and incidence matrix and discuss their usefulness for analysis of networks.

Definition 13.17 ***(Adjacency Matrix)*** Let G be a simple undirected graph of order p with vertices $v_1, v_2, \ldots, v_p$. The *adjacency matrix* $A = A(G) = (a_{ij})$ is the $p \times p$ matrix with $a_{ij} = 1$ if vertices v_i and v_j are adjacent and $a_{ij} = 0$ otherwise.

Example 13.10 A graph G_1 and its associated adjacency matrix A_1 are shown below. Row i of the adjacency matrix corresponds to vertex i and the row entries indicate which other vertices are connected to vertex i.

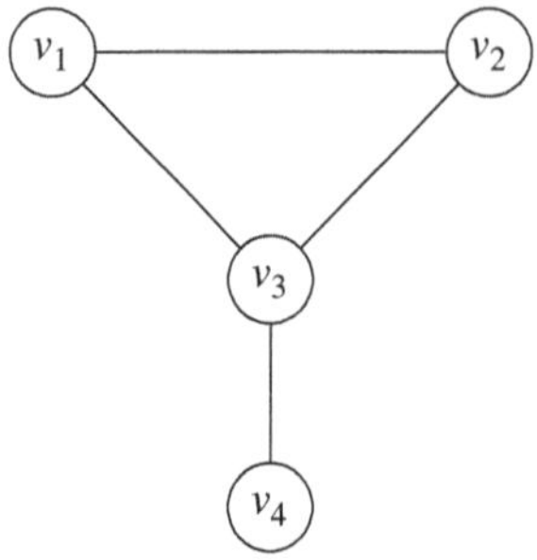

$$A_1 = \begin{bmatrix} 0 & 1 & 1 & 0 \\ 1 & 0 & 1 & 0 \\ 1 & 1 & 0 & 1 \\ 0 & 0 & 1 & 0 \end{bmatrix}$$

Remark 13.2 ***(Properties of the Adjacency Matrix)***

- For an undirected graph, the adjacency matrix is symmetric, i.e., $a_{ij} = a_{ji}$.
- For a simple graph, the diagonal elements of the adjacency matrix are zero and the off-diagonal elements are either 0 or 1.
- The sum of the elements of row i is the degree of vertex i.
- The sum of all the elements of the adjacency matrix is equal to twice the numbers of edges in the graph.

Clearly the adjacency matrix depends on how we label the vertices. The same graph with different labels will generally give rise to a different adjacency matrix.

Example 13.11 Consider the graph G_2 below, which is clearly the same graph as G_1 above. The adjacency matrix A_2 is different from the adjacency matrix A_1 in the previous example.

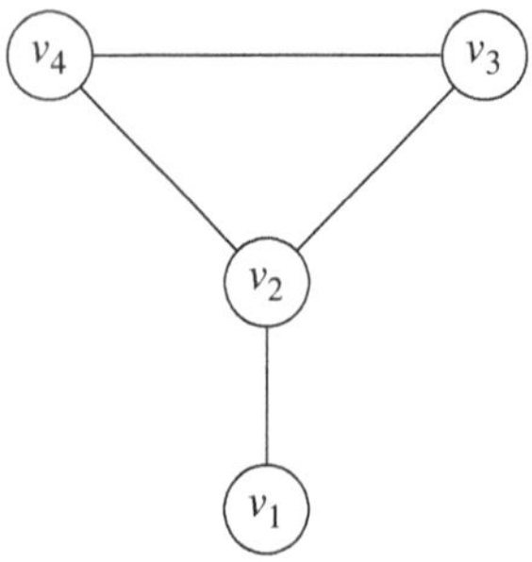

$$A_2 = \begin{bmatrix} 0 & 1 & 0 & 0 \\ 1 & 0 & 1 & 1 \\ 0 & 1 & 0 & 1 \\ 0 & 1 & 1 & 0 \end{bmatrix}$$

Recall that in our discussion of Markov chains in Chapter 10 we said that two Markov chains are isomorphic if they can be relabeled to be identical or, equivalently, if their respective transition probability matrices are the same up to ordering of their rows. This notion of isomorphism extends to general graphs as in the above example. The relation between adjacency matrices for isomorphic graphs is given by the following theorem.

Theorem 13.4 *Suppose two graphs G_1 and G_2 with adjacency matrices A_1 and A_2 are given. Then G_1 and G_2 are isomorphic graphs if and only if there exists a permutation matrix P such that*

$$PA_1P^T = A_2.$$

In the above examples, we can see that the permutation matrix relating the two graphs A_1 and A_2 is given by

$$P = \begin{bmatrix} 0 & 1 & 0 & 0 \\ 0 & 0 & 0 & 1 \\ 1 & 0 & 0 & 0 \\ 0 & 0 & 1 & 0 \end{bmatrix}.$$

Adjacency Matrix for Directed Graphs

If the graph G is directed, we define the (i,j)th entry of the adjacency matrix to be 1 if and only if there is a directed arc from vertex i to vertex j. In this case the adjacency matrix is not necessarily symmetric. The sum of the entries of any row of the adjacency matrix is the out-degree of the vertex, while the sum of the entries of any column is the in-degree of the corresponding vertex. Another useful matrix characterization of a graph is the so-called *incidence matrix*.

Definition 13.18 ***(Incidence Matrix)*** The *incidence matrix* of a graph G of order n and size m is the $n \times m$ matrix $M = (m_{ij})$ with $m_{ij} = 1$ if vertex i is incident with edge j, and zero otherwise.

Example 13.12 Below is an example of a graph and its corresponding incidence matrix. Each column of the incidence matrix has exactly two ones.

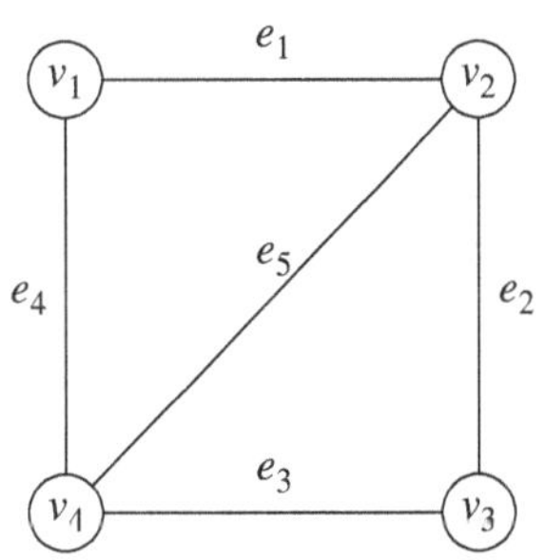

$$M = \begin{bmatrix} 1 & 0 & 0 & 1 & 0 \\ 1 & 1 & 0 & 0 & 1 \\ 0 & 1 & 1 & 0 & 0 \\ 0 & 0 & 1 & 1 & 1 \end{bmatrix}$$

The Graph Laplacian

Next we define the *Laplacian* or *Laplacian matrix* of a graph as follows.

Definition 13.19 ***(Laplacian Matrix)*** Given a simple graph with n vertices, the Laplacian matrix, $L = (\ell_{ij})$ is defined as

$$L = D - A,$$

where D is a diagonal matrix with the vertex degrees on the diagonal and A is the adjacency matrix.

The Laplacian is useful in several ways: for determining the number of connected components of the graph, determining the number of spanning trees, and other applications.

Example 13.13 Consider the graph in Figure 13.20. The Laplacian matrix for this graph is shown in Equation (13.1).

$$L = \begin{bmatrix} 4 & -1 & -1 & -1 & -1 \\ -1 & 2 & -1 & 0 & 0 \\ -1 & -1 & 3 & 0 & -1 \\ -1 & 0 & 0 & 2 & -1 \\ -1 & 0 & -1 & -1 & 3 \end{bmatrix}. \tag{13.1}$$

Properties of the Laplacian Matrix

Since the sum of the elements in each row of the adjacency matrix equals the degree of the node, it follows easily that the sum of the elements in each row of the Laplacian matrix is equal to zero. This means that the vector $x = (1, 1, \dots, 1)^T$ satisfies $Lx = 0$. As a consequence, 0 is always an eigenvalue of the Laplacian, and the Laplacian is therefore never invertible. Since the Laplacian is symmetric for an undirected graph, all of its eigenvalues are real. In fact, they are nonnegative, as we show next using the *oriented incidence matrix*.

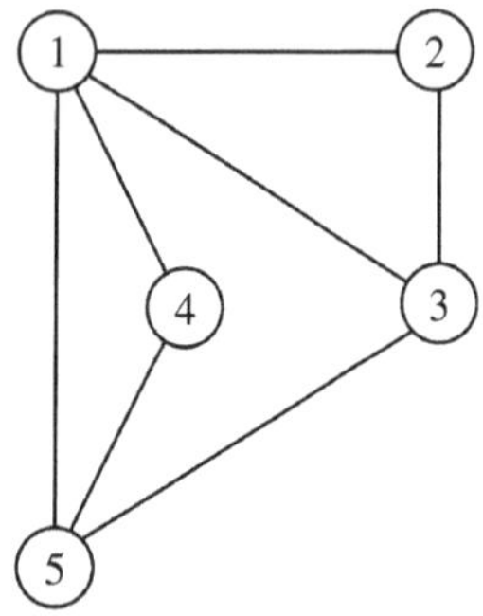

Figure 13.20 A $(5, 7)$ graph example with its associated Laplacian matrix.

Definition 13.20 ***(Oriented Incidence Matrix)*** Given a graph G, of order n and size m, we assign a positive end and a negative end to each edge. For a digraph, one can take the arrow as the positive end. For an undirected graph, one can take either end as positive. Then the *oriented incidence matrix*, $N = (n_{ij})$, is the $n \times m$ matrix defined as

$$n_{ij} = \begin{cases} 1, & \text{if vertex } i \text{ is the positive end of edge } j \\ -1, & \text{if vertex } i \text{ is the negative end of edge } j \\ 0, & \text{otherwise.} \end{cases}$$

Example 13.14 Consider the graph in Figure 13.21.

The oriented incidence matrix is

$$N = \begin{bmatrix} 1 & 1 & 1 & 0 \\ -1 & 0 & 0 & 0 \\ 0 & -1 & 0 & 1 \\ 0 & 0 & -1 & -1 \end{bmatrix},$$

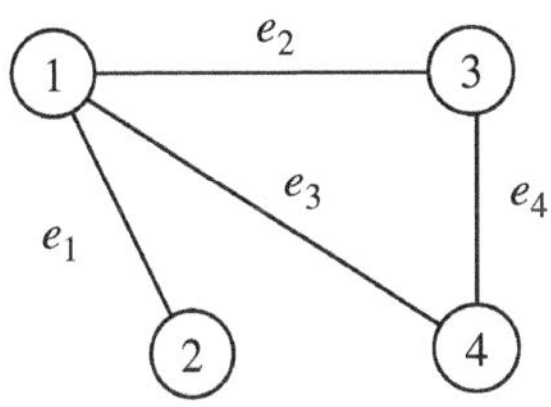

Figure 13.21 With four edges, the oriented incidence matrix will have four columns as shown below.

and one may check that

$$NN^T = \begin{bmatrix} 3 & -1 & -1 & -1 \\ -1 & 1 & 0 & 0 \\ 0 & -1 & 2 & -1 \\ -1 & 0 & -1 & 2 \end{bmatrix},$$

which is equal to the Laplacian of the graph. Note that, if $x = [x_1, x_2, x_3, x_4]^T$, then

$$N^T x = \begin{bmatrix} 1 & -1 & 0 & 0 \\ 1 & 0 & -1 & 0 \\ 1 & 0 & 0 & -1 \\ 0 & 0 & 1 & -1 \end{bmatrix} \begin{bmatrix} x_1 \\ x_2 \\ x_3 \\ x_4 \end{bmatrix} = \begin{bmatrix} x_1 - x_2 \\ x_1 - x_3 \\ x_1 - x_4 \\ x_3 - x_4 \end{bmatrix}.$$

It follows that

$$x^T L x = x^T N N^T x = (x_1 - x_2)^2 + (x_1 - x_3)^2 + (x_1 - x_4)^2 + (x_3 - x_4)^2 \geq 0.$$

This is a general property that holds for all undirected graphs. As a consequence, the Laplacian matrix L is a *positive-semidefinite* matrix. This means that the eigenvalues of L are real and nonnegative.

Further Properties of the Laplacian Matrix

- Let $0 = \lambda_0 \leq \lambda_1 \leq \cdots \leq \lambda_{n-1}$ be the n eigenvalues of L ordered from smallest to largest.
- The multiplicity of the zero eigenvalue is the number of connected components of G.

- If G is connected, therefore, λ_0 is the only zero eigenvalue.
- λ_1 is called the *algebraic connectivity* of G and is a measure of how well the overall graph is connected.

Proposition 13.2 For the complete graph K_n with n vertices, each diagonal entry of the Laplacian matrix is $n - 1$ and all other entries are equal to -1. It follows that the nonzero eigenvalues, $\lambda_1, \dots, \lambda_{n-1}$, are all equal to n, the order of the graph.

Proof: See Appendix D.6.

Kirchhoff's Theorem

In addition to determining the number of connected components of a graph, the Laplacian can also be used to determine the number of spanning trees of a graph. This is known as *Kirchhoff's Theorem.*

Theorem 13.5 ***(Kirchhoff)*** *Let G by a connected graph of order n and let $\lambda_1, \dots, \lambda_{n-1}$ be the nonzero eigenvalues of the Laplacian L(G). Then the number of spanning trees, T(G) of G is given by the product*

$$T(G) = \frac{1}{n}\lambda_1 \cdots \lambda_{n-1}.$$

Equivalently, the number of spanning trees, T(G), is equal to any cofactor of L.

Example 13.15 For the graph in Example 13.13, the Matlab code below computes the number of spanning trees using both methods: as the product of the nonzero eigenvalues of L and by computing the cofactor of $L(1, 1)$. Note that any cofactor of L gives the same result. Running the script below gives the number of spanning trees of G as $T(G) = 21$.

Compute the Number of Spanning Trees of the graph *G*.

```
clear
L = [ 2  -1   0  0  -1  0;
     -1   3  -1  0  -1  0;
      0  -1   3 -1  -1  0;
      0   0  -1  3  -1 -1;
     -1  -1  -1 -1   4  0;
      0   0   0 -1   0  1]; % the graph Laplacian
  x=eig(L);x(1)=[];          % delete the zero eigenvalue
  T=x(1);
  for i=2:length(x)
      T=T*x(i);
  end
  T1=T/(length(x)+1);         % number of spanning trees
```

```
%
Q = L;
Q(:,1)=[];Q(1,:)=[]; % delete first row and column of L
T2=det(Q); % cofactor of L = number of spanning trees
```

13.3 Matlab Graph Functions

Matlab has numerous functions to aid in creating, visualizing, and analyzing networks. The basic command is `graph`, which can be used to create graphs in several ways, for example, from an adjacency matrix or an edge list. A simple, undirected graph is created by the following Matlab script (see Figure 13.22).

Create a Graph from an Adjacency Matrix.

```
clear
A= [0 1 0 0 1 0; 1 0 1 0 1 0; 0 1 0 1 1 0;
    0 0 1 0 1 1; 1 1 1 1 0 0; 0 0 0 1 0 0];
G=graph(A);
%
h=plot(G);
text(h.XData+[0.2 0.2 0 -0.2 0.1 0.2], ...
h.YData+[0 0.1 0.2 0.15 -.21 0],...
h.NodeLabel,'HorizontalAlignment','center','FontSize',14);
h.NodeLabel = {};
h.EdgeColor = 'black';
h.NodeColor = 'black';
h.LineWidth = 2;
h.MarkerSize = 14;
```

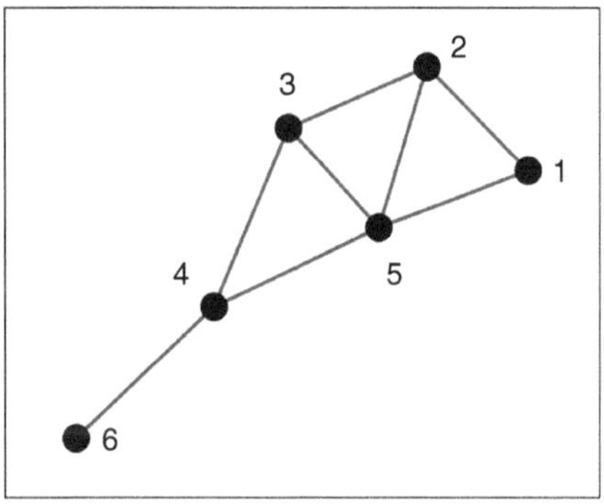

Figure 13.22 A simple undirected graph created by the Matlab `graph` function.

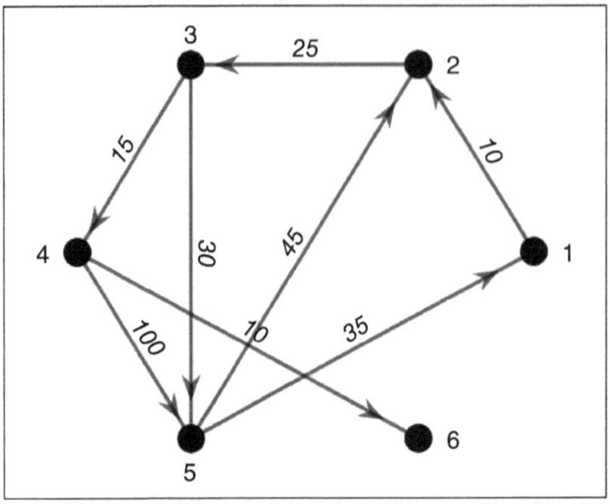

Figure 13.23 A weighted directed network created by Matlab `digraph` function.

The next example is a weighted, directed network defined using an edge list and the Matlab command `digraph` (see Figure 13.23).

Create a digraph from an edge list.

```
clear
s=[1 5 2 5 3 3 4 4]; % edge list
t=[2 1 3 2 4 5 5 6];
G = digraph(s,t);
%
G.Edges.Weight = [10 25 15 30 100 10 35 45]';
h=plot(G,'Layout', 'circle', 'EdgeLabel', G.Edges.Weight);
text(h.XData+[0.15 0.15 0 -0.15 0 0.15],...
 h.YData+[0 0 0.15 0 -0.15 0],...
h.NodeLabel,'HorizontalAlignment','center','FontSize',14);
h.NodeLabel = {};
h.EdgeColor = 'black';
h.NodeColor = 'black';
h.LineWidth = 2;
h.MarkerSize = 14;
h.ArrowSize = 14;
h.ArrowPosition = 0.9;
h.EdgeFontSize = 14;
```

The documentation for the command `graph` lists a number of other available functions for creating, analyzing, and searching on graphs.

13.4 Network Metrics

We next discuss various metrics that can be used for network analysis, data analysis, data mining and other applications. Some important structural questions we may ask for a given network are:

1. Home many connections does the average node have?
2. Which are the most connected nodes?
3. How many links are there between nodes on average?
4. Are there clusters or groupings of nodes?
5. How does one classify different networks?

13.4.1 Degree Distribution

The *degree distribution* of a graph refers to the fraction of links per node. We can think of the degree distribution as a discrete probability distribution where the probability that a randomly chosen node has degree k is proportional to the number of nodes that have degree k. For example, in a railway or highway network, most cities have several links and no city has a huge number of links. The degree distribution will be of the type shown in Figure 13.24.

By contrast, in a typical airline network (Figure 13.25), most airports have very few connections, while the hubs have a huge number of connections. The degree distribution for this network is a *power-law distribution*, which means that the probability that a node has degree k is proportional to $\frac{1}{k^{\alpha}}$. It turns out that the power-law distribution is also present in the internet, WWW, social networks, biological networks, and many others.

In a normal or Gaussian distribution, most values are close to the mean or average value. The tails of the distribution decrease exponentially. For example,

Figure 13.24 A railway network. The degree distribution of this network is approximately a normal distribution.

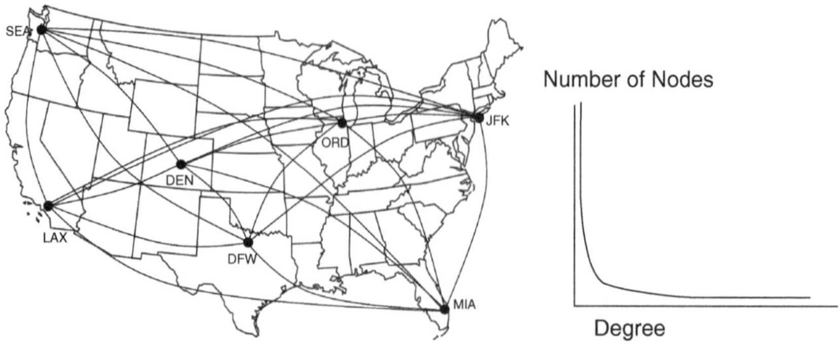

Figure 13.25 An airline network showing several hub airports. Most other airports have only a few connections, mainly to the hubs. A network with a few large hubs and many small-degree nodes approximately follows a power-law distribution.

the average height of males in a give population will follow a normal distribution. Since the number of individuals at a give height falls off exponentially from the mean, the probability of finding very tall (>10 feet) or very short (<1 foot) is essentially zero.

By contrast, a power-law distribution has a heavy tail, which means that the probability falls off slower than exponentially.

Suppose that the distribution of heights of a (very large) population were to follow a power-law distribution. Most individuals (80%) would be below a certain height with a few very tall and many very short. You would not be surprised to occasionally meet someone over 100 feet tall, for example.

It is important to note that, given a power-law distribution $P(k) = \gamma \frac{1}{k^\alpha}$ if we take the logarithm of both sides, we obtain $\log(P) = \log(\gamma) - \alpha \log(k)$ Thus, on a log–log plot, a power-law distribution is a straight line.

Examples of Power-Law Distributions

An example of a power-law distribution is the magnitudes of earthquakes. The vast majority of earthquakes are minor, even going unnoticed without sensitive equipment. However, occasionally a very large earthquake will happen. Plotting the data of earthquake frequency and magnitude on a logarithmic scale, as shown in Figure 13.26, we see that the magnitude data fall on a straight line, which is indicative of a power-law distribution.

Other noteworthy examples of power-law distributions include:

- wealth distribution;
- the internet;
- the World Wide Web;
- airline route networks;

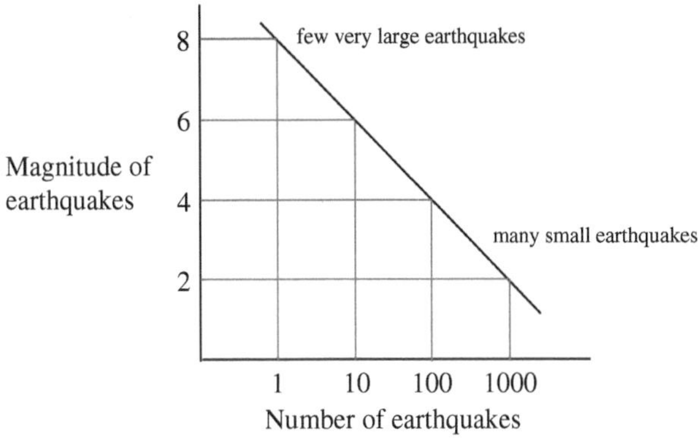

Figure 13.26 Magnitude versus frequency of earthquakes plotted on a log-log scale indicating a power law distribution.

- the Science Citation Index;
- protein–protein reaction networks;
- sexual relationships;
- casualties in wars and battles;
- sizes of cities;
- word frequency in languages;
- family names.

Power-law distributions are also scale-free, which means that the form of the distribution is the same if one zooms in or out. For example, in a group of 1000 individuals, the top 200 will have as much wealth as the bottom 800. Within the top 200, to top 40 will have much wealth as remaining 160. Figure 13.27 shows the income distribution of the top 100 individuals in 2018.

The degree distribution is a useful metric to understand the large-scale structure of networks. For example, if the distribution is uniform the network is regular.

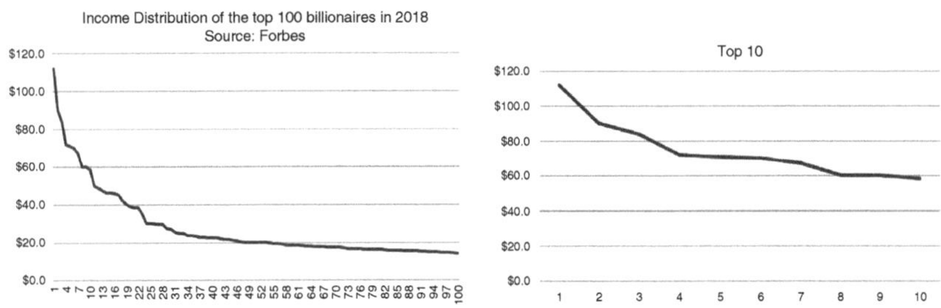

Figure 13.27 Income of the top 100 (left) and the top 10 (right). Despite the small number of individuals in the right-hand plot, the shape of the graph is roughly the same with the top 2 or 3 worth as much as the next 7 or 8.

Removal of a few edges or nodes will not change the large-scale behavior of the network. On the other hand, a network that has a few nodes with large degree (i.e., hubs) and many nodes with small degree has a much different structure. Removal of large-degree nodes may result in a network with unconnected components.

For directed networks the correlation between in-degree and out-degree can be an important indicator; for example, in a network where nodes are people and directed edges indicate listeners (incoming links) and talkers (outgoing links). If one wants a message to spread quickly through the network, then one would want listeners also to be talkers, in order to pass along information. In this case we would want high positive correlation between in-degree and out-degree, i.e., nodes with high in-degree also have high out-degree.

Distance Metrics

The distance between vertices can be thought of as the minimum number of edges among all paths connecting them. Distance can give information about the relative importance of various nodes. Note that distance is generally different than the weight of a path in a network or weighted graph.

Definition 13.21 ***(Geodesic Distance)*** The *geodesic distance* between nodes u and v, denoted $d(u, v)$ in a directed or undirected graph is the length of the shortest path from u to v.

Note that $d(u, v)$ and $d(v, u)$ are the same in an undirected graph but not necessarily equal in a directed graph.

Example 13.16 Consider the graph in Figure 13.28. The geodesic distance between node 5 and node 7 is equal to 3 and there are three distinct paths of length 3 between these nodes. On the other hand, there is only one shortest path (of length 3) between node 4 and node 7.

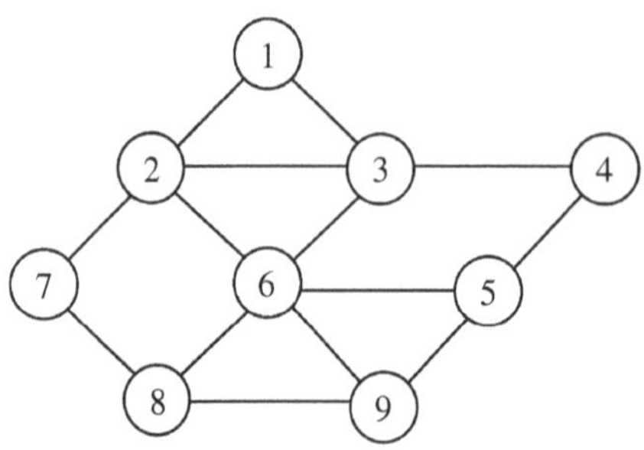

Figure 13.28 The length of a path is the number of 'hops' between nodes. The geodesic distance is the length of the shortest path.

Definition 13.22 ***(Average Path Length)*** Let G be a connected graph with vertex set V. Let $|V|$ denote the order of G and let $\bar{d}(u)$ denote the average length of the shortest paths from vertex u to any other vertex v in G. Then the *average path length* $\bar{d}(u)$ is given by

$$\bar{d}(u) = \frac{1}{|V| - 1} \sum_{v} d(u, v).$$

13.4.2 Centrality

The notion of *centrality* in networks is used to identify the most important vertices in a network, for example the most influential person in a social network, the most likely to spread infections, the most important genes in a gene regulation network, the webpages that best match a given search word. There are several different metrics that have been developed to address the notion of centrality. We will discuss three such measures, namely *degree centrality*, *betweenness*, and *eigenvector centrality*.

Degree Centrality

Definition 13.23 ***(Degree Centrality)*** The *degree centrality*, $c_d(v)$, of a node v is defined as

$$c_d(v) = \frac{\deg(v)}{n - 1},$$

where n is the order of the network.

Degree centrality is the simplest measure of importance and assigns importance solely based on how many neighbors a node has.

Betweenness Centrality

Definition 13.24 ***(Betweenness)*** Let G be a simple and connected graph. Let $S(x, y)$ be the set of shortest paths between x and y and let $S(x, u, y) \subseteq S(x, y)$ denote the set of shortest paths between x and y that pass through u. Then the *betweenness centrality* $c_B(u)$ of the vertex u is defined as

$$c_B(u) = \sum_{x \neq y} \frac{|S(x, u, y)|}{|S(x, y)|}.$$

Betweenness assigns more importance to nodes that lie on many paths, whether or not the nodes themselves have high degree. For example, if removing a node would disconnect the graph, then that node would be important and its importance would be reflected in a larger betweenness centrality.

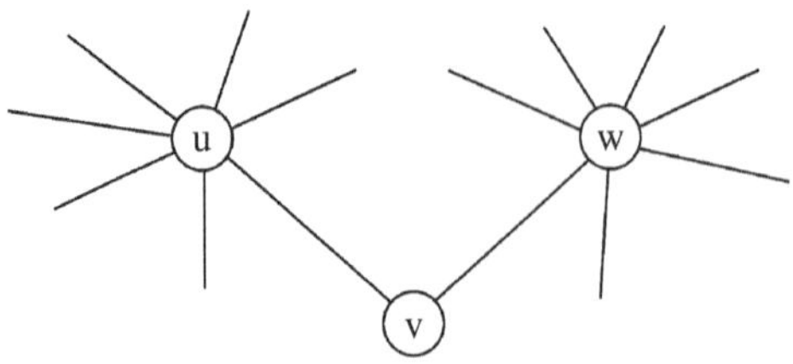

Figure 13.29 A low-degree node may be important if it is adjacent to high degree nodes. Although node *v* only has degree 2, it may have large betweenness centrality and/or large eigenvector centrality.

Eigenvector Centrality

Eigenvector Centrality is another important way to determine the *rank* or *importance* of nodes in a network that is more sophisticated than simply looking at the degree of the node. For example, a node with low degree might be important if it is adjacent to high-degree nodes (Figure 13.29).

To capture this idea, we assign a *weight* x_i to node *i*, that is meant to represent the importance of node *i*. We let x_i be proportional to the *sum of the weights* of all other nodes *adjacent* to x_i,

$$x_i = \alpha \sum_j a_{ij} x_j,$$

where $A = (a_{ij})$ is the adjacency matrix for the graph. Note that in a strongly connected directed graph, we can define x_i to be proportional to the sum of weights of all other nodes that point to x_i. The analysis works for directed as well as undirected graphs.

The above equation can be written in matrix form as

$$Ax = \lambda x, \quad \text{where } \lambda = \alpha^{-1},$$

which we recognize as an *eigenvalue/eigenvector* equation.

Theorem 13.6 *If G is a* (strongly) connected graph *with adjacency matrix A, then there is a* unique *eigenvector* v_0, *with all weights positive, and which corresponds to the* largest *(real) eigenvalue* λ_0 *of A.*

Proof: The proof of the theorem relies on the Perron–Frobenius theorem. We omit the details.

The Matlab script below computes the eigenvector centrality from a given adjacency matrix.

Eigenvector Centrality from Adjacency Matrix.

```
% Input  = Adjacency Matrix A
% Output = Eigenvector Centrality and Maximum Eigenvalue
%
function [V0,lambda]=ECentrality(A)
[V,D] = eig(A,'vector');
[x,ind] = sort(D);
Vs = V(ind,ind);
V0=Vs(:,length(x));
lambda=max(x);
end
```

Example 13.17 Consider the graph G in Figure 13.30.

Computing these three measures of centrality using the Matlab functions shown gives the values shown in Table 13.1.

Three Measures of Centrality

```
A= [0 1 0 0 0 0; 1 0 1 0 0 0; 0 1 0 1 1 1;
    0 0 1 0 1 0; 0 0 1 1 0 1; 0 0 1 0 1 0];
G=graph(A);
CD = centrality(G,'degree'); %degree centrality
CB = centrality(G,'betweenness'); %betweenness centrality
CE = centrality(G,'eigenvector'); %eigenvector centrality
```

Note that there are other measures of centrality that have been defined, such as *Katz centrality* and *closeness centrality*, which we have not covered here. The interested reader is referred to [48, 52] for details of these measures.

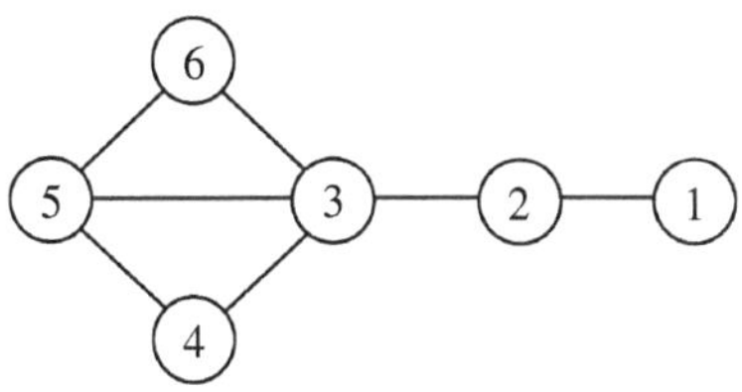

Figure 13.30 A (6, 7) graph with different values of degree centrality, betweenness centrality, and eigenvector centrality. Since these three measures generally give different numbers, it is important to understand how each metric assigns its notion of importance to vertices.

Table 13.1 Three different centrality measures for the graph G in Figure 13.30. In terms of importance, node 3 has the highest degree and is ranked most important in all three measures. Nodes 2, 4, and 6 have the same degree centrality but nodes 4 and 6 have zero betweenness centrality, since the other nodes are reachable without going through 4 or 6. However, nodes 4 and 6 have higher eigenvector centrality since they are connected to higher-degree nodes than is node 2.

Node	Degree centrality	Betweenness	Eigenvector centrality
1	1	0	0.0409
2	2	4.000	0.1108
3	4	6.500	0.2588
4	2	0	0.1803
5	3	0.500	0.2289
6	2	0	0.1803

13.4.3 Clustering

In many networks information is spread among nearest neighbors. Schooling of fish, flocking of birds, spreading of internet viruses, spreading of epidemics and gossip among people, as well as forming opinions are examples of this. To study these networks, it is useful to define a notion of *clustering*, which tells us to what extent neighbors of a given vertex v are also neighbors of each other.

Definition 13.25 ***(Neighbor Set)*** Given a vertex v, the *neighbor set* of v is the set

$$N(v) = \{u \neq v \in G | u \text{ is adjacent to } v\}$$

Intuitively, a maximal notion of clustering would be if all neighbors of v are neighbors of each other, in other words, if the neighbor set, $N(v)$, forms a complete graph. Let $n_v = |N(v)|$ be the order of the neighbor set $N(v)$. Then the maximum number of edges possible for $N(v)$ is $\binom{n_v}{2} = \frac{1}{2}n_v(n_v - 1)$. This motivates the following definition.

Definition 13.26 ***(Clustering Coefficient)*** Let G be a simple, connected, undirected graph and let $v \in V(G)$ be a vertex with neighbor set $N(v)$. Let $n_v = |N(v)|$ and let m_v be the number of edges in the subgraph of G induced by $N(v)$. If the degree $\deg(v) > 1$, then the *clustering coefficient* $cc(v)$ of v is defined as

$$cc(v) = \frac{2m_v}{n_v(n_v - 1)}.$$

If $\deg(v) \leq 1$ the clustering coefficient is not defined.

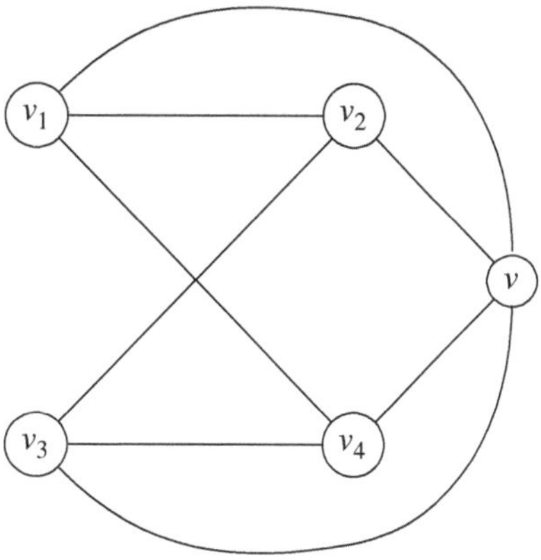

Figure 13.31 The neighbors of v are v_1, v_2, v_3, v_4. The subgraph formed by these nodes has four edges. A complete graph with four nodes has six edges. Therefore, the clustering coefficient of v is 2/3.

Example 13.18 In Figure 13.31, we have that $N(v) = \{v_1, v_2, v_3, v_4\}$ and so $n_v = 4$. The number of edges in the subgraph formed by these four neighbors of v is 4, so $m_v = 4$. Hence the clustering coefficient of v is

$$cc(v) = \frac{2m_v}{n_v(n_v - 1)} = \frac{2}{3}.$$

The intuitive interpretation is that, among the four neighbors of v, there are six possible edges, representing the possible connections among these neighbors, namely, u_1u_2, u_1u_3, u_1u_4, u_2u_3, u_2u_4, u_3u_4. The actual connections are the four u_1u_2, u_1u_4, u_2u_3, u_3u_4, or two-thirds of those possible.

Now we define the clustering coefficient of the entire graph as the average of the clustering coefficients of all vertices in the graph.

Definition 13.27 ***(Clustering Coefficient of a Graph)*** Let G be a simple connected undirected graph. Let V^* be the set of vertices of G with degree greater than 1. The clustering coefficient $CC(G)$ for the entire graph G is defined as

$$CC(G) = \frac{1}{|V^*|} \sum_{v \in V^*} cc(v).$$

Example 13.19 The graph in Figure 13.32 has vertex degrees and clustering coefficients as shown in the Table 13.2.

The clustering coefficient of the graph is therefore $CC(G) = 3/6 = 0.5$.

The previous definition of the clustering coefficient is sometimes called the local view because it is defined in terms of individual vertices. Here we give another definition, which is called the global view.

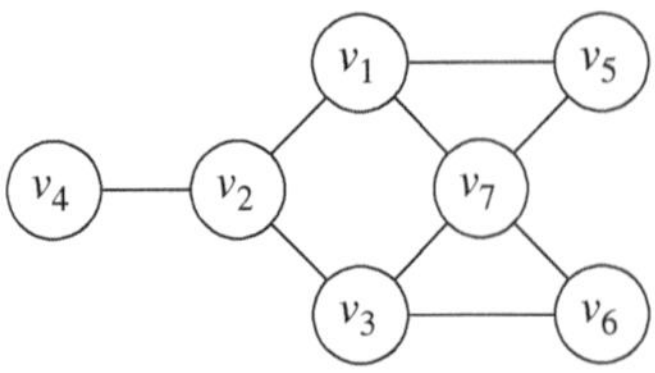

Figure 13.32 Graph clustering example for Example 13.19.

Table 13.2 Clustering coefficients for the graph in Figure 13.32.

Vertex	**1**	**2**	**3**	**4**	**5**	**6**	**7**
n_v	3	3	3	1	2	2	4
m_v	1	0	1	–	1	1	2
$cc(v)$	$\frac{1}{3}$	0	$\frac{1}{3}$	–	1	1	$\frac{1}{3}$

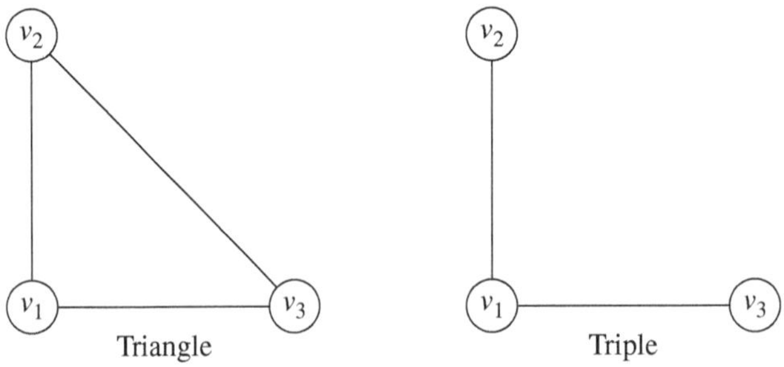

Figure 13.33 A triangle and a triple

Definition 13.28 ***(Triangles and Triples)*** A *triangle* is a complete (sub)graph with exactly three vertices. A *triple* is a (sub)graph with exactly three vertices and two edges (see Figure 13.33).

We first state the following equivalent definition of clustering coefficient for a vertex.

Definition 13.29 ***(Clustering Coefficient)*** Let $n_\Delta(v)$ be the number of triangles at v and $n_\Lambda(v)$ the number of distinct triples at v. Then the clustering coefficient can be expressed as

$$cc(v) = \frac{n_\Delta(v)}{n_\Lambda(v)},$$

which is equivalent to the previous definition.

Definition 13.30 ***(Transitivity)*** Suppose G is a simple, connected graph with $n_\Delta(G)$ distinct triangles and $n_\Lambda(G)$ distinct triples. Then the *network transitivity* is defined as

$$\tau(G) = \frac{n_\Delta(G)}{n_\Lambda(G)}.$$

Note that the network transitivity for the graph is not necessarily the same as the cluster coefficient $CC(G)$, even though for individual vertices the ratio n_Δ/n_Λ is equivalent to $cc(v)$. For the graph in the previous example, we had $CC(G) = 1/2$. One can show (Problem 13.12) for this graph that the transitivity $\tau(G) = 2/17$.

Cliques

Definition 13.31 ***(Clique)*** Let $G = (V, E)$ be a graph. A *clique* is a set of vertices $S \subset V$ such that:

1. The subgraph induced by S is a complete graph.
2. If $\overline{S} \supset S$, there is at least one pair of vertices in S' that is not connected by an edge in E.
3. The *clique number* of a graph G, written $\omega(G)$, is the size of the largest clique in G.

Figure 13.34 shows three separate cliques. Cliques are used in numerous applications:

- in social networks to model groups of people;
- in bioinformatics to model clustering of gene expression data;
- in population dynamics to model ecological niches in food webs;
- in protein–protein interaction, to identify clusters of proteins that interact closely with each other and have few interactions with proteins outside the cluster;
- to design efficient circuits in communication networks.

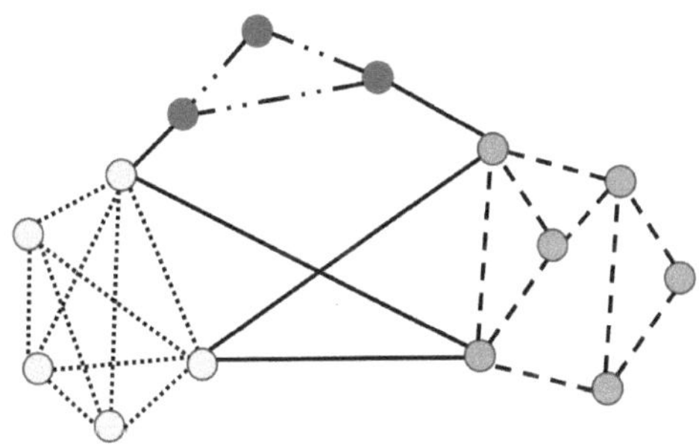

Figure 13.34 A graph with three distinct cliques.

13.5 Random Graphs

In this section we introduce an important class of graphs known as *random graphs*. Random graphs are those in which pairs of vertices are connected according to some probability. Random graphs were defined in the 1950's by Paul Erdős and Alfréd Rényi [22]. The subject was mostly of theoretical interest until recently when they have been applied to a large variety of complex networks. We will discuss three main types of random networks:

1. Erdős–Rényi networks
2. small-world networks
3. scale-free networks

These random networks have important applications in modeling and analyzing real-world phenomena. For example, Erdős–Rényi random networks give us insight into phase transition and percolation on networks, both of which we discuss below. Small-world networks have the property that the average path length tends to be short. Social networks, the internet, and gene regulation networks have this property. Scale-free networks typically have few nodes with very high degree and many nodes with low degree. Airline networks, the World Wide Web, and many other networks have this property.

13.5.1 Erdős–Rényi Networks

Definition 13.32 ***(ER Network)*** An *Erdős–Rényi network* or *ER network*, denoted by $ER(n, p)$, is an undirected graph with n vertices in which each two distinct vertices are connected by an edge with probability p.

With $\delta(u)$ denoting the degree of vertex u, $P[\delta(u) = k]$ denotes the probability that the degree of u is equal to k. Now:

- there are a maximum of $n - 1$ vertices that can be adjacent to u;
- there are $\binom{n-1}{k}$ possibilities for k nodes to connect to u;
- the probability of having exactly one specific set of k neighbors is $p^k(1-p)^{n-1-k}$.

Therefore, we have

$$P(\delta) = P[\delta(u) = k] = \binom{n-1}{k} p^k(1-p)^{n-1-k}.$$

Treating the vertex degree δ as a *discrete random variable* means, therefore, that δ follows a binomial distribution as defined in Appendix C.

Recall that the average (or expected value) of a random variable δ is defined as

$$\overline{\delta} = E[\delta] = \sum_k k \cdot P[\delta = k].$$

Proposition 13.3 The expected value for the vertex degree of an $ER(n, p)$ random graph is

$$\bar{\delta} = E(\delta) = \sum_{k=1}^{n-1} k \cdot P[\delta = k] = p\ (n - 1).$$

The proof is by direct calculation (Appendix D.7).

Proposition 13.4 The expected number of edges, i.e., the expected size of an $ER(n, p)$ random graph is $\binom{n}{2} p = 0.5n(n - 1)p$.

As the number of nodes n in a random graph gets larger and larger it is more likely that the actual number of edges in a particular $ER(n, p)$ graph will be close to the expected value.

Proposition 13.5 For a large $G \in ER(n, p)$, the average path length $\bar{d}(G)$ is approximately equal to $\frac{\log(n)}{\log(\bar{\delta})}$. For example, with $n = 2000$ and $p = 0.015$, the calculation yields an average path length of approximately 2.24.

Clustering Coefficient

For an $ER(n, p)$ random graph, the expected clustering coefficient of every vertex is p. The reasoning is as follows.

- The clustering coefficient is the fraction of edges between neighbors and the maximum number of possible edges.
- The expected number of edges between k neighbors is $\binom{k}{2} p$.
- The maximum number of edges between k neighbors is $\binom{k}{2}$.
- Therefore, the expected clustering coefficient for every vertex is their ratio, p.

Phase Transition and Emergence of a Giant Component

- For an $ER(n, p)$, with $p = 0$ there are no edges and all nodes are completely disconnected, i.e., the number of components equals the number of nodes.
- For $p = 1$ every node is connected to every other node and so there is only one connected component in the network.
- For $p = 0$ the order of the largest component is 1 independent of the number of nodes n, while for $p = 1$ the order of the largest component is equal to the number of nodes n.
- As the probability p increases, more and more nodes are connected. There is a critical value (around $np = 1$) when the size of the largest component undergoes a sudden change to a *giant component*.

For small values of p, the ER random graph has small disconnected components. At the value $np = 1$, the emergence of the giant component is called a *phase transition*. We can calculate the expected size of the giant component from the following argument. We first note that for large n and small probability p, the binomial distribution

$$P(k) = \binom{n-1}{k} p^{n-1}(1-p)^{n-1-k}$$

can be approximated by the *Poisson* distribution

$$P(k) = \frac{\lambda^k e^{-\lambda}}{k!},$$

where $\lambda = p(n-1)$. (See Appendix C.)

Let u be the probability that a vertex chosen uniformly at random from the graph is not in the giant component. Then u is the fraction of vertices on the graph that do not belong to the giant component. The probability that a given vertex does not belong to the giant component is also equal to the probability that none of its neighbors belong to the giant component, which is just u^k if the vertex has degree k. Averaging this expression over the probability distribution of u, we then find the following relation for u in the limit of large graph size:

$$u = \sum_{k=0}^{\infty} p_k u^k = e^{-\lambda} \sum_{k=0}^{\infty} \frac{(\lambda u)^k}{k!} = e^{\lambda(u-1)}.$$

The fraction S of the graph occupied by the giant component is $S = 1 - u$ and hence

$$S = 1 - e^{-\lambda S}.$$

For $\lambda < 1$ the only nonnegative solution of the above equation is $S = 0$, while for $\lambda > 1$ there is also a nonzero solution, which is the size of the giant component. The phase transition occurs at $\lambda = 1$. A giant component is evident in Figure 13.35.

Figure 13.35 ER random graphs showing the emergence of a giant component. Both graphs have 200 nodes. In the graph on the left, $\lambda = 0.9$. In the graph on the right $\lambda = 1.6$.

Generate an ER Random Graph.

```
function G=ERgraph(numnodes,p)
A=zeros(numnodes);
G=graph(A);          % create an empty graph
%
for i=1:numnodes
    for j=i+1:numnodes
X=rand;
if X<p
    G=addedge(G,i,j,1); % add edges with probability p
end
    end
end
h=plot(G,'Layout','force','UseGravity',true);
h.NodeLabel={};
h.EdgeColor = 'blue';
h.NodeColor = 'black';
h.LineWidth = 3;
h.MarkerSize = 3;
```

Percolation

The process of randomly removing some fraction of vertices and their incident edges from a network is called *percolation*. For example, the random failure of routers on the internet can be represented by removing the corresponding vertices and edges of the network representation of the internet. In fact, about 3% of the routers on the internet are not working for one reason or another at any given moment. Percolation helps us understand what effect this might have on the performance of the network.

Another example of a percolation phenomenon is the vaccination or immunization of individuals against the spread of disease. Diseases are spread by contacts between individuals and vaccinated individuals can be represented by removal of vertices. A similar sort of phase transition occurs when vertices are removed at random. Most vertices remain inside a single giant component until a significant fraction of vertices are removed and then the network starts to break up into many smaller connected components, which means that the disease can no longer spread through the network.

Example 13.20 ***(Random Removal of Vertices and Phase Transition)***

Figure 13.36 shows the fraction of nodes outside the giant component in an $ER(2000, 0.015)$ graph as a function of the fraction of nodes removed. Randomly

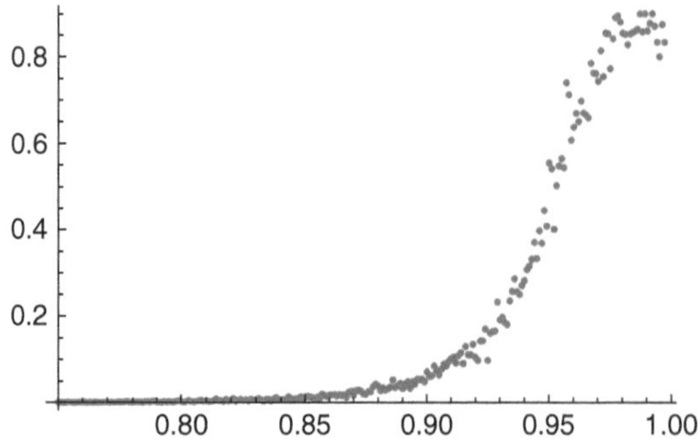

Figure 13.36 Fraction of nodes outside the giant component as a function of the fraction of nodes removed. Figure reprinted from [69] with permission of the author.

removing a large number of nodes does not change the fraction of nodes in the giant component until a critical fraction, around 90%, is reached at which point the network undergoes a phase transition and collapses back to having multiple small components. For some networks, like a network of internet routers, this is good news since the network remains highly connected despite the removal (or failure) of a large number of nodes. In other cases, for example, in epidemic models or models of disease eradication, the robustness of networks to random removal of nodes helps to explain why achieving herd immunity in a population requires vaccination of almost 90% of the population.

13.5.2 Small-World Networks

Many real-world networks show a small average shortest path length and also show a high degree of clustering. ER random networks have a small average shortest path length, but not a high clustering coefficient, which limits their applicability in such networks. The concept of the *small-world network* was introduced by Watts and Strogatz [73] (see also [72]) to address this problem. A Watts–Strogatz (WS) network, denoted by $WS(n, k, p)$, is a network with a fixed order, as in the case of an $ER(n, p)$ network. The construction of a $WS(n, k, p)$ network proceeds as follows.

- Choose n and k such that k is even and $n \gg k \gg \ln(n) \gg 1$.
- Let $V = \{v_1, v_2, \ldots, v_n\}$ be the vertex set with n nodes.
- Order the n vertices in a ring and connect each vertex to its $k/2$ left-hand neighbors, and to its $k/2$ right-hand neighbors.
- With probability p, replace edge $\langle u, v\rangle$ with an edge $\langle u, w\rangle$ where $w \neq u$ is randomly chosen, but such that $\langle u, w\rangle \notin E(G)$.

For many vertices in a WS graph, the distances $d(u, v)$ will be small since each vertex has k nearby neighbors, direct links to other groups or clusters, and long links that cross the ring and shorten the distance to otherwise more distant

Figure 13.37 *WS*(20, 8, 0) network. With probability $p = 0$, the graph is regular with each node connected to exactly $k = 8$ neighbors.

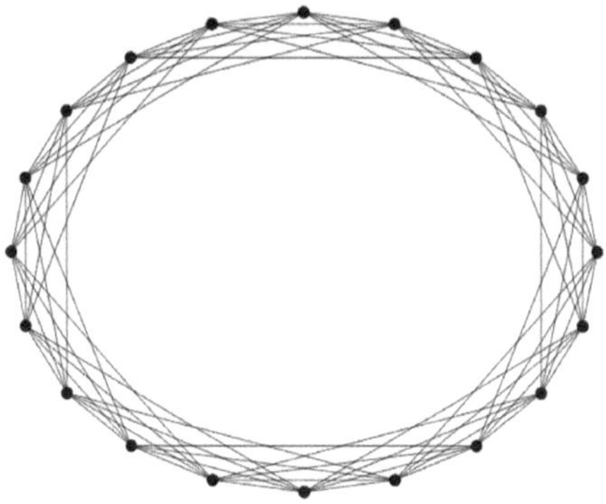

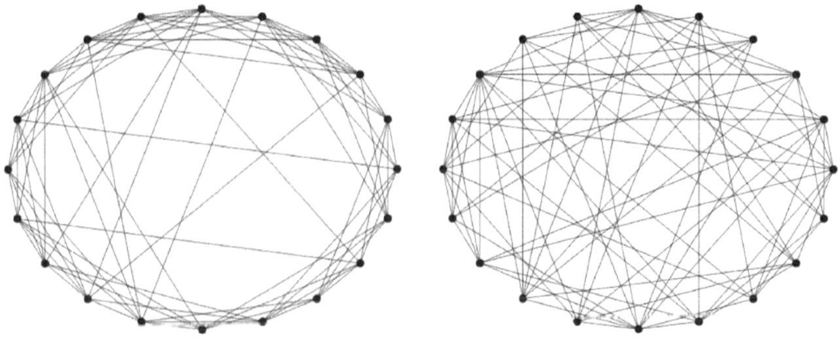

Figure 13.38 *WS*(20, 8, *p*) networks with $p = 0.25$ (top) $p = 0.75$ (bottom). To access Matlab code to construct a WS network, type the command `openExample('matlab/BuildWattsStrogatzSmallWorldGraphModelExample')` in Matlab.

vertices. Figures 13.37 and 13.38 shows three WS networks with three different probabilities.

Theorem 13.7 *For any G from WS(n, k, 0), the clustering coefficient of the graph is*

$$CC(G) = \frac{3}{4}\frac{k-2}{k-1}.$$

The proof is omitted. The clustering coefficient is therefore independent of the order of the graph and is close to $\frac{3}{4}$ for large k.

Theorem 13.8 *For $G \in WS(n, k, 0)$, the average shortest path length $\overline{d}(u)$ from vertex u to any other vertex is approximated by*

$$\overline{d}(u) = \frac{(n-1)(n+k-1)}{2kn}.$$

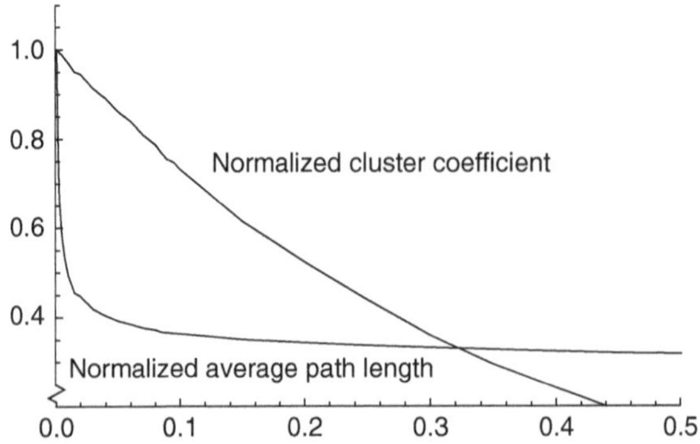

Figure 13.39 Average shortest path length and clustering coefficient versus increasing probability. The average shortest path length is small. The clustering coefficient decreases slowly so that the WS networks can exhibit both short average path length and high clustering. Figure reprinted from [69] with permission of the author.

For example, with $n = 200$ and $k = 10$, we have $\overline{d}(u) = 10$. Therefore, $WS(n, k, 0)$ graphs have high clustering coefficients but not short average path lengths. However, increasing p shows that average path length drops rapidly due to the presence of the long distance links, whereas the cluster coefficient decreases slowly.

13.5.3 Scale-Free Networks

The Watts–Strogatz networks exhibit both high clustering coefficients and short average path length, which overcomes the limitation in the ER graphs. However, the degree distribution of WS random graphs still follows a Poisson distribution. In many real networks, the degree distribution is empirically determined to follow a scale-free or power-law distribution.

Recall that in a power-law distribution the probability that a node u has degree k is given by

$$P(\delta) = P[\delta(u) = k] = \alpha k^{-a},$$

where α and a are parameters to be determined from the data. Whereas ER and WS graphs are constructed from a given set of vertices, scale-free networks result from a growth process combined with *preferential attachment*. The idea of preferential attachment is that the probability of a new node u connecting to an existing node v is proportional to the degree of node v. This is the *rich-get-richer* model. Once a particular node establishes a higher degree relative to other nodes, it then has a higher probability of attracting new nodes and, hence, its degree can increase rapidly. Scale-free networks are therefore characterized by the presence of *hubs*.

We next describe a particular scale-free network called the Barabási–Albert (BA) network [8], which combines growth and preferential attachment. We start with a given network G_0, such as an $ER(n_0, p)$ network or a regular graph with n_0 nodes, and add new nodes and edges sequentially to create networks $G_1, G_2, \dots$, as follows.

1. For each $i = 1, \dots$, add a new vertex v_i forming vertex set $V_i = V_{i-1} \bigcup v_i$.
2. Add $m \leq n_0$ edges incident to v_i and a vertex u from V_{i-1}, with u not chosen previously. Choose u with probability

$$P[\text{select } u] = \frac{\delta(u)}{\sum_{w \in V_{i-1}} \delta(w)}. \tag{13.2}$$

   This means that the probability that a vertex u is selected is proportional to its current degree.
3. Stop when n vertices have been added, otherwise repeat the previous two steps.

The resulting network is called a *Barabási–Albert network*, denoted $BA(n, n_0, m)$. It can be shown that For any $BA(n, n_0, m)$ graph G and $u \in V(G)$, $P[\delta(u) = k]$ is proportional to $\frac{1}{k^3}$.

Figure 13.40 a network grown using the above algorithm. The appearance of hubs in this network, together with many low-degree nodes, is readily apparent.

Figure 13.40 Network grown with preferential attachment with $n = 200$ and $m = 3$ edges added at each step. The presence of hubs is readily apparent in this network. Refer to the Mathworks File Exchange site for programs to generate scale-free networks.

Fitness

The BA model of growth and preferential attachment favors the so-called *early adopters*. It is more or less impossible for lower-degree nodes to overtake higher-degree nodes in this model. However, in practice, it is not uncommon for newcomers to overtake more established agents. For example, the internet search engine Altavista was well established before Google appeared on the scene and yet Google quickly became the dominant search engine on the web. One way to incorporate such phenomena in the BA model is to include a measure of *fitness* so that the probability of a node acquiring new links is a function of both its degree and its fitness.

In this case the probability of selecting node u is modified from Equation (13.2) to

$$P[\text{select } u] = \frac{\eta_u \delta(u)}{\sum_{w \in V_{i-1}} \eta_w \delta(w)}, \tag{13.3}$$

where η_u, and η_w are the fitness coefficients for nodes u and w. Thus, a node with a low degree can increase its probability of acquiring new nodes by increasing its fitness, which is apparent in many phenomena such as startup companies, and biological evolution.

13.6 Synchronization in Networks

In this section we discuss the problem of *synchronization* in networks. Synchronization phenomena appear in many kinds of natural and human-made networks and systems. In nature, important examples include *flocking* of birds, *swarming* of insects and *schooling* of fish (Figure 13.41). These are examples of so-called *emergent behavior*, in which purely local interactions give rise to global behavior. For example, in a school of fish, each fish can only sense a few of its neighbors a short distance from it and yet the entire school is able to move in unison as if it is a single entity. Other examples of the synchronization phenomenon include:

- circadian rhythm;
- contraction of coronary pacemaker cells;
- firing of memory neurons in the brain.

Figure 13.41 Flocking of birds (left), swarming of insects (middle) and schooling of fish (right).

Synchronization is important in engineered systems as well as in nature. Examples include:

- formation flying of autonomous aircraft;
- multiple robot networks;
- design of oscillator circuits;
- supply chains;
- modeling of crowd dynamics.

For problems of synchronization, in addition to modeling a network of agents as vertices on a graph, we also model the dynamics of each agent. The state of each agent may be, for example, a position, velocity, temperature, pressure, or other variable that evolves in time according to some differential or difference equation.

Example 13.21 As a simple example, suppose we have agents $x_1, \ldots, x_4$, each with dynamics

$$\dot{x}_i = u_i, \quad i = 1, \ldots, 4,$$

where u_i represents an input variable, and suppose that the agents are interconnected according to the undirected graph in Figure 13.42.

We assume that each agent knows its own state and receives the state of the agents to which it is connected. Thus we can choose the input u_i according to

$$u_i = \sum_{j=1}^{4} a_{ij}(x_j - x_i) = \sum_{j \in \mathcal{N}_i} (x_j - x_i), \tag{13.4}$$

where a_{ij} is the (i,j)th entry of the adjacency matrix and $\mathcal{N}_i$ is the neighbor set of agent i. This leads to the system

$$\begin{aligned}
\dot{x}_1 &= u_1 = x_2 - x_1 + x_4 - x_1 \\
\dot{x}_2 &= u_2 = x_1 - x_2 + x_3 - x_2 \\
\dot{x}_3 &= u_3 = x_2 - x_3 + x_4 - x_3 \\
\dot{x}_4 &= u_4 = x_3 - x_4 + x_1 - x_4.
\end{aligned}$$

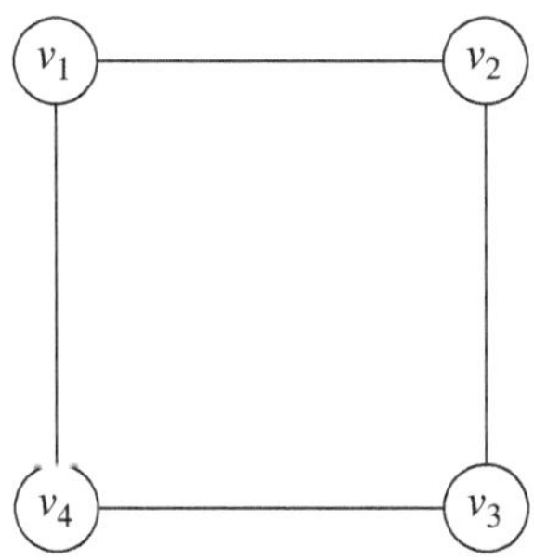

Figure 13.42 Four agents interconnected in a ring.

Writing the above equations in vector-matrix form gives

$$\begin{bmatrix} \dot{x}_1 \\ \dot{x}_2 \\ \dot{x}_3 \\ \dot{x}_4 \end{bmatrix} = \begin{bmatrix} -2 & 1 & 0 & 1 \\ 1 & -2 & 1 & 0 \\ 0 & 1 & -2 & 1 \\ 1 & 0 & 1 & -2 \end{bmatrix} \begin{bmatrix} x_1 \\ x_2 \\ x_3 \\ x_4 \end{bmatrix},$$

which we can write compactly as

$$\dot{x} = -Lx, \tag{13.5}$$

noting that the matrix L is precisely the Laplacian of the graph. Equation (13.5) is a linear system with a nonhyperbolic equilibrium at $x = 0$ since the Laplacian always has an eigenvalue at zero. To analyze the stability of the system, recall that the Laplacian matrix L satisfies

$$L = N^T N,$$

where N is the signed edge adjacency matrix, and let $V = \frac{1}{2}x^T x$ be a Lyapunov function. Then

$$\dot{V} = -x^T Lx = -x^T N^T Nx = -y^T y \leq 0,$$

where $y = Nx$. Now, since

$$y^T y = (x_1 - x_2)^2 + (x_2 - x_3)^2 + (x_3 - x_4)^2 + (x_4 - x_1)^2,$$

it follows from LaSalle's theorem that all solutions converge to the set $x_1 = x_2 = x_3 = x_4$ since the graph is connected.

In other words, we have shown that the agents synchronize, i.e., their states converge asymptotically to the same value. To find out what the limiting value is, we note, from the equations governing this system, that if we add the four equations we get

$$\dot{x}_1 + \dot{x}_2 + \dot{x}_3 + \dot{x}_4 = 0,$$

which implies that $x_1(t) + x_2(t) + x_3(t) + x_4(t) = constant$, for all time t. The only way this can happen is if

$$x_1(t) + x_2(t) + x_3(t) + x_4(t) = x_1(0) + x_2(0) + x_3(0) + x_4(0),$$

the sum of the initial conditions. Since we have shown that the agents synchronize to a common value, it follows that each agent converges to

$$\frac{1}{4}\sum_{i=1}^{4} x_i(0),$$

in other words, to the average value of the initial conditions. Figure 13.43 shows the response of the agents starting at initial conditions $(2, 1, 4, 6)$. The agents converge to the average value $13/4 = 3.25$ as predicted.

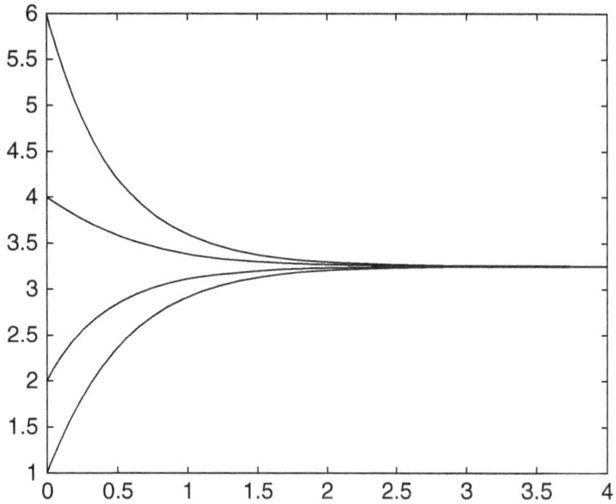

Figure 13.43 Response of the four-agent network showing that the states synchronize to a common value, which is the average of the initial values of the agents at time $t = 0$.

The same procedure can be extended to any connected network. The control given by Equation (13.4) is called the *consensus protocol* and provides a general algorithm for synchronization (and consensus) in a wide variety of networks, such as computer networks, social networks, and mobile robot networks.

Example 13.22 Suppose that we assign a *temperature* T_i to each node in a network. According to *Newton's law of cooling*, the heat transferred between nodes i and j will be proportional to the difference of their temperatures provided that they are connected so that heat can flow between the nodes. This leads to the dynamical system

$$\frac{dT_i}{dt} = r\sum_j a_{ij}(T_j - T_i), \tag{13.6}$$

where $A = (a_{ij})$ is the adjacency matrix and r is called the *coefficient of heat transfer*. Thus, as before, we have

$$\frac{dT}{dt} = -rLT, \tag{13.7}$$

where L is the graph Laplacian. If we let the number of nodes approach infinity, with each node at position (x, y), and replace $-L$ by the Laplacian operator, $\nabla^2 = \frac{\partial^2}{\partial x^2} + \frac{\partial^2}{\partial y^2}$, then Equation (13.7) becomes the two-dimensional heat equation. This is, in fact, the reason why the matrix L is called the *graph Laplacian*.

Remark 13.3 The above example is an example of the more general concept of *diffusion on networks*, which describes many phenomena, such as the spread of gossip, opinions, epidemics, and fads.

Remark 13.4 Another important application of a theory of synchronization, flocking, and/or swarming is in computer animation and the animated film industry. For example, to animate flocking dinosaurs, stampeding zebras, or schooling fish, it is difficult to preprogram the motion of each animal individually. Instead, one programs each animal with dynamics and some rules of behavior, such as move toward your nearest neighbor or move away from predators.

Problems

13.1 Give a one- or two-sentence description in nontechnical terms of the following:

1. Small-world network
2. Hub
3. Rich-get-richer.

13.2 For each of the following networks, how would you define the nodes and edges in order to model the network? Is the network directed or undirected?

1. An ant colony
2. A city road and highway network
3. A VLSI circuit
4. Evolution of species.

13.3 Suppose that you and a date attended a party with three other couples. At the beginning of the party, people shook hands to introduce themselves. No one shook hands with himself or with their own date, and no one shook hands with the same person more than once. Suppose you ask each person, including your own date, how many hands they shook and found that each person gave you a different answer.

1. Draw a graph to represent this example, where the nodes are people and an edge connects two people if they shook hands.
2. How many hands did you shake?
3. How many hands did your date shake?

13.4 Give an example of a graph exhibiting the properties that:

- every vertex is adjacent to two vertices; and
- every edge is adjacent to two edges.

13.5 Let G_1, G_2, and G_3 be simple, undirected $(4, 2)$ graphs. Show that at least two of them must be isomorphic.

13.6 Construct two nonisomorphic simple graphs with six vertices with degrees $[3, 3, 2, 2, 1, 1]$.

13.7 Suppose the graph G is a tree. Prove that Euler's formula is true no matter how many vertices G has.

13.8 Consider the graph below.

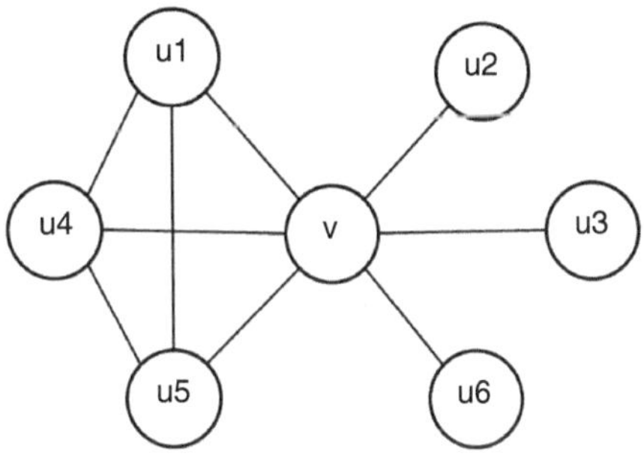

(a) Draw the subgraph induced by the neighbors of the vertex v.
(b) Compute the clustering coefficient of the vertex v.

13.9 Show that the graphs in Example 13.2 are not isomorphic by showing that the adjacency structure of the two graphs are not the same.

13.10 Suppose that the matrix

$$A = \begin{bmatrix} 0 & 1 & 1 & 1 & 1 \\ 1 & 0 & 0 & 0 & 1 \\ 1 & 0 & 0 & 1 & 1 \\ 1 & 0 & 1 & 0 & 0 \\ 1 & 1 & 1 & 0 & 0 \end{bmatrix}$$

is the adjacency matrix of a simple, undirected graph G.

1. What are the degrees of each node?
2. What is the size of G?

13.11 For the graph below find the clustering coefficient of the vertex V. Find the transitivity of the graph.

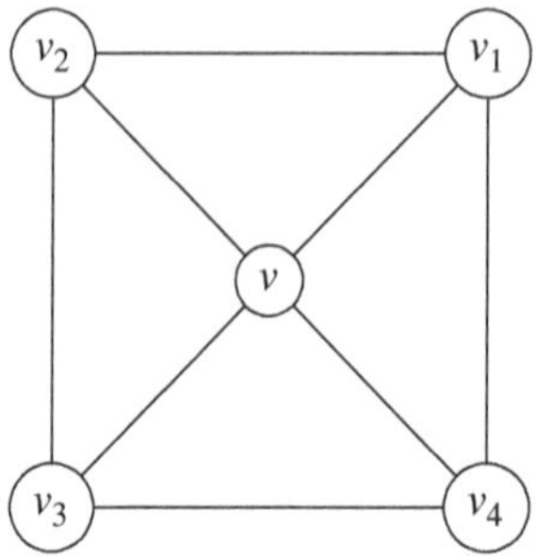

13.12 Verify that the transitivity in the graph in Figure 13.32 is equal to $2/17$.

13.13 Consider the graph below. Write the adjacency matrix A, the incidence matrix N, and the Laplacian matrix L.

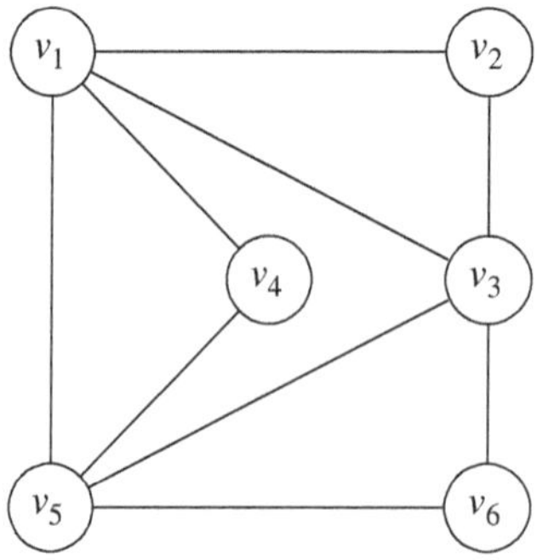

13.14 Compute the Laplacian matrix for the graph G below. How many connected components does G have? Explain why.

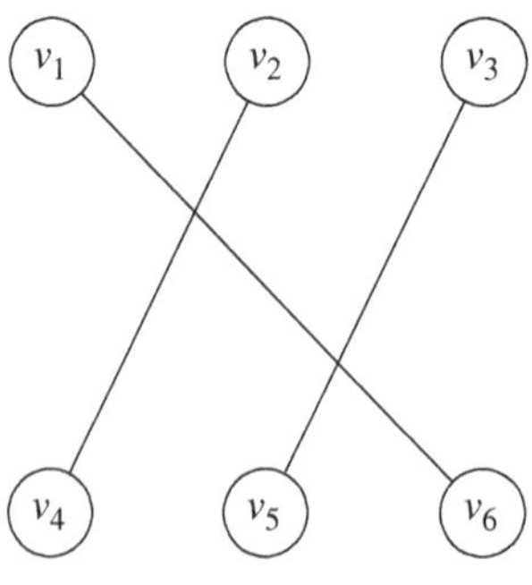

13.15 Explain why the product of the nonzero eigenvalues of the Laplacian of a simple undirected graph is always a whole number and not a fraction.

13.16 Consider the circular graph below.

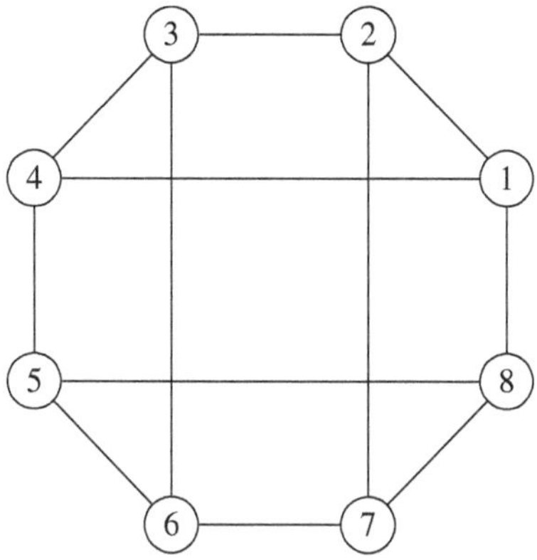

Show that it can be redrawn as the bipartite graph below.

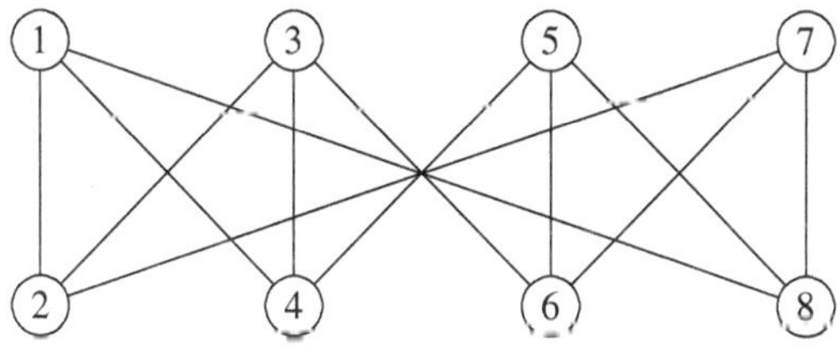

13.17 Show that a complete graph of order p is $(p-1)$-regular. Show that the complete graph K_n has $\binom{n}{2}$ edges.

13.18 Given an ER-random graph $G = ER(3, 0.5)$, what is the probability that G has exactly two edges? What is the general formula for the probability that $G = ER(n, p)$ has exactly $m \leq \binom{n}{2}$ edges?

13.19 Given an ER random graph $G = ER(n, p)$, what is the probability that G is a complete graph?

13.20 Let A be the adjacency matrix for an $ER(n, p)$ random graph. What is the expected number of 1s and 0s in A? In other words, on average how many elements $a(i, j)$ of A would be equal to 1?

13.21 Let G be a scale-free graph of order n with degree distribution $P(\delta) = P[\deg(u) = k] = k^{-3}$. What is the expected value $\overline{\delta} = E(\delta)$ for $n = 5$?

13.22 Consider a network of agents interconnected according to the graph below.

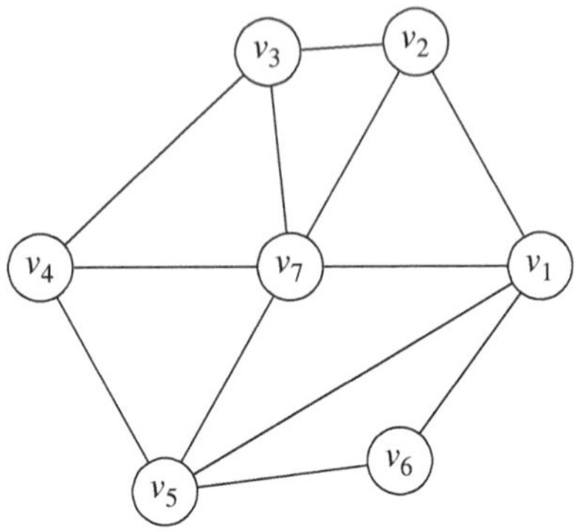

Suppose that the state of each agent evolves according to $\dot{x}_i = u_i$. Write a Matlab script to simulate the evolution of the network with u_i given by the consensus protocol in Equation (13.4). Interpret the results.

Appendix A

Linear Algebra

We assume that the reader has some familiarity with basic properties of vectors and matrices, such as matrix addition, multiplication, and determinants [66]. The symbol $\mathbb{R}$ denotes the set of real numbers and $\mathbb{R}^n$ denotes the set of n-tuples of real numbers. We use lowercase letters a, b, c, x, y, etc., to denote scalars in $\mathbb{R}$ and vectors in $\mathbb{R}^n$. Uppercase letters A, B, C, M, R, etc., denote matrices.

A.1 Vectors

Vectors in $\mathbb{R}^n$ are defined as column vectors

$$x = \begin{bmatrix} x_1 \\ \vdots \\ x_n \end{bmatrix}, \quad \text{with } x_i \in \mathbb{R},\ i = 1, \dots, n.$$

The entries $x_1, \dots, x_n$ are called *components* or *coordinates* of x. The *transpose* of a vector x, denoted x^T, is the row vector

$$x^T = [x_1, \dots, x_n].$$

In this text we will often drop the superscript T when the vector x is not part of a displayed equation and simply write

$$x = [x_1, \dots, x_n] \quad \text{or} \quad x = (x_1, \dots, x_n)$$

to denote x as a row instead of a column. Also, since we do not use bold font or arrows to denote vectors, the difference between, for example, x_i as a vector and x_i as a component of a vector x should be clear from the context.

The *norm* or *length* of a vector $x = (x_1, \dots, x_n)^T$ is denoted $||x||$ and is defined as

$$||x|| = \sqrt{x_1^2 + \cdots + x_n^2}.$$

Introduction to Modeling and Simulation: A Systems Approach, First Edition. Mark W. Spong.

Companion Website: www.wiley.com/go/spong/engineeringmodeling

Linear Independence

A set of vectors $\{x_1, \ldots, x_m\}$ is said to be *linearly independent* if and only if, for arbitrary scalars $\alpha_i \in \mathbb{R}$,

$$\sum_{i=1}^{m} \alpha_i x_i = 0 \text{ implies } \alpha_i = 0 \text{ for all } i;$$

otherwise the vectors $x_1, \ldots, x_m$ are said to be *linearly dependent*. A *basis* of a real vector space is a linearly independent set of vectors $\{e_1, \ldots, e_m\}$ such that every vector x can be written as a linear combination

$$x = x_1 e_1 + \cdots + x_m e_m, \quad x_i \in \mathbb{R},\ i = 1, \ldots, m.$$

The coordinate representation $x = (x_1, \ldots, x_m)$ is uniquely determined by the particular basis $\{e_1, \ldots, e_m\}$. The *dimension* of the vector space is the number of vectors in any basis.

Subspaces

A set of vectors $\mathcal{M}$ contained in a vector space $\mathcal{V}$ is a *subspace* of $\mathcal{V}$ if $\alpha x + \beta y$ belongs to $\mathcal{M}$ for every x and y in $\mathcal{M}$ and every α and β in $\mathbb{R}$. In other words $\mathcal{M}$ is closed under the operations of addition and scalar multiplication. A subspace $\mathcal{M}$ is itself a vector space. The dimension of $\mathcal{M}$ is less than or equal to the dimension of $\mathcal{V}$.

Inner Product

The *inner product* or *dot product* of vectors x and y in $\mathbb{R}^n$ is defined as

$$x \cdot y = x^T y = x_1 y_1 + x_2 y_2 + \cdots + x_n y_n.$$

The inner product satisfies the following properties:

- $x \cdot y = y \cdot x$;
- $(\alpha_1 x + \alpha_2 y) \cdot z = \alpha_1 x \cdot z + \alpha_2 y \cdot z$, where a_1 and a_2 are scalars;
- $x \cdot x \geq 0$ and $x \cdot x = 0$ if and only if $x = 0$.

The inner product is related to the norm as

$$\|x\| = \sqrt{x \cdot x}.$$

Useful properties of the inner product and norm are as follows:

Cauchy–Schwarz inequality

The Cauchy–Schwarz inequality relates the inner product and norms of vectors x and y according to

$$|x \cdot y| \leq \|x\|\ \|y\|,$$

with equality if and only if x and y are linearly dependent.

Homogeneity
If x is a vector and α is a scalar, then $\|\alpha x\| = |\alpha| \|x\|$, where $|\alpha|$ is the absolute value of α.

The Triangle Inequality
The triangle inequality states that, if x and y are vectors, then

$$\|x + y\| \leq \|x\| + \|y\|.$$

This is a generalization of the fact from two-dimensional geometry that the length of the hypotenuse of a right triangle is less than the sum of the lengths of the two sides.

For vectors in $\mathbb{R}^2$ or $\mathbb{R}^3$ the scalar product can be expressed as $x \cdot y = \|x\| \cdot \|y\| \cos\theta$, where θ is the angle between the vectors x and y.

Orthogonal Complement
If W is a subspace of $\mathbb{R}^n$, the *orthogonal complement* of W in $\mathbb{R}^n$ is the subspace

$$W^{\perp} = \{v \in \mathbb{R}^n \mid v^T w = 0 \text{ for all } w \in W\}.$$

The symbol $W^{\perp}$ is read "W perp". Thus $W^{\perp}$ is the set of all vectors $v \in \mathbb{R}^n$ that are orthogonal to all vectors $w \in W$. The orthogonal complement of a subspace W in $\mathbb{R}^n$ satisfies the following:

- $W^{\perp}$ is also a subspace of $\mathbb{R}^n$;
- $(W^{\perp})^{\perp} = W$;
- $\dim(W) + \dim(W^{\perp}) = n$.

A.2 Matrices

Matrices and matrix representations of linear transformations are fundamental to many of the concepts in this text. A *linear transformation* A on a vector space $\mathcal{V}$ is a function $A : \mathcal{V} \to \mathcal{V}$ such that

$$A(\alpha x_1 + \beta x_2) = \alpha A(x_1) + \beta A(x_2), \text{ for } x_1, x_2 \in \mathcal{V} \text{ and } \alpha, \beta \in \mathbb{R}.$$

We will denote $A(x)$ by Ax. We represent linear transformations in coordinate form as matrices by taking a particular basis for $\mathcal{V}$.

An $m \times n$ matrix $A = (a_{ij})$ over $\mathbb{R}$ is an ordered array of real numbers with m row vectors $(a_{i1}, \dots, a_{in})$, for $i = 1, \dots, m$, and n column vectors $(a_{1j}, \dots, a_{mj})$, for $j = 1, \dots, n$, written as

$$A = \begin{bmatrix} a_{11} & a_{12} & \cdots & a_{1n} \\ a_{21} & a_{22} & \cdots & a_{2n} \\ \vdots & \vdots & \ddots & \vdots \\ a_{m1} & a_{m2} & \cdots & a_{mn} \end{bmatrix}. \tag{A.1}$$

The matrix A is a representation of a linear transformation from $\mathbb{R}^n$ to $\mathbb{R}^m$. If $n = m$, i.e., if the number of rows and the number of columns are equal, then the matrix A is said to be *square*. The set of $n \times n$ square matrices, which we denote by $\mathbb{R}^{n\times n}$, is itself a vector space of dimension n^2.

The *rank* of a matrix A is the largest number of linearly independent rows (or columns) of A. Thus, the rank of an $m \times n$ matrix can be no greater than the minimum of m and n.

The *transpose* of a matrix A is denoted A^T and is formed by interchanging rows and columns of A:

$$A^T = \begin{bmatrix} a_{11} & a_{21} & \cdots & a_{m1} \\ a_{12} & a_{22} & \cdots & a_{m2} \\ \vdots & \vdots & \ddots & \vdots \\ a_{1n} & a_{2n} & \cdots & a_{mn} \end{bmatrix}. \tag{A.2}$$

Thus A^T is an $n \times m$ matrix. Some properties of the matrix transpose are as follows:

- $(A^T)^T = A$;
- $(AB)^T = B^T A^T$;
- $(A + B)^T = A^T + B^T$.

A square $n \times n$ matrix A is said to be:

- *symmetric* if $A^T = A$;
- *skew-symmetric* if $A^T = -A$;
- *orthogonal* if $A^T A = AA^T = I$, where I is the $n \times n$ identity matrix.

Matrix Trace

The *trace* of an $n \times n$ matrix A, denoted $\text{tr}(A)$, is the sum of the diagonal entries of A. Thus, if $A = (a_{ij})$, then $\text{tr}(A) = \sum_{i=1}^{n} a_{ii} = a_{11} + \cdots + a_{nn}$.

The Determinant

A *determinant* is a function that assigns a particular scalar to a linear transformation (matrix) $A \in \mathbb{R}^{n\times n}$. For our purposes we may define the determinant, $\det(A)$, of a square matrix A via the recursive formula that holds for any $i = 1, \dots, n$,

$$\det(A) = \sum_{j=1}^{n} a_{ij}(-1)^{i+j} \det(C_{ij}),$$

where C_{ij} is the $(n-1) \times (n-1)$ matrix formed by deleting the ith row and jth column of A. The matrix determinant satisfies the following properties:

- $\det(AB) = \det(A)\det(B)$;
- $\det(\alpha A) = \alpha^n \det(A)$;
- $\det(A^T) = \det(A)$.

If $\det(A) = 0$, the matrix A is called *singular*; otherwise A is *nonsingular* or *invertible*. The *inverse* of a square matrix $A \in \mathbb{R}^{n\times n}$ is a matrix $B \in \mathbb{R}^{n\times n}$ satisfying

$$AB = BA = I$$

where I is the $n \times n$ identity matrix. We denote the inverse of A by A^{-1}. The inverse of a matrix A exists and is unique if and only if A has rank n, equivalently, if and only if the determinant $\det(A)$ is nonzero. The matrix inverse satisfies the following properties:

- $(A^{-1})^{-1} = A$;
- $(AB)^{-1} = B^{-1}A^{-1}$.

If A is an orthogonal matrix, $A^T = A^{-1}$.

Minors

Definition A.1 Let

$$P = \begin{bmatrix} p_{11} & p_{12} & \cdots & p_{1n} \\ p_{21} & p_{22} & \cdots & p_{2n} \\ \vdots & \vdots & \vdots & \vdots \\ p_{n1} & p_{n2} & \cdots & p_{nn} \end{bmatrix}$$

be an $n \times n$ matrix and let P_k be a $k \times k$ submatrix of P formed by deleting $n - k$ rows and $n - k$ columns of P. Then the determinant of P_k is called a *minor* or *minor determinant* of P.

The *principal minors* of P are the determinants of the submatrices P_k formed by deleting rows and columns $k + 1$ to n, i.e., the determinants of

$$P_1 = p_{11}, \quad P_2 = \begin{bmatrix} p_{11} & p_{12} \\ p_{21} & p_{22} \end{bmatrix}, \quad P_3 = \begin{bmatrix} p_{11} & p_{12} & p_{13} \\ p_{21} & p_{22} & p_{23} \\ p_{31} & p_{23} & p_{33} \end{bmatrix}, \quad \ldots, \quad P_n = P.$$

Then the principal minors of P are $M_k = \det(P_k),\ k = 1, \ldots, n$.

A.3 Eigenvalues and Eigenvectors

The *eigenvalues* of a matrix A are the solutions in λ of the equation

$$\det(\lambda I - A) = 0.$$

The function $\det(\lambda I - A)$ is a polynomial of degree n in λ, called the *characteristic polynomial* of A. If λ_e is an eigenvalue of A, an eigenvector of A corresponding to λ_e is a nonzero vector x_e satisfying the system of linear equations

$$Ax_e = \lambda_e x_e.$$

Similarity Transformation

If T is an $n \times n$ nonsingular matrix, then

$$\overline{A} = T^{-1}AT \tag{A.3}$$

is called a similarity transformation. Since T is nonsingular, the column vectors of T are linearly independent, and hence form a basis of $R^{n \times n}$. For this reason, (A.3) is also called a *change of basis*. The matrix $\overline{A}$ represents the same linear transformation as A in the basis defined by T.

Diagonalizing a Symmetric Matrix

If A is a symmetric matrix, then:

- its eigenvalues $\lambda_1, \dots, \lambda_n$ are real;
- eigenvectors corresponding to distinct eigenvalues are orthogonal;
- there exists an $n \times n$ orthogonal matrix T such that

  $$\overline{A} = T^{-1}AT = \text{diag}[\lambda_1, \dots, \lambda_n]$$

  where diag[ ] is a diagonal matrix.

The columns of the matrix T consist of (a basis of) eigenvectors corresponding to the respective eigenvalues $\lambda_1, \dots, \lambda_n$.

Quadratic Forms

Definition A.2 A *quadratic form* on $\mathbb{R}^n$ is a scalar function

$$V(x) = x^T P x = \sum_{i=1}^{n} \sum_{j=1}^{n} p_{ij} x_i x_j,$$

where $P = (p_{ij})$ is an $n \times n$ symmetric matrix. The function V, equivalently, the matrix P, is *positive definite* if:

- $x^T P x > 0$ for all $x \neq 0$;
- all eigenvalues of P are positive;
- all principal minors of P are positive.

If P is positive definite, then we have the useful bounds

$$\lambda_{\min}(P)\|x\|^2 \leq x^T P x \leq \lambda_{\max}(P)\|x\|^2, \tag{A.4}$$

where $\lambda_{\min}(P)$ and $\lambda_{\max}(P)$ are, respectively, the minimum and maximum eigenvalues of P.

Range and Null Space

The *range space*, $\mathcal{R}(A)$, of an $m \times n$ matrix A is the subspace of $\mathbb{R}^m$ defined by

$$\mathcal{R}(A) = \{y \in \mathbb{R}^m : \ y = Ax, \ \text{for some } x \in \mathbb{R}^n\}.$$

The *null space*, $\mathcal{N}(A)$, of an $m \times n$ matrix A is the subspace of $\mathbb{R}^n$ defined by

$$\mathcal{N}(A) = \{x \in \mathbb{R}^n : Ax = 0\}.$$

The range and null spaces of a matrix are related according to

$$\mathcal{N}(A) = \mathcal{R}(A^T)^\perp$$
$$\mathcal{R}(A) = \mathcal{N}(A^T)^\perp.$$

An important property of the null space is that

$$\dim \mathcal{R}(A) + \dim \mathcal{N}(A) = n.$$

An $n \times n$ matrix is invertible if and only if the null space consists of only the zero vector, i.e., $Ax = 0$ implies $x = 0$.

Appendix B

Real Analysis

B.1 Set Theory

Here we give some basic definitions from set theory and topology.

Sets

- Let $x_0 \in \mathbb{R}^n$ be a vector in $\mathbb{R}^n$. An (open) *neighborhood* of x_0, also called an open *ball* of radius ϵ at x_0, is the set

  $$B(x_0, \epsilon) = \{x \in \mathbb{R}^n \mid ||x - x_0|| < \epsilon\}.$$

  A *closed neighborhood* of x_0 is the set

  $$\overline{B}(x_0, \epsilon) = \{x \in \mathbb{R}^n \mid ||x - x_0|| \leq \epsilon\}.$$

- A set $S \subset \mathbb{R}^n$ is an *open set* if, for each $x \in S$, there is a neighborhood $B(x, \epsilon)$ such that $B(x, \epsilon) \subset S$.
  A set $S \subset \mathbb{R}^n$ is a *closed set* if and only if the *complement* of S, $\mathbb{R}^n \backslash S$, meaning the points in $\mathbb{R}^n$ not in S, is open.
- A vector $x \in \mathbb{R}^n$ is a *boundary point* of a set S if and only if any neighborhood $B(x, \epsilon)$ of x contains points in both S and the complement of S. The boundary of S, denoted ∂S, is the set of all boundary points of S.
  If S is an open set then the union, $S \cup \partial S$, of S and its boundary is a closed set.
- A set $S \subset \mathbb{R}^n$ is *bounded* if there exist $r > 0$ and a ball $B(0, r)$ at the origin $x = 0$ such that $S \subset B(0, r)$. In other words, $||x|| < r$ for every $x \in S$.
- A set S is *compact* if and only if it is closed and bounded.

Example B.1 The set consisting of the interval $S = (a, b)$ for a, b real numbers is open. The interval $S = [a, b]$ is closed. The boundary of S is the set $\{a, b\}$ of

Introduction to Modeling and Simulation: A Systems Approach, First Edition. Mark W. Spong.

Companion Website: www.wiley.com/go/spong/engineeringmodeling

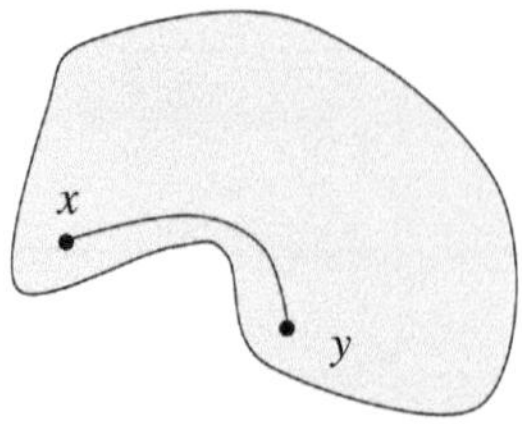

Figure B.1 A connected set. Every pair of points x and y in S can be connected by a continuous curve lying in S.

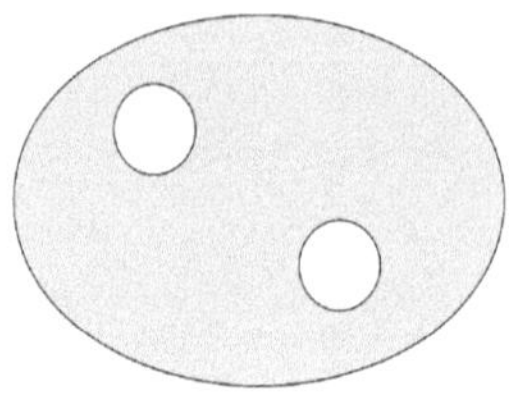

Figure B.2 A simply connected set is one with no holes. The set shown is not simply connected.

endpoints of the interval. The set S is bounded if both a and b are finite. Thus, $S = (0, \infty)$ is open but not bounded. The set $S = [0, 1]$ is closed and bounded, hence it is compact.

Connectedness

Let S be a subset of $\mathbb{R}^n$. Then S is *connected* if any two points x and y in S can be connected by a continuous curve lying entirely in S. This is illustrated by Figure B.1.

A set S is *simply connected* if every simple closed curve can be shrunk to a point without leaving the set. Intuitively, a simply connected set is one with no "holes." The set shown in Figure B.2 is not simply connected.

B.2 Vector Fields

Definition B.1 ***(Vector Field)*** A *vector field* is a function from $\mathbb{R}^n$ to $\mathbb{R}^n$

$$f(x) = \begin{bmatrix} f_1(x_1, \dots, x_n) \\ f_2(x_1, \dots, x_n) \\ \vdots \\ f_n(x_1, \dots, x_n) \end{bmatrix}.$$

We say that $f \in C^k$ if f is at least k-times continuously differentiable in x and we refer to such functions as *smooth functions*.

B.3 Jacobian

Let $f(x) : \mathbb{R}^n \rightarrow \mathbb{R}^n$ be a vector field on $\mathbb{R}^n$. The *Jacobian* of f is the matrix function

$$J(x) = \frac{\partial f}{\partial x} = \begin{bmatrix} \frac{\partial f_1}{\partial x_1} & \frac{\partial f_1}{\partial x_2} & \cdots & \frac{\partial f_1}{\partial x_n} \\ \frac{\partial f_2}{\partial x_1} & \frac{\partial f_2}{\partial x_2} & \cdots & \frac{\partial f_2}{\partial x_n} \\ \vdots & \vdots & & \vdots \\ \frac{\partial f_n}{\partial x_1} & \frac{\partial f_n}{\partial x_2} & \cdots & \frac{\partial f_n}{\partial x_n} \end{bmatrix}.$$

B.4 Scalar Functions

In this section we discuss level sets, the gradient, and the Hessian of scalar functions. Let $V : \mathbb{R}^n \rightarrow \mathbb{R}$ be a *scalar function* on $\mathbb{R}^n$. By "scalar" we mean that V takes a vector x and returns a real number $V(x)$.

Definition B.2 A *level set* or *level curve* of V is the set of points $V^{-1}(c) = \{x \in \mathbb{R}^n \mid V(x) = c\}$, as illustrated in Figure B.3.

Gradient and Hessian

Definition B.3 ***(Gradient and Hessian)*** Let $f : \mathbb{R}^n \rightarrow \mathbb{R}$ be a smooth scalar function. The gradient of $f(x)$ is the column vector

$$\nabla f(x) = \left[\frac{\partial f}{\partial x_1}, \frac{\partial f}{\partial x_2}, \ldots, \frac{\partial f}{\partial x_n}\right]^T.$$

The Hessian, or Hessian matrix, is the $n \times n$ matrix of second partial derivatives of $f(x)$:

$$H(x) = \begin{bmatrix} \frac{\partial^2 f}{\partial x_1^2} & \frac{\partial^2 f}{\partial x_1 \partial x_2} & \cdots & \frac{\partial^2 f}{\partial x_1 \partial x_n} \\ \frac{\partial^2 f}{\partial x_2 \partial x_1} & \frac{\partial^2 f}{\partial x_2^2} & \cdots & \frac{\partial^2 f}{\partial x_2 \partial x_n} \\ \vdots & \vdots & \ddots & \vdots \\ \frac{\partial^2 f}{\partial x_n \partial x_1} & \frac{\partial^2 f}{\partial x_n \partial x_2} & \cdots & \frac{\partial^2 f}{\partial x_n^2} \end{bmatrix}.$$

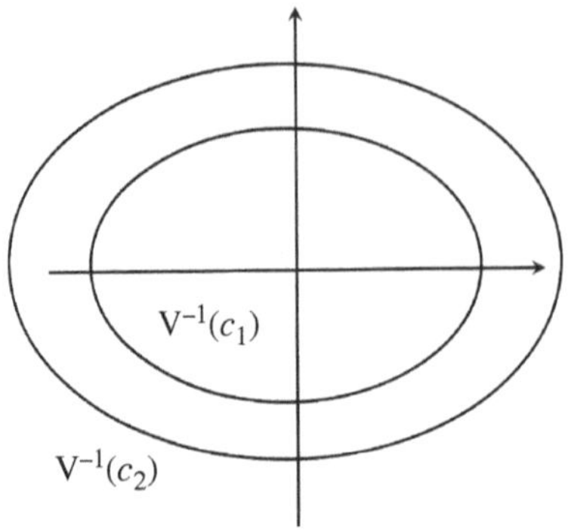

Figure B.3 Level sets of a function $V(x)$ in $\mathbb{R}^2$.

Example B.2 Consider the function $f(x) = x_1^3 + x_1x_2^2 + x_1^2x_2 + x_2^4$. A straightforward calculation shows that the gradient and Hessian are given by

$$\nabla f(x) = \left[3x_1^2 + x_2^2 + 2x_1x_2, 2x_1x_2 + x_1^2 + 4x_2^3\right]^T,$$
$$H(x) = \begin{bmatrix} 6x_1 + 2x_2 & 2x_2 + 2x_1 \\ 2x_2 + 2x_1 & 2x_1 + 12x_2^2 \end{bmatrix}.$$

Definition B.4 ***(Directional Derivative)*** Given a scalar function $f(x) : \mathbb{R}^n \to \mathbb{R}$ and a vector $d \in \mathbb{R}^n$, the *directional derivative* of f in the direction d at a point $x \in \mathbb{R}^n$ is given by

$$D_d(f) = d^T \nabla f(x) = \sum_{i=1}^{n} d_i \frac{\partial f}{\partial x_i}(x).$$

B.5 Taylor's Theorem

Taylor's theorem says, roughly speaking, that a large class of nonlinear functions can be well approximated by polynomials and is one of the most important tools at our disposal for many applications such as numerical solution of differential equations and optimization. In order to quantify these polynomial approximations, we need to introduce some notation.

Order Notation

Definition B.5 ***(Big-O and Little-o Notation)*** Let $h > 0$.

1. We say that a function $R_n = O(h^n)$ if $\frac{|R_n|}{h^n}$ is bounded as $h \to 0$.
2. We say that a function $R_n = o(h^n)$ if $\frac{|R_n|}{h^n} \to 0$ as $h \to 0$.

This notation is read *big O* and *little o*, respectively.

Example B.3 Consider the function $R(x) = 3x^3 + x^2$. Then $\frac{R(x)}{x^2} = 3x + 1$ which approaches 1 as $x \to 0$ and is therefore bounded. Thus, $R(x)$ is $O(x^2)$. On the other hand, $\frac{R(x)}{x} = 3x^2 + x$ approaches zero as $x \to 0$ and hence $R(x)$ is $o(x)$.

A one-dimensional version of Taylor's theorem can be stated as follows. Suppose $f(x) : \mathbb{R} \to \mathbb{R}$ is k times continuously differentiable at a point a. Then $f(x)$ can be approximated in a neighborhood of a as

$$f(x) = f(a) + f'(x)(x-a) + \frac{1}{2}f''(a)(x-a)^2 + \cdots + \frac{1}{(k-1)!}f^{k-1}(a)(x-a)^{k-1} + R_k$$

where the remainder $R_k = O(x-a)^k$.

Example B.4 A straightforward calculation shows that the Taylor series expansions for the exponential function, sine function, and cosine function are:

$$\begin{aligned} e^x &= 1 + x + \frac{1}{2!}x^2 + \frac{1}{3!}x^3 + \cdots + \frac{1}{k!}x^k + \cdots, \\ \sin(x) &= x - \frac{1}{3!}x^3 + \frac{1}{5!}x^5 - \cdots + \frac{(-1)^k}{(2k+1)!}x^{2k+1} + \cdots, \\ \cos(x) &= 1 - \frac{1}{2!}x^2 + \frac{1}{4!}x^4 - \cdots + \frac{(-1)^k}{(2k)!}x^{2k} + \cdots. \end{aligned}$$

B.6 Extreme-Value Theorem

The *extreme-value theorem* is important in optimization theory.

Theorem B.1 *Let $f(x) : \mathbb{R}^n \to \mathbb{R}$ be a continuous real-valued function from $\mathbb{R}^n$ to $\mathbb{R}$. Let S be a compact (i.e., closed and bounded) set in $\mathbb{R}^n$. Then $f(x)$ has a maximum and a minimum value on S.*

Appendix C

Probability

In this appendix we give a short review of basic concepts in discrete and continuous probability.

C.1 Discrete Probability

Discrete probability is largely a matter of counting the number of successful outcomes as a fraction of the total possible outcomes. For example, there are 36 possible outcomes from rolling two dice and there are six ways to roll a 7, namely, $(1,6), (2,5), (3,4), (4,3), (5,2)$ and $(6,1)$. Therefore, the probability of rolling a 7 is $6/36 = 1/6$.

Definition C.1 ***(Random Experiment)*** A *random experiment* is one whose outcome cannot be determined in advance. Examples include:

- tossing a die;
- counting the number of customers that arrive at a store during a fixed time period;
- picking a group of people and noting their birthdays.

Definition C.2 ***(Sample Space)*** The *sample space* Ω of a random experiment is the set of all possible outcomes. Thus, for the problem of rolling two dice, the sample space is

$$\Omega = \{(1,1), (1,2), \dots, (6,6)\}.$$

For the number of customers arriving at a store during a specified time interval, the sample space is

$$\Omega = \{0, 1, 2, \dots\}.$$

Introduction to Modeling and Simulation: A Systems Approach, First Edition. Mark W. Spong.

Companion Website: www.wiley.com/go/spong/engineeringmodeling

For the birthdays of randomly chosen people, the sample space is

$$\Omega = \{1, 2, \ldots, 365\}.$$

Definition C.3 ***(Event)*** An *event* is a subset of the sample space Ω representing one or more outcomes of an experiment. Events are usually denoted by capital letters $A, B, C, \ldots$. For example, the event that the sum of two dice is less than or equal to 3 is

$$A = \{(1,1), (1,2), (2,1)\}.$$

We can apply the usual set-theory notions to events, such as union, intersection, complement, and subset. For example, the event that the sum of the two dice is less than 3 or greater than 10 is the union of $A = \{\text{less than } 3\}$ and $B = \{\text{greater than } 10\}$.

$$A \cup B = \{(1,1)\} \cup \{(5,6), (6,5), (6,6)\} = \{(1,1), (5,6), (6,5), (6,6)\}.$$

Likewise, the event $C = A$ and B is the intersection $A \cap B$. In this case C is equal to the empty set $\emptyset = \{\ \}$. We say that two events A and B are disjoint if and only if $A \cap B = \emptyset$.

Definition C.4 ***(Probability Distribution)*** A *probability distribution* is a function $P : \Omega \to [0, 1]$ such that:

1. $0 \leq P(A) \leq 1$ for any event $A \subset \Omega$;
2. $P(\Omega) = 1$;
3. for any set $\{A_1, A_2, \ldots\}$ of disjoint events, $P\left(\bigcup_i A_i\right) = \sum_i P(A_i)$.

C.2 Conditional Probability

Sometimes we are given partial information on the outcome of an experiment. For example, if we roll two dice, what is the probability of rolling 8 if we know the first die is 2?

The probability of rolling 8 is $5/36$ since the number of ways to roll 8 is $(2,6), (3,5), (4,4), (5,3), (6,2)$. If the first die is 2 then to roll 8 the second die must be 6, which has probability $1/6$, which is greater than $5/36$. Therefore, the additional information that the first die is 2 results in a greater probability. This is called conditional probability.

Definition C.5 ***(Conditional Probability)*** Let A and B be events with $P(B) > 0$. Then the *conditional probability* of A given B is denoted by $P(A \mid B)$ and is defined as

$$P(A \mid B) = \frac{P(A \cap B)}{P(B)}.$$

In the previous example A is the event that the sum of two dice equals 8 and B is the event that the first die equals 2. Then

$$B = \{(2, 1), (2, 2), (2, 3), (2, 4), (2, 5), (2, 6)\}$$

which has probability $1/6$. The event $A \cap B$ is $\{(2, 6)\}$, which has probability $1/36$. Therefore,

$$P(A \mid B) = \frac{P(A \cap B)}{P(B)} = \frac{1/36}{1/6} = 1/6.$$

The conditional probability satisfies the following properties:

1. The conditional probability is nonnegative: $P(A \mid B) \geq 0$.
2. The conditional probability of the entire sample space is $P(\Omega \mid B) = 1$.
3. For any set of mutually exclusive events $A_1, A_2, \ldots,$

$$P\left(\bigcup_i A_i \mid B\right) = \sum_i P(A_i \mid B).$$

Definition C.6 ***(Independent Events)*** Events A and B are *independent* if

$$P(A \mid B) = P(A).$$

In other words, the probability that event B occurs does not effect the probability that event A occurs. This is equivalent to

$$P(B \mid A) = P(B) \quad \text{and} \quad P(A \cap B) = P(A)P(B).$$

We next give some examples of the more common discrete probability distributions.

Uniform Distribution

The *uniform distribution* assigns a probability $P(a) = \frac{1}{n}$ to each element a in a sample space Ω that has n elements.

Binomial Distribution

The *binomial distribution*, also called the *Bernoulli distribution*, arises from a statistical experiment that has the following properties:

1. The experiment consists of n repeated trials.
2. Each trial can result in just two possible outcomes. We call one of these outcomes a success and the other a failure.
3. The probability of success is the same on every trial.
4. The trials are independent, i.e., the outcome on one trial does not affect the outcome on other trials.

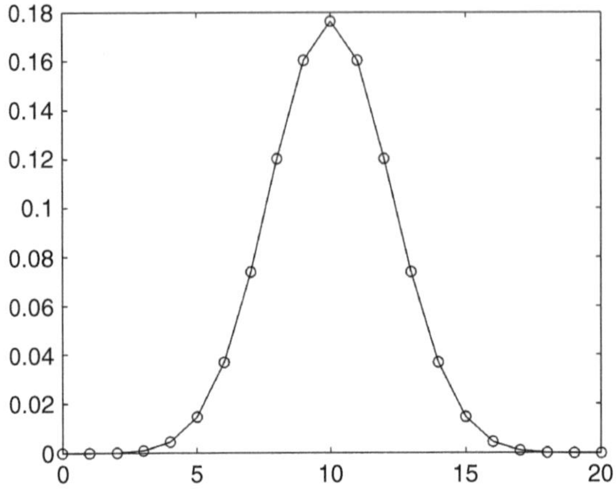

Figure C.1 The binomial distribution with $n = 20$ an $p = 0.5$.

The most common example of a binomial experiment is flipping a coin n times with the probability of heads equal to p, where $0 \leq p \leq 1$. If we identify heads with 1 and tails with 0, the sample space is $\Omega = \{0,1\}^n$ and the probability distribution function is $P(a) = p^k(1-p)^{n-k}$ where k is the number of 1s in the string a.

When $p = \frac{1}{2}$, we have the uniform distribution over $\{0,1\}^n$. If A_k is the event that we get a string of k ones, then

$$P[A_k] = \binom{n}{k} p^k(1-p)^{n-k},$$

where $\binom{n}{k} = \frac{n!}{k!(n-k)!}$ is the binomial coefficient.

Figure C.1 shows the binomial distribution for $n = 20$ trials with $p = 0.5$.

Poisson Distribution

The *Poisson distribution* is used to determine probabilities of events occurring in fixed time intervals, assuming that the events are independent and the mean rate of occurrence is known. For example, suppose a large convention center has infinitely many light bulbs. For each integer k, let $P(k)$ denote the probability that k bulbs will burn out during the next 24 hours. Then the Poisson distribution with mean rate λ is

$$P(k) = \frac{\lambda^k}{k!}e^{-\lambda},$$

Note that the mean rate λ is not required to be an integer. Figure C.2 shows the Poisson distribution for $\lambda = 4$.

For large n and small probability $p = \lambda/n$, the binomial distribution

$$P(k) = \binom{n}{k} p^k(1-p)^{n-k}$$

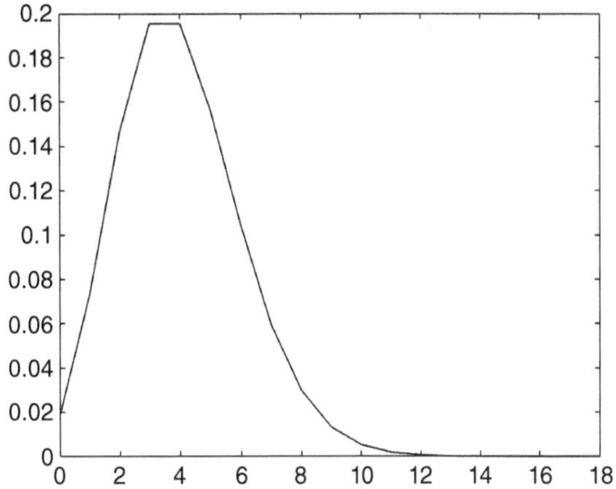

Figure C.2 The Poisson distribution with mean $\lambda = 4$.

can be approximated by the Poisson distribution

$$P(k) = \frac{\lambda^k e^{-\lambda}}{k!},$$

where $\lambda = np$. Roughly speaking, the approximation is valid for $n > 100$ and $np < 1$.

C.3 Random Variables

Definition C.7 ***(Random Variable)*** A *random variable* X is a function $X : \Omega \to V$, where Ω is a sample space and V is called the range of the random variable.

The notation $X = v$ for $v \in V$ denotes the event $\{a \in \Omega\ :\ X(a) = v\}$ and so the expression $P[X = v]$ is well defined.

Example C.1 Consider again the experiment of rolling a pair of dice with $\Omega = \{1, \dots, 6\}^2$, and for $(a, b) \in \Omega$ we have $P(a, b) = 1/36$. Suppose the random variable X is given by $X(a, b) = a + b$. Then the range $V = \{2, \dots, 12\}$. Thus, we can compute $P[X = v]$ for every $v \in V$ as $P[X = 2] = 1/36$, $P[X = 3] = 2/36$, etc.

Definition C.8 ***(Expectation)*** If X is a random variable, the *expectation* of X, also called the *expected value* of X, is defined as

$$E[X] = \sum_{v \in V} v P[X = v],$$

where V is the range of X.

Example C.2 Suppose a certain game has three outcomes, win (W), lose (L), or draw (D), with $P(W) = 2/3$, $P(L) = 1/6$, $P(D) = 1/6$. A win generates \$1.00, a loss forfeits \$2.00 and a draw does not win or lose anything. Let $\Omega = \{W, L, D\}$ and define the random variable X to be the dollar amount of winnings. Specifically, $X(W) = 1, X(L) = -2, X(D) = 0$. The expected value of X is therefore

$$E[X] = \frac{2}{3} \cdot 1 + \frac{1}{6} \cdot (-2) + \frac{1}{6} \cdot 0 = \frac{1}{3}.$$

Remark C.1 The expectation is linear in the sense that if $X_1, \dots, X_n$ are random variables over the same sample space, and $\alpha_1, \dots, \alpha_n$ are constants, then

$$E[\alpha_1 X_1 + \cdots + \alpha_n X_n] = \alpha_1 E[X_1] + \cdots + \alpha_n E[X_n].$$

Example C.3 If we flip a coin n times, what is the expected number of heads? Since $p = \frac{1}{2}$ we have that $P[X = k] = \binom{n}{k} p^k (1-p)^{n-k} = \binom{n}{k} 2^{-n}$. Thus the expected value is

$$E[X] = \sum_{k=0}^{n} k \binom{n}{k} 2^{-n}.$$

This sum is not easy to compute. However, we can we think of X as the sum of random variables $X = X_1 + \cdots + X_n$, where X_i equals 1 or 0 depending on whether or not the ith coin flip is 1 or 0, respectively. Then clearly, for each i,

$$E[X_i] = \frac{1}{2} \cdot 0 + \frac{1}{2} \cdot 1 = \frac{1}{2}$$

and so

$$E[X] = E[X_1 + \cdots + X_n] = E[X_1] + \cdots + E[X_n] = \frac{n}{2}.$$

Definition C.9 ***(Variance)*** The *variance* $V[X]$ of a random variable X is defined as

$$V[X] = E[X^2] - E[X]^2.$$

The standard deviation of X is

$$\sigma[X] = \sqrt{V[X]}.$$

If X has the binomial distribution over $\{0, \dots, n\}$ with probability p then

$$E[X] = np, \quad V[X] = np(1-p).$$

If X has the Poisson distribution with rate λ, then

$$E[X] = \lambda, \quad V[X] = \lambda.$$

C.4 Continuous Probability

For continuous variables, the probability distribution is described by a *probability density function* $p(x)$ that gives the probability of all outcomes with $p(x)dx$ equal to the probability that an outcome X is in the interval $x - dx < X \leq x$.

The probability density function $p(x)$ is nonnegative and satisfies

$$\int_{-\infty}^{\infty} p(x)dx = 1.$$

We also define the cumulative distribution function according to

$$F(x) = \int_{-\infty}^{x} p(u)du.$$

$F(x)$ is equal to the probability that X is less than or equal to x.

The expected value of a function $f(x)$ with respect to the probability density function $p(x)$ is given by

$$E[f] = \int_{-\infty}^{\infty} f(x)p(x)dx.$$

The mean and variance are defined, respectively, as

$$E[X] = \mu = \int_{-\infty}^{\infty} xp(x)dx$$

$$\mathrm{Var}[X] = \sigma^2 = E[(x-\mu)^2] = \int_{-\infty}^{\infty} (x-\mu)^2 p(x)dx.$$

Uniform Distribution

The uniform distribution $p_u(x)$ on $[a, b]$ is

$$p_u(x) = \begin{cases} 1/(b-a) & a \leq x \leq b, \\ 0, & \text{otherwise,} \end{cases}$$

with mean $\mu = (a+b)/2$ and variance $\sigma^2 = (b-a)^2/12$.

Gaussian Distribution

The *Gaussian* or *normal* distribution with mean μ and variance σ^2 is given by

$$p(x) = \frac{1}{\sqrt{2\pi\sigma^2}} \exp \frac{(x-\mu^2)}{2\sigma^2}.$$

Figure C.3 shows the normal distribution with mean $\mu = 0$ and variance $\sigma = 1$, and its cumulative distribution function.

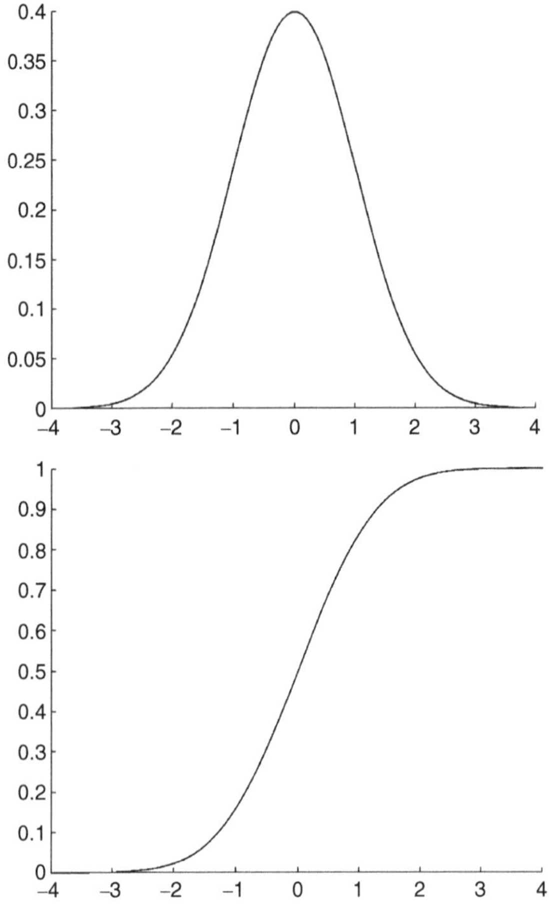

Figure C.3 A normal or Gaussian distribution and cumulative distribution function with mean $\mu = 0$ and variance $\sigma = 1$.

Exponential Distribution

The *exponential* distribution with mean $\mu = 1/\lambda$ and variance $\sigma^2 = 1/\lambda^2$ is given by

$$p_e(x) = \begin{cases} \lambda \exp(\lambda x), & x \geq 0 \\ 0, & x < 0. \end{cases}$$

Figure C.4 shows the exponential distribution and its cumulative distribution function.

Figure C.4 The exponential distribution and cumulative distribution function with $\mu = 1$.

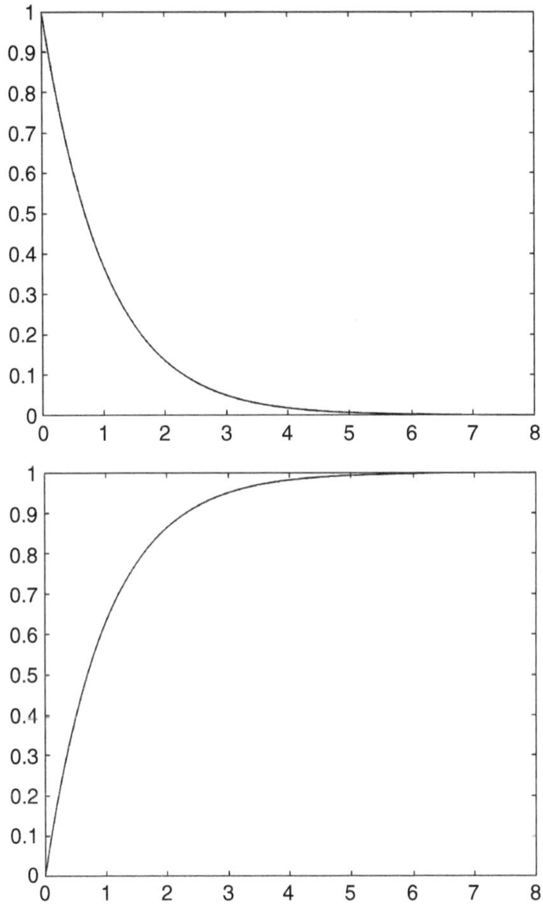

Power-Law Distribution

A power-law distribution assigns a probability

$$P(x) = a\left(\frac{1}{x}\right)^{\alpha}.$$

1. Power-law distributions have a well-defined mean only if the exponent $\alpha > 2$.
2. Power-law distributions have a finite variance only if the exponent $\alpha > 3$.
3. Most of the systems that we will discuss later have exponents satisfying $2 < \alpha < 3$ and so have finite mean and infinite variance. This means that many standard statistical tools do not apply (like the central limit theorem).

Figure C.5 shows a power-law distribution for $\alpha = 2.5$.

Power-Law distributions are also scale-free.

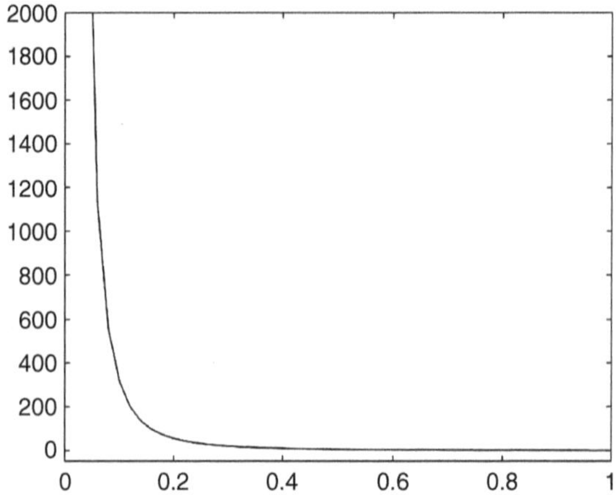

Figure C.5 A power-law distribution for $\alpha = 2.5$.

Definition C.10 ***(Scale Invariance)*** A function $F(x)$ is *scale-free* if there exist a function $g(\cdot)$ and x_0 such that

$$F(\lambda x) = g(\lambda)F(x) \quad \text{for } \lambda x > x_0.$$

This is easy to see. If $P(x) = ax^{-\alpha}$, then

$$\begin{aligned} P(\lambda x) &= a(\lambda x)^{-\alpha} \\ &= a\lambda^{-\alpha}x^{-\alpha} = g(\lambda)P(x), \quad \text{where } g(\lambda) = \lambda^{-\alpha}. \end{aligned}$$

Appendix D

Proofs of Selected Results

D.1 Proof of Theorem 2.2

Suppose J is a closed trajectory in D. Then the inner product $f(x) \cdot n(x) = 0$, where $f(x)$ is the vector field defining (2.27) and $n(x)$ is the outward normal to J. Consequently, the line integral around the closed orbit satisfies $\oint_J f \cdot n d\ell = 0$. Therefore, by Green's theorem we have

$$\int \int_S \nabla \cdot F dA = \oint_J f \cdot n ds = 0,$$

where S is the region enclosed by J. Therefore, either $\nabla \cdot f$ is identically zero in S, or it changes sign in S.

D.2 Proof of Theorem 5.1

The proof closely follows Khalil [44]. See also [33]. Given $\epsilon > 0$, choose $r \in (0, \epsilon)$ such that

$$B_r = \{x \in \mathbb{R}^n \mid ||x|| \leq r\} \subset D.$$

Let

$$\alpha = \min_{||x||=r} V(x).$$

Then $\alpha > 0$ by condition (a) in the statement of the theorem. Let

$$\Omega = \{x \in B_r \mid V(x) \leq \alpha\}.$$

Then Ω is positively invariant for (5.1) since, from condition (b),

$$V(x(t)) \leq V(x(0)) \leq \alpha.$$

Introduction to Modeling and Simulation: A Systems Approach, First Edition. Mark W. Spong.

Companion Website: www.wiley.com/go/spong/engineeringmodeling

Now choose $\delta > 0$ so that the ball $B_\delta \subset \Omega$. Then, if $x(0) \in B_\delta$ we have $x(t) \in \Omega \in B_\epsilon$ for all $t > 0$. This shows stability. Note that invariance of Ω guarantees the existence of a unique solution for all $t > 0$ since there can be no finite escape.

To show asymptotic stability, if $\dot{V}(x(t)) < 0$, let $x(t)$ be a solution starting in $\Omega \backslash \{0\}$ and suppose that $x(t_n) \to z_0 \in B_r$ for some sequence $t_n \to \infty$. Such a sequence exists by compactness of B_r. The claim is that $z_0 = 0$. To see this observe that

$$V(x(t)) > V(z_0), \quad \text{for all } t \geq 0,$$

and

$$V(x_n(t)) \to V(z_0).$$

If $z_0 \neq 0$, let $z(t)$ be the solution starting at z_0. For any $s > 0$ we have

$$V(z(s)) < V(z_0).$$

Hence, for any solution $y(s)$ starting sufficiently near z_0 we have

$$V(y(s)) < V(z_0).$$

Setting $y(0) = x(t_n)$ for sufficiently large n yields

$$V(x(t_n + s)) < V(z_0),$$

which is a contradiction. Therefore, $z_0 = 0$, and, therefore, $x = 0$ is the only limit point of $\{x(t) \mid t \geq 0\}$.

D.3 Proof of Theorem 5.5

We have already shown sufficiency, i.e., if Q is positive definite, then A is Hurwitz. To show the converse, suppose A is Hurwitz and Q is given. Define P as

$$P = \int_0^\infty e^{A^T t} Q e^{At} dt.$$

The matrix P is clearly symmetric. We claim that P is positive definite, i.e., $x^T P x > 0$ for all $x \neq 0$. To see this, compute

$$x^T P x = \int_0^\infty x^T e^{A^T t} Q e^{At} x dt = \int_0^\infty y^T Q y dt \geq 0$$

since Q is positive definite, where $y = e^{At} x$. Now $x^T P x = 0$ if and only if $y = 0$, which implies $x = 0$ since e^{At} is nonsingular. Therefore, P is positive definite.

Next, we compute

$$\begin{aligned} A^T P + PA &= \int_0^\infty A^T e^{A^T t} Q e^{At} dt + \int_0^\infty e^{A^T t} Q e^{At} A dt \\ &= \int_0^\infty \frac{d}{dt}(e^{A^T t} Q e^{At}) dt \end{aligned}$$

$$= e^{A^T t} Q e^{At} |_0^\infty = -Q,$$

which shows that P is a solution to the Lyapunov equation. It only remains to show that P is unique. To show uniqueness, suppose that there is another matrix $\tilde{P}$ satisfying

$$A^T \tilde{P} + \tilde{P} A = -Q.$$

Then a straightforward calculation shows that

$$A^T (P - \tilde{P}) + (P - \tilde{P}) A = 0,$$

which implies that

$$e^{A^T t} [A^T (P - \tilde{P}) + (P - \tilde{P}) A] e^{At} = 0,$$

which reduces to

$$\frac{d}{dt} [e^{A^T t} (P - \tilde{P}) e^{At}] = 0.$$

This means that $e^{A^T t}(P - \tilde{P})e^{At}$ is constant, which holds only if $P = \tilde{P} = 0$.

D.4 Proof of Theorem 13.3

The proof is by induction on r, the number of regions. If $r = 1$, then there can be no regions enclosed by edges of G. Thus G must be acyclic, in which case $e = v - 1$. Thus $v - e + r = v - (v - 1) + 1 = 2$. Therefore, Euler's formula is true for $r = 1$.

Now assume that the formula is true for all planar graphs with fewer than r regions and choose a graph with $r > 1$. Choose an edge e, which is not a cut edge, and let $G_1 = G - e$, the graph obtained by removing edge e. Since e is part of a cycle, removing e results in two merged regions, reducing the number of regions by 1. For G_1, therefore, we the formula is true by the induction hypothesis. Therefore,

$$V(G_1) - E(G_1) + (r - 1) = 2$$

since $V(G_1) = V(G)$ and $E(G_1) = E(G) - 1$. Substituting these into the above equation gives the result.

D.5 Proof of Corollary 13.2

Consider a region r in any planar graph G. For any interior region, let $B(r)$ denote the number of edges enclosing r. Clearly, $B(r) \geq 3$. However, with the number of vertices, $v \geq 3$, we also have that the exterior region is bounded by at least three edges. Therefore, if there are a total of r regions, then $\sum B(r) \geq 3r$. On the other hand, it is easy to see that $\sum B(r)$ counts every edge in G once or twice and hence $\sum B(r) \leq 2e$ so that $3r \leq \sum B(r) \leq 2e$ and so $r \leq \frac{2}{3}e$. From Euler's formula, therefore, we have $e = v + r - 2 \leq v + \frac{2}{3}e - 2$ so that $e \leq 3v - 6$.

D.6 Proof of Proposition 13.2

The eigenvectors associated with eigenvalues $\lambda_1, \ldots \lambda_{n-1}$ form an $(n-1)$-dimensional subspace orthogonal to the vector $x = [1, 1, \ldots, 1]$. Let v be any vector orthogonal to x. Then

$$v^T x = \sum_{i=1}^{n} v_i = 0.$$

Computing the jth entry of Lv gives

$$(Lv)_j = (n-1)v_j - \sum_{i \neq j} v_i = nv_j$$

Therefore, n is an eigenvalue and any vector orthogonal to x is an eigenvector for n.

D.7 Proof of Proposition 13.3

The expected value of a random variable δ, is defined as

$$\overline{\delta} = E[\delta] = \sum_k k \cdot P[\delta = k].$$

Therefore, substituting the expression for $P[\delta = k]$ and rearranging the summation yields

$$\begin{aligned}
\sum_{k=1}^{n-1} kP[\delta = k] &= \sum_{k=1}^{n-1} \binom{n-1}{k} k\, p^k\, (1-p)^{n-1-k} \\
&= \sum_{k=1}^{n-1} \frac{(n-1)!}{k!(n-1-k)!} k\, p^k\, (1-p)^{n-1-k} \\
&= p\,(n-1) \sum_{k=1}^{n-1} \frac{(n-2)!}{(k-1)!(n-k-1)!} p^{k-1}\, (1-p)^{n-1-k} \\
&= p\,(n-1) \sum_{l=0}^{n-2} \frac{(n-2)!}{l!(n-l-2)!} p^l\, (1-p)^{n-l-2} \\
&= p\,(n-1) \sum_{l=0}^{m} \binom{m}{l} p^l\, (1-p)^{m-l} \\
&= p\,(n-1).
\end{aligned}$$

Appendix E

Matlab Command Reference

Math	Matlab

$A = \begin{bmatrix} a & b \\ c & d \end{bmatrix}$	`A = [a b; c d];`
	`A(1,1) = a; A(1,2) = b;`
	`A(2,1) = c; A(2,2) = d;`

$I = \begin{bmatrix} 1 & 0 \\ 0 & 1 \end{bmatrix}$	`I = eye(M,N); or`
	`I = eye(N); if square matrix`
	`M = number of rows`
	`N = number of columns`

$Q = \begin{bmatrix} 1 & 1 \\ 1 & 1 \end{bmatrix}; R = \begin{bmatrix} 0 & 0 \\ 0 & 0 \end{bmatrix};$	`Q = ones(M,N); R = zeros(M,N);` `or Q = ones(N), R = zeros(N);`
$A + B$	`A + B;` The sizes of A and B must be compatible.
$A - B$	`A - B;` A and B must be compatible.
AB	`A*B;` Matrix multiplication. The number of columns of A must equal the number of rows of B.
$A * B$	`A.*B;` element-by-element product. A and B must be the same size unless one is scalar.
$A^{-1}B$	`A\B;` Matlab left divide.
BA^{-1}	`B/A;` Matlab right divide.
A^T	`A';` Matrix transpose.
A^n	`A^n;` Raises the matrix A to the power n.

Introduction to Modeling and Simulation: A Systems Approach, First Edition. Mark W. Spong.

Companion Website: www.wiley.com/go/spong/engineeringmodeling

$v = \begin{bmatrix} a \\ b \\ c \end{bmatrix}$	`v = [a;b;c];` column vector.
	`v(1) = a; v(2) = b; v(3) = c;`
$v = [a, b, c]$	`v = [a b c];` or `v = [a,b,c];` row vector.
	v′ changes a row vector to a column vector and vice versa.
$p(s) = s^3 + 3s^2 + 2s + 1$	`p = [1 3 2 1];`
	Define cubic polynomial coefficients.
$p(3) = 3^3 + 3 * 3^2 + 2 * 3 + 1 = 36$	`polyval(p,3);` evaluate $p(s)$ at $s = 3$.
$p(s) = 0$ find roots of p.	`roots(p)` or `roots([1 3 2 1]);`

Symbolic Math	Matlab
$\lim_{n\to\infty}(1 + \frac{x}{n})^n$	`syms n x; limit((1+x/n)^n, n, inf)`
$f = \exp x \cos(x)$ $g_1 = \frac{df}{dx}; \; g_2 = \frac{d^2f}{dx^2}$	`syms x;`
	`f = exp(x)*cos(x);`
	`g1 = diff(g); g2 = diff(g,2);`
$\int x^3 dx; \; \int e^{-t^2} dt$	`syms x t;`
	`int(x^3); int(exp(-t^2));`
$S = \sum_{n=0}^{n-1} \frac{a^n}{n!} f^{(n)(a)}$	`syms x;`
	`f = cos(2*x);`
	`S = taylor(f,n,a);` order up to $n = 8$ allowed.

Useful Matlab Commands	**Description**
`help(command)`	Displays Matlab help for `command`.
`doc(command)`	Opens a separate documentation page for `command`.
`clear`	Clears all variables in the workspace.
`clc`	Clears the command window.
`length`	Returns the length of the vector x.
`size`	Returns the size of an array.
`zeros`	Creates an array of zeros.
	Useful for preallocating memory to speed up calculations.
`ones`	Creates an array of ones.
`rand`	Create an array of random numbers.
`input`	Prompts for user-defined input.
`plot` and `subplot`	Extensive features for plotting 2D and 3D figures.

Symbolic Math	Matlab
Matlab ODE solvers	
`ode45`	This is the most commonly used solver
`ode23`	Better for stiff equations (defined later)
`ode113`	Can be more efficient than `ode45` at problems with strict tolerances
`ode15s`	Try this one first for stiff equations
`ode23s`	Can be more efficient than odes15s at problems with crude tolerances
`ode23t`	Can solve DAEs
`ode23tb`	Also can be efficient at problems with crude error tolerances
`ode15i`	For fully implicit problems $f(t, y, \dot{y}) = 0$.

Control System Toolbox	Description
`tf`	`H=tf(q,p)` calculates the transfer function $H(s) = \frac{q(s)}{p(s)}$.
`ss`	Creates a state-space model. `sys = ss(A,b,c,d)` creates the system $\dot{x} = Ax + bu;\ y = cx + du$.
`tf2ss`	Converts a transfer function description to state space.
`ss2tf`	Converts a state-space description to a transfer function
`step`	`step(sys)` computes the step response of the system `sys`.
`lqr`	Linear-quadratic regular design `[k,P,poles] = lqr(A,b,Q,r)` calculates the optimal gain k the solution of the Riccati equation P and the closed-loop pole locations `poles`
`icare`	Solves the continuous-time algebraic Riccati equation $A^TP + PA - Pbb^TP + Q = 0$.
`lyap(A',Q)`	Solves the continuous Lyapunov equation $A^TP + PA + Q = 0$
`dlyap(A',Q)`	Solves the discrete Lyapunov equation $APA^T - P + Q$
`pidtune`	Tunes gains in a PID controller
`ctrb`	`C=ctrb(A,b)` returns the controllability matrix $[b, Ab, A^2b, \dots, A^{n-1}b]$.
`obsv`	`O = obsv(A,c)` returns the observability matrix $[c^T, A^Tc^T, \dots, A^{T^{n-1}}c^T]$.

References

1 D. Abramovitch. The outrigger: a prehistoric feedback mechanism. *IEEE Control Systems Magazine*, 25(4):57–72, 2005.

2 J. Ackermann. Der Entwurf linearer Regelungssysteme im Zustandsraum. *At-Automatisierungstechnik*, 20:1–12, 1972.

3 R. L. Ackoff. Towards a system of systems concepts. *Management Science*, 17(11), 1971.

4 G. B. Airy. On the intensity of light in the neighbourhood of a caustic. *Transactions of the Cambridge Philosophical Society*, 6:379–402, 1838.

5 K. J. Åström and R. Murray. *Feedback Systems: An Introduction for Scientists and Engineers*. Princeton University Press, Princeton, NJ, 2008.

6 K. J. Åström and T. Hägglund. *PID Controllers: Theory, Design, and Tuning*. Instrument Society of America, Research Triangle Park, NC, 1995.

7 N. Bacaër. Lotka, Volterra and the predator-prey system (1920–1926). In *A Short History of Mathematical Population Dynamics*, pages 71–76. Springer, London, 2011.

8 A. L. Barabási and R. Albert. The emergence of scaling in random networks. *Science*, 286:797–817, 1999.

9 J. Bechhoefer. Feedback for physicists: A tutorial essay on control. *Reviews of Modern Physics*, 77(3):783–835, 2005.

10 S. Bennett. A brief history of automatic control. *IEEE Control Systems Magazine*, 16(3):17–25, 1996.

11 F. Black and M. Scholes. The pricing of options and corporate liabilities. *Journal of Political Economy*, 81(3):637–654, 1973.

12 D. J. Block, K. J. Åström, and M. W. Spong. *The Reaction Wheel Pendulum*. Morgan & Claypool Publishers, San Rafael, CA, 2007.

13 F. Brauer and C. Castillo-Chavez. *Mathematical Models in Population Biology and Epidemiology*. Springer, New York, 2001.

14 V. Capasso. *The Mathematical Structure of Epidemic Systems*. Springer-Verlag, Berlin, 1993.

Introduction to Modeling and Simulation: A Systems Approach, First Edition. Mark W. Spong.

Companion Website: www.wiley.com/go/spong/engineeringmodeling

15 P. B. Checkland. *Systems Thinking, Systems Practice*. John Wiley & Sons, Chichester, 1999.

16 E. K. P. Chong and S. H. Zak. *An Introduction to Optimization*. John Wiley & Sons, Hoboken, NJ, fourth edition, 2013.

17 C. W. Churchman, R. L. Ackoff, and E. L. Arnoff. *Introduction to Operations Research*. John Wiley & Sons, New York, 1950.

18 P. G. L. Dirichlet. Sur la convergence des series trigonométriques qui servent à représenter une fonction arbitraire entre des limites donnés. *Journal of Mathematics*, 4:157–169, 1829.

19 J. R. Dormand and P. J. Prince. A family of embedded Runge-Kutta formulae. *Journal of Computational and Applied Mathematics*, 6(1):19–26, 1980.

20 D. G. Duffy. *Advanced Engineering Mathematics with Matlab*. Chapman and Hall/CRC, Boca Raton, FL, 2003.

21 P. Ehrenfest. Über zwei bekannte Einwände gegen das Boltzmannsche H-Theorem. *Physikalische Zeitschrift*, 8:311, 1907.

22 P. Erdős and A. Rényi. On random graphs. *Publicationes Mathematicae*, 6:290–297, 1959.

23 R. FitzHugh. Mathematical models of threshold phenomena in the nerve membrane. *Bull. Math. Biophysics*, 17:257–278, 1955.

24 H. Freedman. *Deterministic Mathematical Models in Population Ecology*. Marcel Dekker, New York, 1980.

25 F. Galton. Kinship and correlation (reprinted 1989). *Statistical Science*, 4(2):80–86, 1989.

26 M. Gautier and W. Khalil. On the identification of the inertial parameters of robots. In *IEEE Conf. on Decision and Control*, pages 2264–2269, Austin, TX, 1988.

27 E. Gilbert. Controllability and observability in multivariable control systems. *SIAM Journal of Control*, 1(1):128–151, 1963.

28 K. Godfrey. *Compartment Models and Their Application*. Academic Press, New York, 1983.

29 H. Goldstein. *Classical Mechanics*. Addison-Wesley, Reading, MA, 1974.

30 G. H. Golub and C. F. Van Loan. *Matrix Computations*, 3rd Edition. Johns Hopkins University Press, Baltimore, MD, 1983.

31 R. M. Goodwin. A growth cycle. In C. H. Feinstein, editor, *Socialism, Capitalism and Economic Growth*. Cambridge University Press, Cambridge, 1967.

32 A. Greenbaum and T. P. Chartier. *Numerical Methods: Design, Analysis, and Computer Implementation of Algorithms*. Princeton University Press, Princeton, NJ, 2012.

33 W. Hahn. *Stability of Motion*. Springer-Verlag, Berlin, 1967.

34 D. Halliday, R. Resnick, and J. Walker. *Fundamentals of physics*. John Wiley & Sons, Hoboken, NJ, 2004.

35 R. L. Harrison, C. Granja, and C. Leroy. Introduction to monte carlo simulation. In *AIP Conference Proceedings*, volume 1204, pages 17–21, 2010.

36 H. W. Hethcote. The mathematics of infectious diseases. *SIAM Review*, 42(4):599–653, 2000.

37 M. W. Hirsch and S. Smale. *Differential Equations, Dynamical Systems, and Linear Algebra*. Academic Press, New York, 1974.

38 A. L. Hodgkin and A. F. Huxley. A quantitative description of membrane current and its application to conduction and excitation in nerve. *Journal of Physiology*, 117(4):500–544, August 1952.

39 Y. Ji and J. Shen. Turing instability of Brusselator in the reaction-diffusion network. *Complexity*, 2020:12, 2020.

40 R. E. Kalman. Contributions to the theory of optimal control. *Boletin de la Sociedad Matématica Mexicana*, 5:102–119, 1960.

41 R. E. Kalman. On the general theory of control systems. In *Proc. IFAC Congress on Automatic Control*, volume 1, pages 481–492, Moscow, 1961.

42 O. D. Kellogg. *Foundations of Potential Theory*. Dover Books, New York, 2010.

43 W. O. Kermack and A. G. McKendrick. A contribution to the mathematical theory of epidemics. *Proceedings of the Royal Society of London*, 115:700–721, 1927.

44 H. K. Khalil. *Nonlinear Systems*, 3rd ed. Macmillan, New York, 2001.

45 F. G. Kilgour. *Engineering in History*. Dover Publications, New York, 1990.

46 P. Kokotović, H. Khalil, and J. O'Reilly. *Singular Perturbation Methods in Control: Analysis and Design*. Academic Press, London, 1986.

47 J.P. LaSalle. Some extensions of liapunov's second method. *IRE Transactions on Circuit Theory*, CT-7:520–527, 1960.

48 T. G. Lewis. *Network Science: Theory and Practice*. John Wiley & Sons, Hoboken, NJ, 2009.

49 D. G. Luenberger and Y. Ye. *Linear and Nonlinear Programming*. Springer, Cham, fourth edition, 2016.

50 A. M. Lyapunov. The general problem of the stability of motion. *International Journal of Control*, 55(3):531–534, 1992.

51 J. C. Maxwell. On governors. *Proceedings of the Royal Society of London*, 16:270–283, 1868.

52 M. E. J. Newman. *Networks: An Introduction*. Oxford University Press, New York, 2010.

53 Y. Nievergelt. A tutorial history of least squares with applications to astronomy and geodesy. *Journal of Computational and Applied Mathematics*, 121:37–72, 2000.

54 A. V. Oppenheim, R. W. Schafer, and J. R. Buck. *Discrete-Time Signal Processing*. Prentice Hall, Upper Saddle River, NJ, 1998.

55 R. Pearl and L. J. Reed. On the rate of growth of the population of the United States and its mathematical representation. *Proceedings of the National Academy of Sciences*, 6:275–288, 1920.

56 A. R. Plastino and J. C. Muzzio. On the use and abuse of newton's second law for variable mass problems. *Celestial Mechanics and Dynamical Astronomy*, 53(3), 1992.

57 H. Poincaré. L'équilibre du'ne masse fluide animée d'un mouvement de rotation. *Acta Mathematica*, 7:259–380, 1885.

58 H. Rodrigues, N. Panza, D. Portes, and A. Soares. The motion of a leaking oscillator: a study for the physics class. *Physics Education*, 49(5), 2014.

59 F. Scheid. *Numerical Analysis*. McGraw-Hill, New York, 1968.

60 L. F. Shampine. *Numerical Solution of Ordinary Differential Equations*. Chapman & Hall, New York, 1994.

61 L. F. Shampine, I. Gladwell, and S. Thompson. *Solving ODEs with MATLAB*. Cambridge University Press, Cambridge, 2003.

62 L. F. Shampine and M. K. Gordon. *Computer Solution of Ordinary Differential Equations: the Initial Value Problem*. W. H. Freeman, San Francisco, 1975.

63 L. F. Shampine and M. W. Reichelt. The MATLAB ODE suite. *SIAM Journal on Scientific Computing*, 18:1–22, 1997.

64 M. W. Spong, S. Hutchinson, and M. Vidyasagar. *Robot Modeling and Control*. John Wiley & Sons, New York, second edition, 2020.

65 N. C. Stenseth, W. Falck, O. N. Bjørnstad, and C. J. Krebs. Population regulation in snowshoe hare and Canadian lynx: Asymmetric food web configurations between hare and lynx. *Proceedings of the National Academy of Sciences*, 94(10):5147–5152, 1997.

66 G. Strang. *Introduction to Linear Algebra*. Wellesley-Cambridge Press, Wellesley, MA,1993.

67 W. Thomson. Mechanical integration of the general linear differential equation of any order with variable coefficients. *Proceedings of the Royal Society of London*, 24:262–275, 1876.

68 B. van der Pol. On oscillation hysteresis in a triode generator with two degrees of freedom. *Philosophical Magazine*, 6(43):700–719, 1922.

69 M. van Steen. *Graph Theory and Complex Networks: An Introduction*. Maarten van Steen, Amsterdam, 2010.

70 P.-F. Verhulst. Notice sur la loi que la population sui dans son accroissement. *Correspondance Mathématique et Physique, publiée par A. Quetelet*, 10:113–120, 1838.

71 M. Vidyasagar. *Hidden Markov Models: Theory and Applications to Biology*. Princeton University Press, Princeton, NJ, 2014.

72 D. J. Watts. *Small Worlds*. Princeton University Press, Princeton, NJ, 1999.

73 D. J. Watts and S. H. Strogatz. Collective dynamics of "small-world" networks. *Nature*, 393:440–442, 1998.

74 D. B. West. *Introduction to Graph Theory*. Prentice Hall, Upper Saddle River, NJ, 1996.

75 E. P. M. Widmark and J. Tandberg. Über die Bedingungen für die Akkumulation indifferenter Narkotika. *Biochemische Zeitung*, 148:358–389, 1924.

Index

Introduction to Modeling and Simulation: A Systems Approach, First Edition. Mark W. Spong.

Companion Website: www.wiley.com/go/spong/engineeringmodeling

t